Advances in

MICROBIAL
PHYSIOLOGY

Advances in

MICROBIAL
PHYSIOLOGY

Edited by
A. H. ROSE

School of Biological Sciences
Bath University
England

and

D. W. TEMPEST

Microbiological Research Establishment
Porton, Salisbury
England

VOLUME 7

1972

ACADEMIC PRESS · LONDON and NEW YORK

ACADEMIC PRESS INC. (LONDON) LTD.
24/28 Oval Road,
London NW1

United States Edition published by
ACADEMIC PRESS INC.
111 Fifth Avenue
New York, New York 10003

Copyright © 1972 by
ACADEMIC PRESS INC. (LONDON) LTD.

All Rights Reserved

No part of this book may be reproduced in any form by photostat, microfilm, or any
other means, without written permission from the publishers

Library of Congress Catalog Card Number: 67-19850
ISBN: 0-12-027708-5

PRINTED IN GREAT BRITAIN BY
WILLIAM CLOWES AND SONS LIMITED
LONDON, COLCHESTER AND BECCLES

Contributors to Volume 7

R. E. BURGE, *Physics Department, University of London, Queen Elizabeth College, London, England.*

U. EBERHARDT, *Institute for Microbiology, University of Göttingen, West Germany.*

D. C. ELLWOOD, *Microbiological Research Establishment, Porton, Nr. Salisbury, Wiltshire, England.*

H. GEST, *Department of Microbiology, Indiana University, Bloomington, Indiana 47401, U.S.A.*

J. R. QUAYLE, *Department of Microbiology, University of Sheffield, Sheffield, England.*

D. A. REAVELEY, *Physics Department, University of London, Queen Elizabeth College, London, England.*

H. G. SCHLEGEL, *Institute for Microbiology of the Gesellschaft für Strahlenforschung m.b.H., München, Göttingen, West Germany.*

D. W. TEMPEST, *Microbiological Research Establishment, Porton, Nr. Salisbury, Wiltshire, England.*

Contents

Contributors to Volume 7 v

Walls and Membranes in Bacteria.   D. A. REAVELEY and R. E. BURGE

I. Introduction 2
  A. Protoplasts and Sphaeroplasts 3
  B. Division of Protoplasts 4
  C. Reversion of L Forms of *Bacillus subtilis* to the Bacillary Form 5

II. Intracytoplasmic Membranes 6
  A. Occurrence and Appearance 6
  B. Extrusion of Mesosomal Vesicles and Tubules from Protoplasts 10
  C. Mesosomes and Septum Formation in Gram-Positive Bacteria 12
  D. Mesosomes and the Nucleus in Gram-Positive Bacteria . 12
  E. Intracytoplasmic Membranes in *Escherichia coli* . . . 13
  F. Localization of Respiratory Enzymes in Bacteria . . 14
  G. Other Proposed Functions of Mesosomes 15

III. The Cytoplasmic Membranes of Gram-Positive Bacteria . . 16
  A. Introduction 16
  B. Morphology 17
  C. Isolation of the Cytoplasmic Membranes of Gram-Positive
     Bacteria and Requirement for Magnesium 18
  D. Association of Ribosomes with the Cytoplasmic Membrane . 19
  E. Yield and Composition of Cytoplasmic Membranes . . 20
  F. Disaggregation and Re-aggregation of Cytoplasmic Mem-
     branes 25
  G. Enzymes of the Cytoplasmic Membrane 26
  H. Structure of the Cytoplasmic Membrane 28

IV. Cell Walls of Gram-Positive Bacteria 30
  A. Introduction 30
  B. Morphology 31
  C. Peptidoglycans 35
  D. Teichoic and Teichuronic Acids 42
  E. Other Cell-Wall Polymers 46

V. The Cell Envelopes of Gram-Negative Bacteria . . . 47
  A. Introduction 47
  B. Morphology 49
  C. Peptidoglycans of Gram-Negative Bacteria . . . 52
  D. The Lipoprotein of the Rigid Layer of the Cell Envelope of
     Escherichia coli 55

viii CONTENTS

E. Lipopolysaccharides 56
F. Lipids 60
G. Proteins 63
H. Effect of EDTA on Gram-Negative Bacteria . . . 64
I. Osmotic Shock 66
J. Structure of the Outer Membrane of the Gram-Negative Cell
Envelope 67
K. The Cell Envelope of Halophilic Bacteria 68
VI. Structure Arrangements of Polymeric Components Within
Bacterial Cell Walls and Cytoplasmic Membranes . . . 69
VII. Acknowledgements 71
References 71

Effects of Environment on Bacterial Wall Content and Composition. D. C. ELLWOOD and D. W. TEMPEST

I. Introduction 83
II. Bacterial Wall Content 84
III. Teichoic Acids and Teichuronic Acids 86
A. Effects of Specific Nutrient Limitations 88
B. Effects of Growth Rate 95
C. Effects of Ionic Environment 95
D. Effects of Medium pH Value 98
E. Control of Synthesis of Teichoic Acid and Teichuronic Acid . 100
IV. Intracellular Teichoic Acids 102
V. Mucopeptide 102
A. Comparative Effects of Lysozyme on *Bacillus subtilis* . . 105
B. Effect of Growth Condition 106
VI. Lipopolysaccharide 110
VII. Protein 113
VIII. Lipids 113
IX. Concluding Remarks 114
References 115

The Metabolism of One-Carbon Compounds by Micro-Organisms. J. R. QUAYLE

I. Introduction 119
II. Description and Physiology of One-Carbon-Utilizing Micro-
Organisms 120
A. Aerobic Organisms 120
B. Photosynthetic Bacteria 138
C. Non-Photosynthetic Anaerobic Bacteria 140

III.	Energy Metabolism	143
	A. Aerobic Oxidation	143
	B. Anaerobic Dismutation	161
	C. Photometabolism	168
IV.	Carbon Assimilation	169
	A. Assimilation of Carbon Dioxide	169
	B. Assimilation of Reduced One-Carbon Compounds	177
V.	Acknowledgements	197
	References	197

Regulatory Phenomena in the Metabolism of Knallgasbacteria.
H. G. SCHLEGEL and U. EBERHARDT

I.	Introduction	205
II.	Chemolithotrophic Metabolism	206
	A. General Survey	206
	B. Hydrogen Oxidation	207
	C. Hydrogenase Synthesis	210
	D. Carbon Dioxide Fixation	215
	E. Ribulose Diphosphate Carboxylase and Phosphoribulokinase Synthesis	216
III.	Chemo -Organotrophic and Mixotrophic Metabolism	219
	A. Enzymes Involved in Hexose and Gluconate Utilization	219
	B. Effect of Hydrogen on Hexose Degradation	222
	C. Formation and Function of the Tricarboxylic Acid Cycle Enzymes	229
	D. Utilization of Nitrogenous Compounds	230
IV.	Biosynthesis of Amino Acids.	234
V.	Final Considerations	238
VI.	Acknowledgements	239
	References	240

Energy Conversion and Generation of Reducing Power in Bacterial Photosynthesis.   HOWARD GEST

I.	Introduction	243
II.	On the "Eras" of Photosynthesis; Primary versus Secondary Processes	245
III.	Photophosphorylation	248
IV.	Generation of Net Reducing Power	251
	A. Molecular Hydrogen as an Accessory Electron Donor	253
	B. Reduction of NADP by $NADH_2$ (Transhydrogenase)	258
	C. Succinate as a Hydrogen Donor for Photoreduction of NAD	260
	D. Another Way in Which ATP May "Drive" Nicotinamide Nucleotide Reduction	263

CONTENTS

E.	Photoproduction of Molecular Hydrogen	.	.	264
F.	General Comments on Proposed Non-Cyclic Electron-Flow Mechanisms .	.	.	268
V.	A Comparison of Energy Metabolism and Electron-Transfer Patterns in Photosynthetic Bacteria and Clostridia	.	.	269
VI.	Regulatory Mechanisms	.	.	273
VII.	Epilogue	.	.	277
VIII.	Acknowledgements	.	.	278
	References	.	.	278

Author Index 283

Subject Index 295

Walls and Membranes in Bacteria

D. A. REAVELEY* and R. E. BURGE

*Physics Department, University of London,
Queen Elizabeth College, London, England*

I. Introduction	. . . .	2
A. Protoplasts and Sphaeroplasts	. . . . .	3
B. Division of Protoplasts	. . . . . .	4
C. Reversion of L Forms of *Bacillus subtilis* to the Bacillary Form		5
II. Intracytoplasmic Membranes	. . . . . .	6
A. Occurrence and Appearance	. . . . . .	6
B. Extrusion of Mesosomal Vesicles and Tubules from Protoplasts		10
C. Mesosomes and Septum Formation in Gram-Positive Bacteria .		12
D. Mesosomes and the Nucleus in Gram-Positive Bacteria	. .	12
E. Intracytoplasmic Membranes in *Escherichia coli*	. . .	13
F. Localization of Respiratory Enzymes in Bacteria	. . .	14
G. Other Proposed Functions of Mesosomes	. . . .	15
III. The Cytoplasmic Membranes of Gram-Positive Bacteria .	. .	16
A. Introduction	. . . .	16
B. Morphology	. . . .	17
C. Isolation of the Cytoplasmic Membranes of Gram-Positive Bacteria and Requirement for Magnesium	. . .	18
D. Association of Ribosomes with the Cytoplasmic Membrane	.	19
E. Yield and Composition of Cytoplasmic Membranes	. .	20
F. Disaggregation and Re-aggregation of Cytoplasmic Membranes		25
G. Enzymes of the Cytoplasmic Membrane	. . . .	26
H. Structure of the Cytoplasmic Membrane	. . . .	28
IV. Cell Walls of Gram-Positive Bacteria	. . . . .	30
A. Introduction	. . . .	30
B. Morphology	. . . .	31
C. Peptidoglycans	. . . .	35
D. Teichoic and Teichuronic Acids	. . . . .	42
E. Other Cell-Wall Polymers	. . . . . .	46
V. The Cell Envelopes of Gram-Negative Bacteria	. . .	47
A. Introduction	. . . .	47
B. Morphology	. . . .	49
C. Peptidoglycans of Gram-Negative Bacteria	. . .	52
D. The Lipoprotein of the Rigid Layer of the Cell Envelope of *Escherichia coli*	. . . . .	55
E. Lipopolysaccharides	. . . . . .	56

* Member of the Medical Research Council External Staff.

F. Lipids 60
G. Proteins 63
H. Effect of EDTA on Gram-Negative Bacteria 64
I. Osmotic Shock 66
J. Structure of the Outer Membrane of the Gram-Negative Cell
Envelope 67
K. The Cell Envelope of Halophilic Bacteria 68
VI. Structure Arrangements of Polymeric Components Within Bacterial
Cell Walls and Cytoplasmic Membranes 69
VII. Acknowledgements 71
References 71

I. Introduction

Almost one hundred years ago, Cohn (1872) suggested that the resistance of bacteria to attack by acids and alkalies was due to the presence of a rigid structure surrounding the cell. The phenomenon of plasmolysis, first observed twenty years later by Fischer (1891), produced the first visual evidence for a cell wall. During plasmolysis the cytoplasm loses water and retracts from an apparently rigid cell wall. The suggestion that the retracted cytoplasm is itself surrounded by a membrane responsive to osmotic changes followed naturally from Fischer's observations. Although some evidence for this membrane was obtained from observations with the light microscope, conclusive visual evidence came only with the advent of electron microscopy and the allied thin-section techniques (Kellenberger and Ryter, 1958; Glauert, 1962). This membrane has been variously termed a "cytoplasmic membrane", a "plasma membrane", a "protoplast membrane" and more recently a "plasmalemma". In this review it is proposed to use the term cytoplasmic membrane.

A correlation between reaction to the Gram stain and ease of plasmolysis was quickly established and it appears that in general only Gram-negative bacteria are plasmolysed. Mitchell and Moyle (1956) cited this observation as evidence for a stronger link between the cell wall and the cytoplasmic membrane in Gram-positive bacteria. Connections between specific regions of the cytoplasmic membrane and the inner layer of the cell wall can be seen in thin-sections of plasmolysed cells of *Escherichia coli* (Bayer, 1968). Photographs of replicas of freeze-etched bacilli (Holt and Leadbetter, 1969; Nanninga, 1968) indicate that strands or fibres connect the inner layer of the cell wall with the outer layer of the cytoplasmic membrane. The relationship between these fibres and the finger-like protrusions of material observed by negative staining to extend from the cytoplasmic membrane to the cell wall in many Gram-positive bacteria is not clear (Hurst and Stubbs, 1969). Bridges extending from the outer regions of the cytoplasmic membrane are clearly visible in thin sections of *Bacillus licheniformis* (Rogers, 1970)

WALLS AND MEMBRANES IN BACTERIA 3

and have been reported to be present in other Gram-positive bacteria (Leadbetter and Holt, 1968; Ghosh and Murray, 1967; Edwards and Stevens, 1963). An intimate connection between cell wall and cytoplasmic membrane is not unexpected in view of recent findings that the cytoplasmic membrane is involved in the biosynthesis of some cell-wall components (Ellar, 1970). Preparations of Gram-negative "cell walls" produced by mechanical disintegration often contain cytoplasmic membranes retained within the walls. Such preparations are better regarded as cell envelopes (Salton, 1967a).

Cell walls of Gram-positive bacteria consist principally of an insoluble polymer, the peptidoglycan (synonyms—murein, mucopeptide, glycosaminopeptide) which in some cases accounts for as much as 90% of the dry weight of the cell wall, together with one or more other macromolecular components which may be protein, polysaccharide or teichoic acid. The cell walls of Gram-negative bacteria are more complex containing only relatively small amounts (5–20%) of peptidoglycan and large amounts of protein, lipid and lipopolysaccharide (Salton, 1964; Rogers and Perkins, 1968). Isolated cell walls retain the shape of the cell from which they were obtained and, because of this retention of shape, they have often been described as rigid structures. Recent physico-chemical experiments (Ou and Marquis, 1970) on the cell walls of Gram-positive bacteria have demonstrated that the walls behave like polyelectrolyte gels expanding and contracting in response to changes in environmental pH value and ionic strength. Since the cell walls are also known (Gerhardt and Judge, 1964) to function as heteroporous molecular sieves, such a contraction or expansion may be important in controlling the size of molecules passing through the cell wall. Binding studies with monovalent and divalent cations show that cell walls also behave as weak ion-exchange resins (Cutinelli and Galdiero, 1967).

A. PROTOPLASTS AND SPHAEROPLASTS

Treatment of many bacilli with lysozyme in hypertonic or isotonic media transforms the cells from rods into osmotically fragile spheres. These spheres, termed protoplasts by Weibull (1953), are free of cell-wall material and are surrounded simply by the cytoplasmic membrane. Comparable methods have been used to prepare protoplasts from other Gram-positive bacteria (Gooder, 1968; Weibull, 1968). The cytoplasmic membrane behaves as a porous differential dialysis membrane and the pore size can be increased by osmotic swelling of the protoplast (Corner and Marquis, 1969). The structure is highly extensible and enormous protoplasts may be prepared by slowly dialysing away the stabilizing solutes. The lysis of protoplasts appears to be caused by a rapid influx

of solutes when the pores reach a critical size leading to very rapid local extensions of the membrane and brittle fracture.

Historically, the absence of cell-wall material, spherical shape and osmotic sensitivity were taken as essential criteria for protoplasts (Brenner *et al.*, 1958). Rod-like forms, otherwise similar to protoplasts, are produced from *B. megaterium* (Op den Kamp *et al.*, 1967) and *B. subtilis* (Van Iterson and Op den Kamp, 1969) when cells previously subjected to acidic pH (5) values are treated with lysozyme. These forms which do not contain any residual cell-wall components are more resistant to osmotic lysis than spherical protoplasts produced from cells harvested at neutral pH values. The change in shape is correlated with the replacement of phosphatidylglycerol by glucosaminylphosphatidylglycerol in the cytoplasmic membrane of the former organism, and by the lysyl derivative in the latter. Bizarre shaped protoplasts of bacilli are produced when the cell walls are removed in the presence of high concentrations of magnesium ions (Rogers *et al.*, 1967).

The complete removal of the chemically more complicated Gram-negative cell wall to leave the organism surrounded simply by the cytoplasmic membrane is technically very difficult, and has only been achieved in one specialized case. A marine pseudomonad is converted into a wall-less form by lysozyme after prior removal of the outer layers of the cell wall by washing with $0 \cdot 5$ M-sodium chloride and resuspension in $0 \cdot 5$ M-sucrose. Similar forms may be produced by the action of lysozyme and EDTA in tris-HCl buffer made $0 \cdot 5$ M with respect to sucrose (Costerton *et al.*, 1967; De Voe *et al.*, 1970).

Spherical, osmotically sensitive forms of Gram-negative bacteria, termed sphaeroplasts, may be obtained by penicillin treatment or by the action of lysozyme (McQuillen, 1960). Gram-negative bacteria are more resistant to lysozyme than Gram-positive organisms and are only susceptible to the action of this enzyme if previously sensitized. Cells may be sensitized in several ways, which include heat treatment (Myerholtz and Hartsell, 1952), freezing and thawing (Kohn, 1960), plasmolysis (Birdsell and Cota-Robles, 1967) or by the action of EDTA in tris-HCl buffer (Repaske, 1956). Sphaeroplasts possess a weakened cell wall caused by the loss of the peptidoglycan (Salton, 1964).

B. Division of Protoplasts

Dumbell-shaped protoplasts reminiscent of division forms have been observed when *B. megaterium* is incubated in a stabilized medium for 8–9 hr at 28° (McQuillen, 1960) but, generally, true division of protoplasts in liquid media has yet to be observed (Weibull, 1968). Landman (1968) and Gooder (1968) have reported the propagation of protoplasts of *B. subtilis* and streptococci respectively on soft agar. The colonies

consist of wall-less forms of irregular size and shape which are morphologically indistinguishable from L forms. Such forms may be described as stable L forms since the organism will persist in the wall-less state. Lysozyme-produced protoplasts of *B. subtilis* still containing small amounts of cell wall revert to the bacillary state when plated on suitable media (Miller *et al.*, 1967). Apparently, for this organism, commitment to the L state is initiated by complete (or almost complete) loss of cell wall (Landman, 1968). However, cell wall damage rather than complete loss of cell wall is necessary for efficient propagation of lysozyme-treated cells of streptococci as L forms (King and Gooder, 1970).

C. Reversion of L Forms of *Bacillus subtilis* to Bacillary Forms

Stable L forms of *B. subtilis* propagated on hard agar (2·5%) or a medium containing 25% gelatin revert to the bacillary form (Landman *et al.*, 1968; Landman and Forman, 1969). In addition to hard agar and gelatin, reversion is stimulated by growth on certain membrane filters and by the addition of cell walls of whole organisms (Clive and Landman, 1970). The stimulation by whole cell walls is non-specific since similar stimulation is obtained by the addition of a wide range of autoclaved bacteria and yeasts. The factors involved in reversion have little in common chemically, and reversion must be dependent on the physical properties of the surface provided adjacent to the naked protoplast. How the barrier acts is not known. Synthesis of 2,6-diaminopimelic acid, a constituent of peptidoglycan is repressed in L forms and protoplasts (Bond and Landman, 1970), although cytoplasmic membranes of L forms retain the ability to synthesize peptidoglycan intermediates when supplied with appropriate substrates (Chatterjee *et al.*, 1967).

The process of reversion has been monitored by electron microscopy (Landman *et al.*, 1968). In the initial stages of reversion, the chromosome of the protoplast becomes more condensed and then a thin layer of cell wall appears around the cytoplasmic membrane. It is not clear whether this cell wall originates at specific regions on the cytoplasmic membrane; during the process cell shape is extremely variable. Mesosomes of an unusual compartmented type not linked to the nucleus appear in about 25% of the cells but only after the cell wall has been laid down. Control experiments with bacilli suggest that these mesosomes are induced by gelatin and not by the reversion process.

Biochemically the process of reversion is divided into three stages (Landman and Forman, 1969). In the first stage, primed by casein hydrolysate, synthesis of both protein and RNA takes place. In the second stage, experiments with lysozyme and penicillin suggest that cell-wall synthesis is taking place. During this stage the cells must be continuously surrounded by the solid medium; heating the gelatin briefly

at 40° severely retards the reversion process. The third stage confers osmotic stability on the cells and is inhibited by actinomycin D and chloramphenicol.

II. Intracytoplasmic Membranes

A. OCCURRENCE AND APPEARANCE

Classical intracellular organelles, such as mitochondria and chloroplasts, existing in the cytoplasm and carrying out specific cellular functions are not present in prokaryotic organisms. Such functions are performed in bacteria by the cytoplasmic membrane and intracytoplasmic membranes of varying complexity (Gel'man et al., 1967). The intracytoplasmic membranes appear to be formed almost universally by invaginations of the cytoplasmic membrane and, as a result, they maintain contact with the extracytoplasmic space. A possible exception is the complex intracellular membrane system present in the nitrifying bacterium, Nitrosocystis oceanus (Remsen et al., 1967); these membranes are apparently not formed by invagination of the cytoplasmic membrane and are distributed between daughter cells on division. Possibly this system represents an evolutionary intermediate in the formation of a true intracellular organelle. The photosynthetic organism, Ectothiorhodospira mobilis, contains an exceedingly complex internal membrane system comprising one to eight stacks of lamellar membranes (Remsen et al., 1968). The stacks are penetrated by negative staining showing contact with the external environment and are formed initially by infolding of the cytoplasmic membrane. The membrane may infold again on itself at any point to produce disc-like structures which are reminiscent of chloroplasts in plants. Complicated internal membrane systems have also been demonstrated in the methane-utilizing bacteria (Davies and Whittenbury, 1970).

Simpler intracytoplasmic membranes systems are found in Gram-positive bacteria. Detailed studies of such systems, termed mesosomes (Fitz-James, 1960), have in the main been confined to bacilli, particularly B. subtilis (Ryter, 1968) and B. licheniformis (Rogers, 1970). The mesosome takes the form of a sac-like invagination of the cytoplasmic membrane into the cytoplasm. Examination of the cells, after fixation and thin-sectioning in the electron microscope, reveals that the sac may be filled with lamellar, tubular or vesicular structures (Salton, 1967a; Rogers and Perkins, 1968). In order to prevent confusion at later stages, it is proposed that the following terminology be used when discussing the mesosomes of Gram-positive bacteria. It is suggested that the entire invagination of the cytoplasmic membrane together with the contents of the invagination be called a "mesosome". The part of the cytoplasmic

WALLS AND MEMBRANES IN BACTERIA 7

membrane which invaginates will be called the "mesosomal sac" and the contents of the invagination (or sac) will be referred to as "mesosomal tubules", "mesosomal vesicles" or "lamellar mesosomal membranes" as the situation demands.

The appearance in thin section (Fig. 1) of the structures within the mesosomal sac apparently depends upon the fixation procedure (Highton, 1969, 1970a, b; Burdett and Rogers, 1970); thus lamellar mesosomal membranes are seen only very rarely after prefixation with glutaralde-hyde (Ryter, 1968; Burdett and Rogers, 1970) but are present in cells fixed at 0° using the standard Ryter-Kellenberger procedure (Highton, 1969; Fitz-James, 1968). When fixed at room temperature by the standard Ryter-Kellenberger procedure, the mesosomal sac of *B. licheni-formis* is seen to contain inflated tubules constricted at regular points along their length (Burdett and Rogers, 1970). After using a fixing solution containing a lower concentration of Ca^{2+} or a lower ionic strength the sacs appear to be filled with lamellar mesosomal membranes. Vesicles, 300–400 Å in diameter, are observed in the sacs if the ionic strength of the fixative is increased or alternatively the cells are fixed with a solution of unbuffered osmium tetroxide containing sucrose $(0·05–0·3\ M)$.

Better understanding of the nature of mesosomes under physiological conditions is obtained from freeze-etch studies. This procedure does not permanently damage living cells and the rapid freezing used might be expected to preserve cells in an *in vivo* condition. The mesosomes of *B. subtilis* appeared as collections of vesicles (size 300–1000 Å) implanted into or attached to the cytoplasmic membrane (Remsen, 1968); in contrast to the cytoplasmic membrane both inner and outer surfaces of the vesicles were smooth. Comparable pictures have been obtained by Nanninga (1968). A lamellar-type of mesosome covered with particles similar to those present on the inner side of the cytoplasmic membrane appeared prior to spore formation. Remsen suggests that this mesosome is functionally different from the type containing vesicles.

A disorganization of mesosomes occurs when cells are chilled to 15° (Fitz-James, 1965; Neale and Chapman, 1970). The extent of the disorganization is probably dependent on the nature, function and position of the mesosome in the cell. The mesosomes associated with growing septa are severely disorganized whereas those associated with the nucleus appear intact. Changes in mesosomal structure have frequently been observed when cells are subjected to low doses of antibiotics such as streptomycin, cycloserine (Fitz-James, 1967) or bacitracin (Rieber *et al.*, 1969). A rapid increase in the areas of mesosomes, as observed in thin section, occurred when cells of *Streptococcus faecalis* were starved of valine or threonine for 1–2 hr. (Higgins and Shockman, 1970b).

8

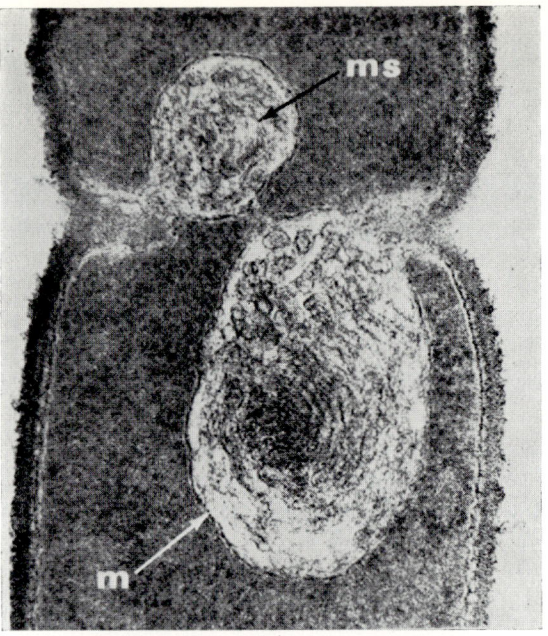

(a)

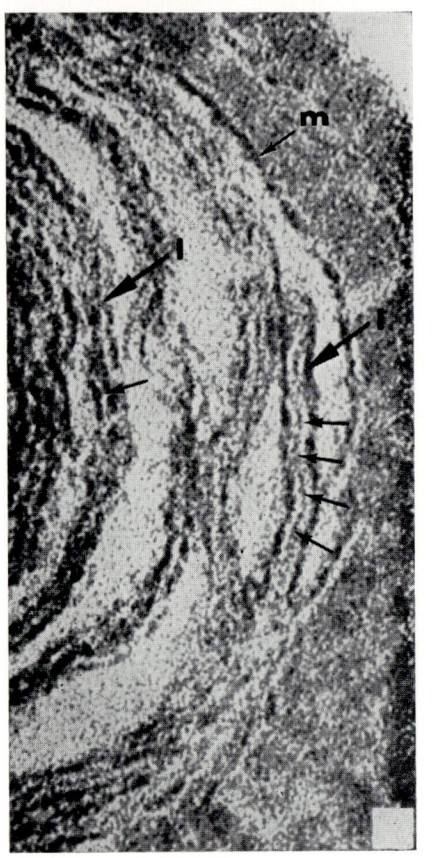

(b)

Fig. 1

WALLS AND MEMBRANES IN BACTERIA 9

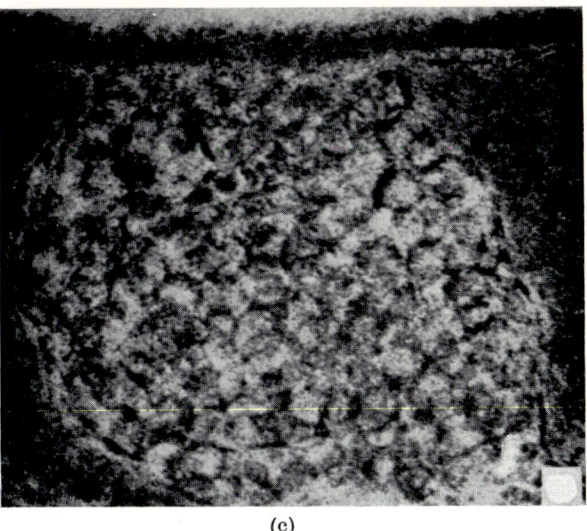

(c)

FIG. 1. Mesosomes of *Bacillus licheniformis*. Photographs are reproduced from Burdett and Rogers (1970). (a) shows a thin section through a cell fixed by the standard Ryter-Kellenberger procedure. Mesosomes (ms) are associated with a forming septum and are filled with tubular structures. The membrane (m) is continuous with the cytoplasmic membrane, and is termed the mesosomal sac. Magnification × 104,800. (b) shows the appearance of a mesosome when the ionic strength of the Ryter-Kellenberger fixative was decreased. Lamellar membranes (l) are present in the mesosome. The smaller arrows point to a fine densely stained line present in the centre of the sheets of membrane and which may be composed of globular units. Magnification × 372,000. (c) shows a thin section through a mesosome when the ionic strength of the fixative is increased by addition of sodium chloride (0.1 *M*). The mesosome appears to be filled with vesicles. Magnification ×184,000.

B. Extrusion of Mesosomal Vesicles and Tubules from Protoplasts

Observations of thin sections of bacilli revealed that, when the cells are placed in a hypertonic medium, the mesosomal sac evaginates (Ryter, 1968; Fitz-James, 1964; Weibull, 1965) to become continuous with the cytoplasmic membrane. The contents of the sac in the form of tubules or vesicles (Ryter, 1968) are extruded into the region between the cell wall and the cytoplasmic membrane. A similar extrusion of mesosomal vesicles was observed on chilling cells (Fitz-James, 1965; Neale and Chapman, 1970) or during cell lysis (Silva, 1967). The phenomenon

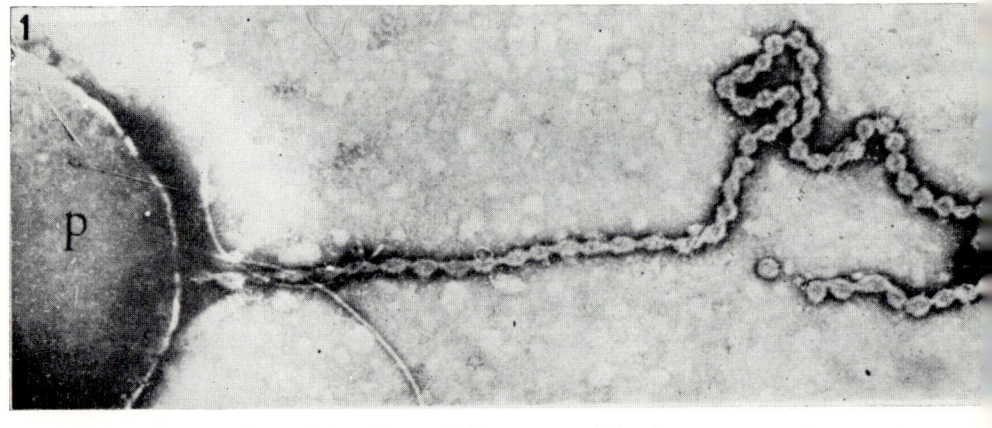

Fig. 2. A protoplast of *Bacillus subtilis* prepared in the presence of magnesium ions (0·2 mM) and examined by negative staining. A long, uncoiled mesosomal tubule remains attached to the protoplast (p) ($\times$45,000). Photograph taken from Ryter *et al.* (1967).

of mesosomal evagination has been demonstrated in several other Gram-positive organisms including *Listeria monocytogenes* (Ghosh and Murray, 1969), *Staphylococcus aureus* (Popkin *et al.*, 1970), *Micrococcus lysodeikticus* (Ellar and Freer, 1969) and *Lactobacillus plantarum* (Kakefuda *et al.*, 1967).

Whether the mesosomal vesicles or tubules remain attached (Fig. 2) to the protoplast after removal of the cell wall or pass into the surrounding medium depends very critically on the concentration of Mg^{2+} in the protoplasting medium (Ryter *et al.*, 1967; Rogers *et al.*, 1967). Low concentrations of Mg^{2+} favour release of mesosomal material. In practice it is necessary to preserve a fine balance between a magnesium ion concentration low enough to ensure the release of the bulk of the mesosomal material and a concentration high enough to prevent lysis of the

protoplast (Reaveley, 1968). The concentration of Mg^{2+} appears to be very dependent on the organism and possibly on the nature of the protoplasting fluid. The mesosomal material can be recovered from the external medium by ultracentrifugation (100,000 g, two hours) after removal of the protoplasts by low-speed centrifugation and purified by density-gradient centrifugation. With the mesosomal material from

TABLE 1. A Comparison of the Chemical and Enzymic Compositions of Mesosomal Material and Cytoplasmic Membranes for several Gram-positive Bacteria.

Organism	Nature of preparation	Composition relative to the cytoplasmic membrane
Bacillus subtilis	Tubules	Much richer in cytochromes (Ferrandes *et al.*, 1966).
Bacillus subtilis	Vesicles	Much richer in phospholipid (Fitz-James, 1967).
Bacillus licheniformis	Vesicles	About the same content of protein, lipid and cytochromes, but complement of proteins different. Much lower in succinate dehydrogenase (5.6) and $NADH_2$ oxidase (27) (Reaveley and Rogers, 1969).
Bacillus megaterium	Vesicles and tubules	Richer in lipid. Lower in succinate dehydrogenase (6.0), malate dehydrogenase and $NADH_2$ dehydrogenase (67). (Ellar, 1969; D. J. Ellar, quoted in Rogers, 1970).
Micrococcus lysodeikticus	Vesicles and tubules	Richer in lipid. Lower in succinate dehydrogenase (3.1) and $NADH_2$ dehydrogenase (48) (Ellar and Freer, 1969; D. J. Ellar, quoted in Rogers, 1970).
Listeria monocytogenes	Vesicles and tubules	About the same content of lipid and protein. Lower in $NADH_2$ oxidase (55) and ATPase (60), higher in nitroblue tetrazolium chloride reductase using succinate (183), $NADH_2$ (350) or lactate (275) as substrate (Ghosh and Murray, 1969).

The figures in brackets express the enzymic activities of mesosomal preparations when the activities of the cytoplasmic membranes are set at 100 units/mg. membrane.

B. licheniformis, two bands each consisting of vesicles 300–3000 Å in diameter surrounded by unit membranes, are produced on caesium-chloride gradients run to equilibrium. Two similar fractions have been isolated from *S. aureus* (Theodore *et al.*, 1970). The chemical and enzymic compositions of some preparations of mesosomal vesicles and tubules produced by comparable methods are given in Table I. Protoplasts prepared from cells of *B. megaterium* (Van Iterson and Op Den Kamp, 1969) and *B. subtilis* (Op Den Kamp, 1967) harvested at acid pH values

12 D. A. REAVELEY AND R. E. BURGE

retain coiled intracellular mesosomal material. Possibly the membranes of these cells are more rigid than those of cells harvested at neutral pH values.

C. MESOSOMES AND SEPTUM FORMATION IN GRAM-POSITIVE BACTERIA

At least one mesosome is present in exponentially growing cells (Rogers, 1970; Ryter, 1968; Highton, 1970a, b; Ellar *et al.*, 1967; Higgins and Shockman, 1970a, b) and a mesosome is always present at the site of a forming septum. The precise position of the septum appears to be dependent on the location of the septal mesosome (Higgins and Shockman, 1970a; Ellar *et al.*, 1967). When septum formation is highly irregular, as in some mutants of *B. subtilis* grown in the absence of sodium chloride (Rogers *et al.*, 1970) or reverting L forms, the mesosomes are of a very unusual type or absent altogether; thus it seems that mesosomes are involved in ordered septum formation. A detailed electron microscope study of synchronously dividing cells of *B. megaterium* (Ellar *et al.*, 1967) reveals that in the initial stages the mesosome develops as a concentric fold of the cytoplasmic membrane at the future site of the septum. The invagination then fills with small vesicles 300–500 Å in diameter and one large vesicle of diameter around 2000 Å. This latter vesicle is characteristic of mesosomes involved in septum formation. During the septum formation the mesosome moves towards the centre of the cell whilst still retaining connection with the extracytoplasmic space by means of a narrow tube extending from the base of the mesosome. The width of this tube narrows as the mesosome moves across the cell, and the cell-wall septum appears to originate in the tube from the original wall by addition of material to the free-growing septum. Prior to completion of the septum the membranes bounding the mesosome fuse to form a large mesosome which is partitioned on either side of the septum. Electron micrographs of other bacilli (Rogers, 1970; Highton, 1970a, b) also show a separation of the mesosome into the two daughter cells on division. With *L. plantarum*, however, two mesosomes, each close to the newly forming septa and in opposite halves of the cell, are required for cell division (Kakefuda *et al.*, 1967). After cell division, the polar mesosome present in each daughter cell of *B. licheniformis* (Highton, 1970a) and *B. subtilis* (Highton, 1970b) travels by a combination of cell growth and mesosomal movement to the new site of septum formation.

D. MESOSOMES AND THE NUCLEUS IN GRAM-POSITIVE BACTERIA

Many electron micrographs of bacteria show that the nucleus is joined to the cytoplasmic membrane and that, in a large number of

cases, this linkage is mediated by a mesosome (Ellar *et al.*, 1967; Van Iterson, 1965; Ryter, 1968; Kakefuda *et al.*, 1967). Frequently, both a polar mesosome and a mesosome present close to the site of a newly formed septum are in contact with the nucleus (Ellar *et al.*, 1967; Kakefuda *et al.*, 1967). The significance of this connection is not clear at the present time, although again it may be that ordered cell division is involved. Ryter and Jacob (1964a) suggest that two separate mesosomes are involved in separating sister nuclei after DNA replication.

In protoplasts, the DNA appears to be linked directly to the cytoplasmic membrane (Ryter and Jacob, 1964b; Ryter, 1968) suggesting that the nucleus is actually linked to the cytoplasmic side of the mesosomal sac. Using an ingenious gradient technique involving the detergent sarkosyl, Trembley *et al.* (1969) were able to isolate from *B. megaterium* a small piece of membrane, presumably the mesosomal sac, linked to virtually all the cellular DNA. This membrane fraction accounted for 10–30% of the total cellular membranes.

E. Intracytoplasmic Membranes in *Escherichia coli*

The demonstration of intracytoplasmic membranes in *E. coli* similar in morphology to the mesosomes present in Gram-positive bacteria has proved particularly difficult.

The difficulty of observing these structures appears to be due to their nature and form which is a delicate and intricate folding of the cytoplasmic membrane (Ryter, 1968). Such folds, which may be coiled or twisted (Pontefract *et al.*, 1969; Fig. 3) are only visible when the section is cut in a plane perpendicular to the fine folds of the membrane. The position of these structures close to the nuclei and in the region of the division constriction makes it seem likely that they are functionally similar to the mesosomes seen in Gram-positive bacteria. Thus they are generally described as mesosomes (Ryter, 1968; Pontefract *et al.*, 1969). On plasmolysis of cells, vesicles are not ejected into the cell wall-membrane interspace (Weibull, 1965) but rather the coils of membrane unfold to become continuous with the cytoplasmic membrane (Ryter, 1968). There seems little doubt that the presence of mesosomes in *E. coli* is dependent on the growth medium (Ryter, 1968). Cota-Robles (1966) observed that cells grown aerobically in a glucose-salts minimal medium contained more invaginations of the cytoplasmic membrane than when glycerol was used as carbon and energy source. Mesosomes are also induced by starvation of cells for magnesium (Fiil and Branton, 1969; Cota-Robles, 1966). The relationship between the intracytoplasmic membranes discussed and the membranous structures which appear when certain thermosensitive mutants of *E. coli* (Hirota *et al.*, 1968;

Weigand *et al.*, 1970) are grown at temperatures in the range of 35–40° is not clear. Possibly the latter structures may arise due to overproduction of cytoplasmic membrane.

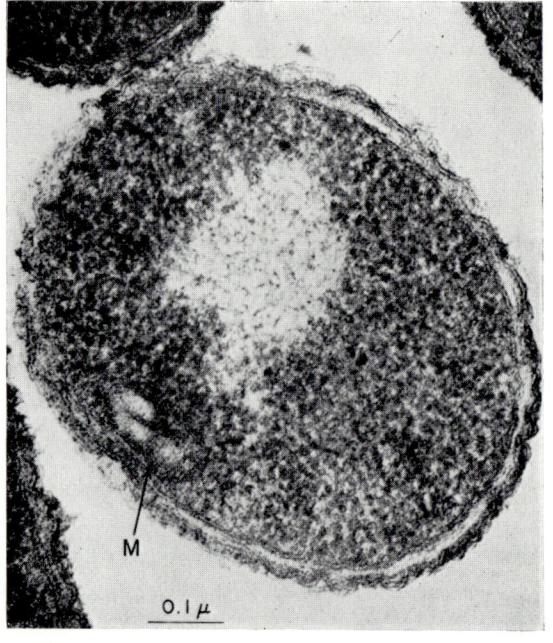

FIG. 3. Mesosomes in *Escherichia coli*. A coiled infolding (**M**) of the cytoplasmic membrane lies very close to nuclear material: ×72,000. The photograph was kindly provided by Dr. R. D. Pontefract.

F. LOCALIZATION OF RESPIRATORY ENZYMES IN BACTERIA

One of the first suggested roles for mesosomes was that they were the sites of electron transport in bacteria. The presence of an energy-producing system close to a newly forming septum or to replicating DNA would be advantageous to the cell. Many cytochemical studies of cells using oxidation-reduction indicators suggest that the enzymes of the electron-transport system, in particular succinate dehydrogenase, are located preferentially in the region of the mesosome (Sedar and Burde, 1965; Leene and Van Iterson, 1965). However the use of oxidation-reduction indicators to localize respiratory-chain enzymes is still a matter of discussion (Gel'man *et al.*, 1967). That mesosomes have other functions is indicated by the observation that mesosomes of *L. plantarum*, an organism which obtains its high-energy compounds from glycolysis, possesses mesosomes similar in type and number to bacilli (Kakefuda

et al., 1967). The number of mesosomes is similar in cells of *S. epidermidis* grown aerobically or anaerobically. The size and appearance of these mesosomes does not change despite the fact that anaerobically growing cells respire at only 10% of the rate of aerobically growing cells and contain less than 15% of the cytochrome content (Conti *et al.*, 1968).

Studies on isolated mesosomal vesicles and tubules have also provided confusing results about the location of the electron-transport system in bacteria. Mesosomal material isolated from *B. licheniformis*, *B. megaterium* and *M. lysodeikticus* is much diminished in succinate dehydrogenase activity compared to the cytoplasmic membrane (Table I, p. 11) and is also lower in $NADH_2$ oxidase or dehydrogenase activities. $NADH_2$ oxidase is also concentrated in the cytoplasmic-membrane fraction of *Listeria monocytogenes* whereas nitroblue-tetrazolium chloride reductase, using $NADH_2$, succinate or lactate as substrate, is found preferentially in the mesosomal material. A preparation of mesosomal tubules from *B. subtilis* was considerably enriched in cytochromes relative to the cytoplasmic membrane (Ferrandes *et al.*, 1966); however comparable preparations from *B. licheniformis* (Reaveley and Rogers, 1969) and *S. aureus* (Theodore *et al.*, 1970) contained roughly the same amounts of cytochromes as the cytoplasmic membrane. Studies by Fitz-James (1968) have shown that ^{59}Fe is preferentially incorporated into mesosomal material of *B. megaterium*, suggestive evidence that mesosomes are the sites of cytochrome synthesis.

G. OTHER PROPOSED FUNCTIONS OF MESOSOMES

Electron-microscope observations have implicated mesosomes in spore formation (Fitz-James, 1960; Mandelstam, 1969) although little or no direct chemical evidence is available to support these suggestions.

The proposal that mesosomes are the sites of synthesis of some components of the cytoplasmic membrane is supported by some radioactive labelling experiments. The evidence for cytochrome synthesis within the mesosome has already been given. Pulse-labelling experiments with exponential-phase cells of *B. megaterium* using ^{14}C-acetate and ^{32}P as lipid labels demonstrated that the lipid label first appears in the mesosomal fraction and may then be "chased" into the cytoplasmic membrane (Fitz-James, 1968). Ghosh and Murray (1969) also showed that the phospholipids present in mesosomal vesicles and tubules of *L. monocytogenes* are preferentially labelled during the exponential phase of growth. Radioactive amino acids, on the other hand, are not preferentially taken up by mesosomal vesicles and tubules (Fitz-James, 1967). These results could be interpreted as showing that mesosomes are the major sites in the cell for the synthesis of membrane phospholipid,

but not for membrane protein. However, more detailed studies involving rates of turnover of mesosome and cytoplasmic membrane components are required.

That mesosomes play a part in the secretion of enzymes from bacterial cells has been suggested by Lampen (1965). Typical of such enzymes is penicillinase, a protein of molecular weight 28,000 which may be secreted from some strains of *B. licheniformis*. A penicillinase-inducible strain (749) normally only possesses a single septal mesosome; however when penicillinase production is induced, complex membranous systems of vesicles and tubules are produced (Ghosh *et al.*, 1968). Similar, but more extensive, structures are present in a magnoconstitutive strain (749/C) derived from the 749 strain. On induction with cephalosporin, cells of strain 749 first produce intracellular vesicles; the attachment of these vesicles to the surface membrane is only rarely seen. Subsequently the cytoplasmic membrane invaginates and the invagination fills with vesicles and tubules. These vesicles and tubules are extruded on proto-plast formation. About 60% of the cell-bound penicillinase is liberated from strain 749/C on protoplast formation (Sargent *et al.*, 1968), approximately one half of which can be deposited by centrifugation (65,000 *g*, 20 hr.) in a fraction consisting exclusively of tubules and vesicles. The penicillinase activity of the tubules and vesicles is six times greater than in the protoplast fraction. Thus, it seems likely that the vesicles and tubules are the principal site of bound penicillinase in the cell. The role which these structures play in the actual secretion process is not clear. Growing protoplasts, devoid of vesicles and tubules, will synthesize and secrete penicillinase (Sargent *et al.*, 1969). The isolated vesicles and tubules from penicillin-secreting cells are morphologically very similar to mesosomal material isolated from other cells. Whether the invaginations of the cytoplasmic membranes filled with vesicles and tubules in the penicillinase-secreting cells should be regarded as mesosomes is a matter for conjecture. That they are functionally different from polar and septal mesosomes is also evident from the impossibility of demonstrating a connection with the nucleus (Sargent *et al.*, 1968).

III. The Cytoplasmic Membranes of Gram-Positive Bacteria

A. INTRODUCTION

A principal function of the cytoplasmic membrane is to control the internal environment of the cell. This is achieved, as with other plasma membranes, by preventing the diffusion of certain solutes between the cytoplasm and the extracellular region and by catalysing the trans-location of solutes and chemical groups from one side of the membrane to the other. The absence of intracellular organelles in bacteria dictates

WALLS AND MEMBRANES IN BACTERIA 17

that the functions undertaken by the membranes of such organelles must be carried out by the bacterial cytoplasmic membrane or mesosomes. As was apparent when discussing the role of mesosomes in electron transport, it is not possible at present to allocate specific roles to mesosomes and cytoplasmic membranes, but clearly the cytoplasmic membrane must be a multifunctional structure. In thickness and overall composition, in particular lipid composition (Korn, 1969), the cytoplasmic membrane of Gram-positive bacteria bears a striking resemblance to the inner mitochondrial membrane.

B. Morphology

In thin section, the cytoplasmic membrane of the Gram-positive bacterium normally presents the characteristic trilamellar appearance typical of all natural membranes (Korn, 1969), although in a few instances the structure appears five-layered (Robinow, 1962; Suganuma, 1966). The profile depends on the growth phase and may be symmetric or asymmetric (Glauert, 1962). It is not clear whether the demonstration of an asymmetric profile in thin section is indicative of an asymmetry of composition. A change from an asymmetric to a symmetric profile is associated with cell lysis (Silva, 1967). Frequently the inner densely stained layer is masked by the cytoplasm (Burdett and Rogers, 1970; Kakefuda et al., 1967); this layer is seen more clearly in cells when the wall and membrane are separated. As with mitochondrial membranes (Fleischer et al., 1966) the trilamellar appearance is retained after almost complete extraction of the membrane lipid (Grula et al., 1967).

When estimated from electron micrographs of thin sections of bacteria, the thickness of the cytoplasmic membrane varies from 65–70 Å in *Bacillus licheniformis* (Highton, 1969; Burdett and Rogers, 1970) to 85–100 Å in *Listeria monocytogenes* (Ghosh and Carroll, 1968). The thickness of the cytoplasmic membrane of *Streptococcus faecalis* increases when the cells are starved of amino acids (Higgins and Shockman, 1970b). Thick cytoplasmic membranes (about 100 Å) are also present in some morphological mutants of *B. subtilis* (Cole et al., 1970; Rogers et al., 1970). Thickness values obtained by X-ray diffraction of wet suspensions of the cytoplasmic membranes of *B. licheniformis* were considerably greater than values obtained from electron micrographs of thin sections (105 Å against 65–70 Å; D. A. Reaveley and R. E. Burge, unpublished results), indicative of a shrinkage of membrane during fixing and sectioning for electron microscopy.

Structures akin to the stalked particles seen by negative staining on the surfaces of mitochondrial cristae (Fernández-Moran et al., 1964) are present on the cytoplasmic side of the membrane of several bacilli

(Abram, 1965) and *Micrococcus lysodeikticus* (Munoz *et al.*, 1968b). The stalked particles of *M. lysodeikticus*, like the analogous particles of the mitochondrion, appear to be associated with an ATPase. The headpiece is approximately 100 Å in diameter and consists of a central unit surrounded by six additional units. The particles of bacilli are somewhat smaller (65–85 Å in diameter) and loosely attached to the membrane by stalks 40–60 Å long.

Such units are not seen on the inner surface of the cytoplasmic membrane of *B. subtilis* as revealed by freeze-etching (Nanninga, 1968). Instead the surface is sparsely covered with particles of diameter about 50 Å; other freeze-etch studies of bacilli (Holt and Leadbetter, 1969) reveal that the inner surface is sparsely covered with particles of diameter 120 Å. The outer surface (Nanninga, 1968) is covered with 50 Å diameter granules and many short strands of material, the latter occasionally entering the cell wall. Caution must be used at the present time in interpreting freeze-etch micrographs especially when, as here, there is an obvious discrepancy with other electron-microscope techniques. Both Nanninga (1968) and Holt and Leadbetter (1969) have interpreted their micrographs by assuming that fracturing occurs along the inner and outer surfaces of the cytoplasmic membrane; thus they assumed that the surface structures of the cytoplasmic membrane were revealed. In contrast, Fiil and Branton (1969), who studied the cytoplasmic membrane of *Escherichia coli*, were not able to distinguish clearly the fracture plane. They considered the additional possibility that fracture might occur at a plane within the membrane rather than at a membrane surface. It seems clear from ferritin-labelled antibody experiments with red-blood cells that, in this case at least, the fracturing takes place within the plane of the membrane (Da Silva and Branton, 1970).

C. Isolation of the Cytoplasmic Membranes of Gram-Positive Bacteria and Requirement for Magnesium

Intact cytoplasmic membranes, due to their delicate nature, may only be produced from organisms whose cell wall may be removed enzymically. The methods employed basically take the form devised by Weibull (1953) for *B. megaterium*; the cell wall is removed enzymically in hypertonic medium and the resulting protoplast gently lysed by a decrease in tonicity. Weibull (1956) reported that intact cytoplasmic membranes are only obtained when magnesium ions are present in the lysing solution. This requirement for magnesium ions, though depending absolutely on the organism, has been confirmed many times (Rogers *et al.*, 1967; Ghosh and Carroll, 1968; Bodman and Welker, 1969; Fitz-James, 1968).

With the discovery of the mesosomal membrane system in bacteria it became clear that many of the early preparations of cytoplasmic membranes were contaminated with mesosomal vesicles, tubules and lamellar membranes (Salton, 1967a). Since mesosomes have been ascribed specific cellular functions, and from freeze-etch pictures appear to be morphologically different from cytoplasmic membranes, it became increasingly important to separate the two systems. Homogeneous preparations of mesosomal material can be easily obtained but the complete removal of adhering mesosomal tubules and vesicles from the cytoplasmic membrane has proved difficult if not impossible (D. A. Reaveley, unpublished results; Ghosh and Murray, 1969; Fitz-James, 1968). Another complicating factor is the mesosomal sac, which although probably functionally different from the rest of the cytoplasmic membrane (Nanninga, 1968), becomes continuous with this membrane on protoplast formation.

Purified preparations of intact cytoplasmic membranes are heterogeneous when examined by density-gradient centrifugation (Salton et al., 1968; Reaveley and Rogers, 1969). The membranes in the various fractions differ slightly in their lipid:protein ratios. Such variations can be effected by changes in the composition of the growth medium or by altering the concentration of magnesium ions in the solutions used either to lyse the protoplasts or wash the cytoplasmic membranes (Reaveley and Rogers, 1969).

D. Association of Ribosomes with the Cytoplasmic Membrane

The bulk of the RNA associated with the bacterial cytoplasmic membrane resembles ribosomal-RNA (Yudkin and Davis, 1965; Abrams et al., 1964) and electron micrographs show clusters of ribosomes attached to the membrane surface (Pfister and Lundgren, 1967; Abrams et al., 1964). It has been frequently suggested that membrane-bound polyribosomes are the major sites of protein synthesis in bacteria. The close proximity of an energy-generating system in the cytoplasmic membrane coupled with the possible presence of membrane-bound stimulating factors might be advantageous for protein synthesis. Whether the association is confined to specific regions of the membrane is not known, but disaggregated membranes bind ribosomes as efficiently as intact cytoplasmic membranes (Aronson, 1966).

A review of the literature reveals that there are large differences in the RNA contents of isolated cytoplasmic membranes (Bishop et al., 1967a; Salton and Freer, 1965; Coleman, 1969). Part of the explanation for these differences lies in the compositions of the solutions used to prepare and lyse protoplasts (Van Dijk-Salkinoja et al., 1970; Patterson

et al., 1970) and to wash the cytoplasmic membranes. Magnesium ions are critical for the binding of ribosomes to membranes for they serve as bridges between negatively charged groups on the two fractions (Coleman, 1969). A lowering of the magnesium-ion content from 20 mM to zero of solutions used to wash the cytoplasmic membranes of *Bacillus licheniformis* decreases the RNA content of the membranes from 13–15% to 2–3% (Rogers *et al.*, 1967). The effect of magnesium ions has been shown to be antagonized by potassium ions (Coleman, 1969); an increase in the potassium-ion concentration from zero to 100 mM in the presence of 10 mM magnesium ions decreases the amount of membrane-bound ribosomal material seven-fold in *B. amyloliquefaciens*. A further complicating factor has been a tendency to compare RNA contents of membranes from cells in different phases of growth. There seems little doubt that cytoplasmic membranes prepared from cells harvested in the exponential-phase of growth have more associated ribosomes than membranes from stationary-phase cells.

E. Yield and Composition of Cytoplasmic Membranes

The cytoplasmic membrane-mesosome system of Gram-positive bacteria normally accounts for 16–30% of the cell dry weight depending upon the organism (Salton and Freer, 1965). Bacterial cytoplasmic membranes consist almost entirely of protein and lipid (Table 2). The total lipid content of the cytoplasmic membrane is affected by the growth phase; membranes prepared from cells of *S. faecalis* harvested in the exponential phase of growth contain 28% lipid compared to 40% present in cells from the stationary phase (Shockman *et al.*, 1963; Higgins and Shockman, 1970b).

Recent gel electrophoresis experiments have shown that many different proteins are present in the cytoplasmic membrane (Salton *et al.*, 1967; Reaveley, 1968) but studies on bacterial membrane proteins have been few in number; even an accurate assessment of the number of different proteins present in a cytoplasmic membrane has yet to be made. Part of the difficulty is caused by the lack of techniques available at the present time to investigate proteins which are insoluble in an aqueous environment. Treatment of membranes with a detergent such as sodium dodecyl sulphate is normally necessary before it is possible to use such techniques as gel electrophoresis. Mirsky (1969) counted 12 distinct bands in gels of proteins from *B. megaterium*, but this figure has been exceeded by a factor of three for *B. licheniformis* by use of a more sophisticated electrophoresis procedure (Fig. 4). A rather interesting organism derived from *B. megaterium* has recently been investigated by Patterson and Lennarz (1970). Over 90% of the membrane protein of this bacillus consists of a

TABLE 2. Composition of the Cytoplasmic Membranes of some Gram-positive Bacteria.

Organism	Per cent protein[a]	Per cent lipid	Per cent composition of lipid			Principal phospholipids	Reference
			neutral lipid	glyco-lipid	phospho-lipid		
Bacillus licheniformis[b]	43–49 (L)	18–25	30	14	40	Phosphatidylethanolamine	Rogers et al. (1967) Reaveley (1968)
Bacillus subtilis[b]	55–64 (L)	10–19	11	12	77	Diphosphatidylglycerol (49%), phosphatidylglycerol (11%), phosphatidylethanolamine (34%), lysophosphatidylglycerol (6%)	Bishop et al. (1967a)
Streptococcus faecalis[b]	65–68 (B)	23–26	22	23	54	Diphosphatidylglycerol	Ibbott and Abrams (1964)
Micrococcus lysodeikticus[b]	49 (L)						Salton (1967b)
Sarcina lutea[b]	57 (B)	23					
Staphylococcus aureus H[c]	63–67 (B)	22–25	15	8	69	Phosphatidylglycerol and diphosphatidylglycerol	Salton (1967b) Ward and Perkins (1968)
Staphylococcus aureus 100[c]	62–71 (B)	22–25	18	10	68	Phosphatidylglycerol and diphosphatidylglycerol	Ward and Perkins (1968)
Listeria monocytogenes[b]	55–60 (L)	30–35					Ghosh and Carroll (1968)

[a] Protein estimated by the method of Lowry et al. (1951; L) or the biuret method (B).
[b] Culture harvested in the stationary phase.
[c] Culture harvested in the late exponential phase.

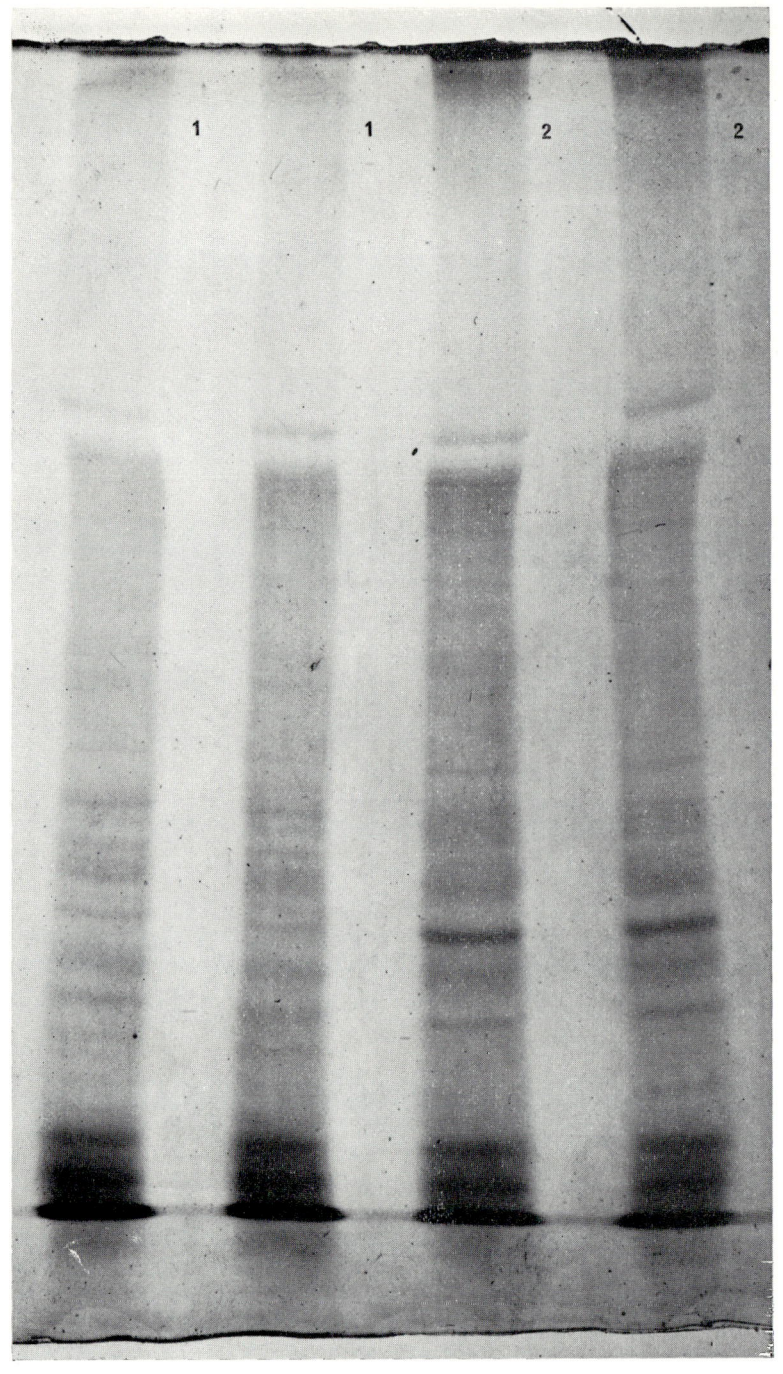

WALLS AND MEMBRANES IN BACTERIA 23

single protein. This protein is present as a minor component in the cytoplasmic membrane of the parent *B. megaterium*; otherwise the protein
composition of the two membranes is similar. The molecular weights of
the proteins and polypeptides of the cytoplasmic membrane of *B.
licheniformis* range from 20,000 to 160,000 with principal components
having molecular weights around 50,000 (D. A. Reaveley, unpublished
results). Therefore, the molecular weights are similar to those of proteins
present in other natural membranes and a protein fraction has been
isolated from the bacterial cytoplasmic membrane which is similar to
the structural protein fraction of the mitochondrion (Mirsky, 1969).

There have been relatively few detailed lipid analyses of the cytoplasmic membranes of Gram-positive bacteria (Table 2). This lack of
information is partially compensated by information available on the
lipid compositions of whole cells, since it is clear that in many cases
virtually all the cell lipid is present in the cell membranes (Vorbeck and
Martinetti, 1965; Rogers *et al.*, 1967). Unfortunately mesosomal lipids
are included in total cell lipids and the possibility that mesosomes have a
lipid composition different from that of cytoplasmic membranes has yet
to be investigated. The non-nitrogenous lipids, diphosphatidylglycerol and phosphatidylglycerol, generally predominate in Gram-
positive bacteria with the exception of the bacilli (and clostridia) which
contain appreciable amounts of phosphatidylethanolamine (Table 2).
Diphosphatidylglycerol appears to be rather tightly bound to the
membrane in Bacillacae (Yudkin, 1966; Bertsch *et al.*, 1969; Lang and
Lundgren, 1970). Glycosyl diglycerides are widely distributed in Gram-
positive bacteria where their presence is of some taxonomic value (Shaw
and Baddiley, 1968). They are also present in mycoplasmas but are
invariably absent from Gram-negative organisms. So far only three
sugars (glucose, galactose and mannose) have been positively identified
in the carbohydrate moiety, which is principally a disaccharide.

The levels of phospholipids in a cell depend upon the phase of growth
in which the cells are harvested (Lang and Lundgren, 1970; Card *et al.*,
1969; Kocun, 1970) and on the composition of the growth media and the
growth conditions (Card, 1969; Frerman and White, 1967). The appearance of the glucosaminyl derivative of phosphatidylglycerol when cells

---

Fig. 4. Polyacrylamide gel electrophoresis of cytoplasmic-membrane proteins of
Bacillus licheniformis. The membranes were disaggregated by sodium dodecyl
sulphate (0·1%, pH 7·6) and dialysed overnight against the disaggregating solution.
Sodium dodecyl sulphate (0·1%) was incorporated into the gel (7% acrylamide)
and the running buffer (pH 8·3). Protein bands were visualized by means of
amido black. Preparations 1 and 2 were two batches of membranes prepared from
stationary-phase cells. At least 30 separate protein bands are present in each
preparation (D. A. Reaveley, unpublished results).

of *B. megaterium* are harvested at pH 5 has already been discussed. Similar experiments have been performed with *Streptococcus faecalis* and *Staphylococcus aureus* (Houtsmuller and Van Deenen, 1964, 1965); in these cases the lysyl ester of phosphatidylglycerol accumulates at acid pH values. The amount of lysyl derivative of phosphatidylglycerol accumulating when *Bacillus subtilis* is grown in a glucose-containing medium is less than for *S. aureus*. This is attributed to a smaller decrease of pH value of the medium (7·0 to 5·5 compared to 7·0 to 4·8) during the growth of *B. subtilis* (Op Den Kamp *et al.*, 1969). The process of formation of aminoacyl esters of phosphatidylglycerol in the cell membrane may be more complex than a simple response of the cell to a more acidic pH value. *Streptococcus faecalis*, when grown by Kocun (1970), contained an alanyl ester of phosphatidylglycerol as the principal phospholipid irrespective of pH value, but a lysyl ester was formed as a response to a decrease in pH value. [^{14}C]lysine was incorporated into this lipid but, rather surprisingly, [^{14}C]alanine was not incorporated into the alanyl ester.

The change in phospholipid composition of *Staphylococcus aureus* has been followed during the transition from anaerobic to aerobic growth. The change is accompanied by the formation of a membrane-bound electron transport system (Frerman and White, 1967). As a consequence of the shift from anaerobic growth the amount of phosphatidylglycerol in the cell increases two-fold whereas diphosphatidylglycerol increases 1·6-fold and the amount of lysylphosphatidylglycerol remains constant. Apparently the cytoplasmic membrane and other cell membranes are able to carry out specific functions, such as transport and cell-wall synthesis, with widely varying ratios of individual phospholipids. In this context it is interesting to note that a mutant of *B. subtilis* has been isolated which is deficient in phosphatidylethanolamine but contains the other lipids characteristic of the organism (Beebe, 1970). The ratios of these lipids (phosphatidylglycerol, diphosphatidylglycerol, lysylphosphatidylglycerol and diglycosyldiglyceride) differ from those of the parent strain presumably to compensate for the loss of phosphatidylethanolamine. The mutant organism appeared to possess different rates of uptake for the amino acids aspartic acid and lysine compared to the parent organism.

The principal fatty acids present in the cytoplasmic membranes of *B. licheniformis*, *M. lysodeikticus*, *Sarcina lutea* (Cho and Salton, 1966), *Staphylococcus aureus* (Ward and Perkins, 1968), *B. subtilis* (Bishop *et al.*, 1967a), and *B. megaterium* (Yudkin, 1966), are C_{15} branch-chain acids. In the cases quoted the anteiso acid predominated, but in other instances the iso acid may be in excess (Kaneda, 1967; Moss *et al.*, 1969; Shen *et al.*, 1970); the ratio of anteiso to iso depends on the availability

WALLS AND MEMBRANES IN BACTERIA

of precursors in the growth medium (Kaneda, 1967; Tornabene et al., 1967). The predominant fatty acids of thermophilic bacilli have slightly longer chain lengths than those of the mesophiles (Cho and Salton, 1966; Daron, 1970; Shen et al., 1970); yet increasing the growth temperature for a particular organism does not significantly increase the average chain length of the fatty acids, although the amount of branching decreases as the temperature increases (Daron, 1970). Lactobacilli and streptococci contain high proportions of normal even-numbered saturated and unsaturated acids and low amounts of odd-numbered branch-chain acids (Veerkamp, 1970).

F. Disaggregation and Re-aggregation of Cytoplasmic Membranes

With appropriate detergents in relatively high concentrations, complete or almost complete solubilization of most cytoplasmic membranes can easily be achieved. Sodium dodecyl sulphate generally appears to be the most effective detergent for bacilli (Yamaguchi et al., 1967; D. A. Reaveley, unpublished results). The non-ionic detergent, Nonidet P40 (a polyoxyethylated alkyl phenol), is more effective (95% solubilization) than sodium dodecyl sulphate (75% solubilization) in solubilizing the cytoplasmic membrane of M. lysodeikticus (Salton and Netschey, 1965). Cationic detergents are ineffective in solubilizing membranes (Yamaguchi et al., 1967; Salton and Netschey, 1965).

The process of solubilization of components of a cytoplasmic membrane by detergents is not random (Bishop et al., 1967b; Yu and Wolin, 1970). Working with concentrations of sodium dodecyl sulphate (0·05–0·10%) insufficient to give complete disaggregation of the cytoplasmic membranes of B. subtilis, Bishop et al. (1967b) demonstrated that, at the lower detergent concentration, the greater part of the membrane $NADH_2$-2,6-dichlorophenolindophenol oxido-reductase is solubilized whereas the succinate dehydrogenase remains insoluble. In the detergent concentration range 0·05–0·10%, protein is preferentially solubilized from the membrane to leave a lipid-rich residue. On the contrary, when the cytoplasmic membranes of M. lysodeikticus are treated with deoxycholate, a lipid-depleted residue (95–97% protein) is produced which accounts for 15% of the membrane (Salton et al., 1968). Virtually all the succinate dehydrogenase and cytochromes of the membranes are present in this residue which is in the form of membranous sheets.

Membranes disaggregated with detergents sediment in the analytical ultracentrifuge as though they consisted of a single homogeneous component (Salton and Netschey, 1965). The sedimentation coefficients of the complexes are in the range 0·7–3·1, and depend upon the detergent used for disaggregation and also to a lesser extent on the organism from

26   D. A. REAVELEY AND R. E. BURGE

which the membranes were obtained. When disaggregated ultrasonically, material with a sedimentation coefficient of 4·2 is produced from the membranes of *M. lysodeikticus*. The obvious conclusion that these peaks contain homogeneous lipoprotein subunits from which the cytoplasmic membrane is constructed was disproved when it was shown that lipid and protein are separable in sodium dodecyl sulphate-membrane mixtures by polyacrylamide gel electrophoresis (Salton and Schmitt, 1967; Rottem *et al.*, 1968) or by density-gradient centrifugation (Engelman *et al.*, 1967; Rodwell *et al.*, 1967).

The requirement of cytoplasmic membranes for magnesium ions has already been mentioned. When membranes are washed with solutions preferably of low ionic strength, not containing magnesium ions, various membrane proteins may be solubilized (Munoz *et al.*, 1968a; Abrams and Baron, 1968). Some examples are given in Section III, G, p. 27. Another way in which selected components may be solubilized from membranes is to resuspend the membranes in buffers of alkaline pH values. Thus, little material is removed from the cytoplasmic membrane of *B. licheniformis* at pH 5·8 but at pH 9·0 38% of the preparation is not deposited at 50,000 *g*. The material sedimented at 50,000 *g* has the form of the intact membranes but possesses a smaller complement of proteins (D. A. Reaveley, unpublished results). Broadly similar techniques have been used to solubilize proteins from *B. megaterium* (Eisenberg *et al.*, 1970a; Mizushima *et al.*, 1966).

Cytoplasmic membranes of *M. lysodeikticus* disaggregated with sodium dodecyl sulphate will re-aggregate on removal of detergent in the presence of magnesium ions. Reaggregation is maximal at pH 7·4, low ionic strength and 37°. An examination by negative staining and thin-section techniques in the electron microscope reveals that the re-aggregated membranes are very similar to whole membranes (Butler *et al.*, 1967). However, unlike whole membranes, the re-aggregated membranes are disaggregated to a significant extent by cationic detergents, suggesting a structural difference between the two membranes. Lipid-depleted membranes, disaggregated by sodium dodecyl sulphate, will re-aggregate into membranous sheets giving a characteristic three-layered appearance in thin section, but the process is considerably enhanced by the presence of phospholipids (Grula *et al.*, 1967). Not all membrane proteins take part in the re-aggregation of lipid-depleted membranes since the sheets produced have a significantly different amino-acid composition from sheets produced from unextracted membranes.

G. ENZYMES OF THE CYTOPLASMIC MEMBRANE

Although enzymes involved in such important cellular functions as oxidative phosphorylation, active transport and cell-wall biosynthesis

WALLS AND MEMBRANES IN BACTERIA

(Rogers and Perkins, 1968) are present in the cytoplasmic membrane, remarkably little is known as to how these enzymes fit into the membrane and perform their specific roles. Perhaps the most intensely studied membrane-bound enzyme is ATPase which in many cases can be rather easily detached from the membrane. A large part of the enzyme is released from the cytoplasmic membrane of *B. megaterium* by adjusting the pH value of a membrane suspension to 8·5 and dialysing against water (Mizushima *et al.*, 1966); better results are obtained if dialysis is against 10 mM-tris-HCl buffer at pH 7·2 (Ishida and Mizushima, 1969a). The properties of the solubilized ATPase differ from the membrane-bound enzyme particularly with respect to calcium activation and cold inactivation. Most of the ATPase activity in the crude extract will combine with depleted membranes in the presence of magnesium or calcium ions. Upon recombination, part of the activity (10–40%) is once again protected from cold inactivation (Ishida and Mizushima, 1969b). Two ATPases, both stimulated by calcium and magnesium, are apparently present in the crude extract. Calcium ions are more effective than magnesium ions for recombination of the purified enzymes (Ishida and Mizushima, 1969b).

The cytoplasmic membranes of *M. lysodeikticus* contain a latent ATPase which is activated by trypsin treatment (Munoz *et al.*, 1969) and possesses properties very reminiscent of mitochondrial ATPase. The enzyme, which accounts for approximately 10% of the total membrane protein, is released in an active calcium-dependent form by washing the membranes with solutions of low ionic strength (3 mM-tris-HCl, pH 7·6) not containing magnesium ions. It appears to have a subunit structure. Both the *M. lysodeikticus* ATPase and the ATPases from *B. megaterium* are associated with spherical particles having diameters of approximately 100 Å. Neither ATPase is stimulated by sodium or potassium. The ATPase of the fermentative organism, *S. faecalis*, remains attached to the membrane after 6–7 washes with tris buffer, but during the next two washes there is a sudden almost quantitative release of the enzyme which is believed to be caused by a leaching out of magnesium or other cations from the membrane (Abrams and Baron, 1968). The enzyme, which accounts for about 2% of the total membrane protein, is acidic, has a molecular weight of 350,000, and consists of non-identical subunits. The recombination of the enzyme with depleted membranes depends upon the concentration of free ATPase, and reaches a maximum in the presence of a large excess of the enzyme. Magnesium ions bind the ATPase tightly to the membrane assuring that the complex is stable in the absence of free ATPase; presumably two membrane-binding sites are involved. Native membranes will not bind the enzyme and their activity is similar to reconstituted membranes. The ATPase

28 D. A. REAVELEY AND R. E. BURGE

is believed to be involved in the active transport of cations and it is rather interesting that this ATPase will, in the presence of magnesium, sodium and potassium ions at concentrations optimum for enzyme activity, interact with artificial lipid bilayers to produce a 10^2–10^4 increase in electrical conductance (Redwood *et al.*, 1969). Such systems should prove useful for examining structure-function relationships in membranes.

Both $NADH_2$ oxidase and succinate dehydrogenase are membrane-bound enzymes, and their specific activities may be increased by low concentrations of sodium deoxycholate (Eisenberg *et al.*, 1970a; Owen and Freer, 1970). $NADH_2$ oxidase activity is unstable and depends on the state of the membrane particularly with respect to the magnesium ion concentration used for preparation (Eisenberg *et al.*, 1970a; Reaveley and Rogers, 1969). Part of the oxidase system, a $NADH_2$-2,6-dichloro-phenolindophenol reductase, is solubilized from the cytoplasmic membrane of *B. megaterium* with 0·2%-sodium deoxycholate, 8 M-urea or at pH 9·0 in the presence of EDTA; such treatments inactivate the $NADH_2$ oxidase. The oxidase activity can be restored (Eisenberg *et al.*, 1970b) by dilution of the deoxycholate-treated membranes in the presence of divalent cations. When deoxycholate-treated membranes are centrifuged at 100,000 g for one hour, all the $NADH_2$ oxidase which can be activated by cations is present in the supernatant fluid and can be recovered by dilution in an insoluble form. Recovery of activity from the supernatant fluid is not good, but nonetheless it is clear that at least part of the entire electron-transport system from $NADH_2$ to oxygen can be solubilized, re-activated and isolated in an active form.

The procedure used to solubilize ATPase from the cytoplasmic membrane of *B. megaterium* solubilizes two-thirds of the membrane $NADH_2$-2,6-dichlorophenolindophenol oxido-reductase (Mizushima *et al.*, 1966); the remaining one third is more firmly bound to the membrane. The purified enzyme will not recombine with the depleted membrane even in the presence of magnesium ions (Mizushima, 1968) until the membrane suspension has been adjusted to pH 4·0. Treatment of membranes with dilute hydrochloric acid is also effective in ensuring recombination, but this destroys the residual enzyme activity of the membrane. Rather surprisingly the activity of the solubilized enzyme is not affected by hydrochloric acid even when bound to the membrane, an indication that there may be two different $NADH_2$-2,6-dichloro-phenolindophenol oxido-reductases in the membrane.

H. STRUCTURE OF THE CYTOPLASMIC MEMBRANE

Structural studies on biological membranes have generally been confined to membranes amenable to study by physical methods;

WALLS AND MEMBRANES IN BACTERIA 29

typical of such membranes are the myelin membranes and the membranes of the retinal rod outer segments. Other membranes which have attracted attention are those of unusual biochemical interest as for instance the mitochondrial and chloroplast membranes. Structural information about the membranes of micro-organisms has tended to be produced as an offshoot of such studies. Recently however an increased interest has been taken in the cytoplasmic membranes of mycoplasmas. These organisms which are sensitive to osmotic lysis are surrounded simply by the cytoplasmic membrane and possess no intracellular membranes. In this respect they resemble bacterial protoplasts although they are more resistant to osmotic lysis. There is a great resemblance between the cytoplasmic membranes of Gram-positive bacteria and the membranes of mycoplasmas; the phospholipids are very similar and both contain glycolipids of the diglycosyldiglyceride type. Also, both the membranes of Gram-positive bacteria and mycoplasmas are disaggregated by sodium dodecyl sulphate and will re-aggregate on dialysis in the presence of magnesium ions into membranous material. Unlike bacteria, most mycoplasmas require cholesterol for growth and this is incorporated unchanged into the cytoplasmic membrane. *Mycoplasma laidlawii* does not require cholesterol for growth and incorporates only small amounts into the cytoplasmic membrane. For a review on the membranes of mycoplasmas, see Razin (1969).

Mycoplasmas will incorporate fatty acids added to the growth medium in an unchanged state into the membrane lipids. Consequently it is possible to produce mycoplasmas with membranes containing large amounts of specific fatty acids. Differential thermal analysis shows that the membranes of *M. laidlawii* (Steim *et al.*, 1969) exhibit an endothermic transition at a temperature dependent on the fatty acid present in the membrane lipids. Similar transitions, at the same temperature, are observed with aqueous dispersions of membrane lipids and in viable organisms (Reinert and Steim, 1970). These observations have been interpreted in terms of a change of packing of the fatty acids within a lipid bilayer from a crystalline to a more fluid state. The thermal transition has also been detected by X-ray diffraction (Engelman, 1970) and was explained in terms of a transition of fatty acyl chains from a closely packed hexagonal array to a more fluid state. Engelman suggests that, below the transition temperature, the lipids are present in a monolayer or bilayer, and that this general configuration is retained at the higher temperature. The 'melting' of the fatty acid chains is not observed in the red-blood cell membrane. It will be recalled however that these membranes contain cholesterol in a roughly equimolar amount to phospholipid, which in some indefined way freezes the fatty-acid chains (Glaser *et al.*, 1970).

The simple membrane model proposed by Danielli and Davson (1935), in which a lipid bilayer is covered with protein which is bound ionically to the polar headpieces of the lipid molecules, fails to account for much recent data. Amongst these data is the known importance of protein-lipid hydrophobic interactions in membranes (Green and Perdue, 1966; Vanderkooi and Green, 1970), the observations that, in mitochondrial (Vanderkooi and Green, 1970) and red-blood cell membranes, phosphatidylcholine is attacked by phospholipase C (Glaser *et al.*, 1970) without degradation of the membrane, and the repeated extraction of lipids from many membranes without change in thickness or appearance of the membrane in thin section. To explain these observations, Vanderkooi and Green (1970) propose that a membrane consists of spherical protein units packed in two layers with only few points of

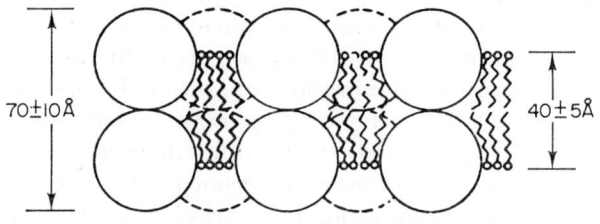

FIG. 5. Cross section of a membrane (after Vanderkooi and Green, 1970). The gaps between the protein molecules (large circles) are filled with phospholipid in the form of a bilayer with the polar headpieces located on the membrane surfaces. The dashed circles represent proteins lying behind the plane of the section and which are in contact with the solid circle proteins.

contact between adjacent units. In the cavities between the protein units, which are believed to be lined predominately with non-polar amino acids, lipid molecules are situated in the form of a bilayer with the polar headpieces on the surface of the membrane (Fig. 5). Such a double layer of globules has been detected occasionally in thin sections of mitochondrial membrane (Crane and Hall, 1969). The essentially static model of Vanderkooi and Green (1970) has been produced in a dynamic form for the retinal rod outer segment membrane (Vanderkooi and Sundaralingam, 1970). Here the protein molecules are thought to float, half submerged in a sea of lipid, and thermal motion takes place in the plane of the membrane.

IV. Cell Walls of Gram-Positive Bacteria

A. INTRODUCTION

The cell walls of several Gram-positive bacteria have been intensively studied over the past ten years (for reviews see Rogers, 1970; Ghuysen, 1968; Osborn, 1969). Peptidoglycan is always present in Gram-positive

cell walls and in some cases may comprise as much as 90% of the dry weight of the cell wall. The rest of the cell wall is usually made up of protein, polysaccharide and teichoic acid although large quantities of lipid are present in mycobacterial cell walls. The biosynthesis in the cell of two of these polymers, peptidoglycan and teichoic acid, has been shown to be co-ordinated (Rogers and Garrett, 1965; Boylen and Ensign, 1968). Both components turnover extensively in exponential-phase cells but at different rates (Mauck and Glaser, 1970). The proteins of the Gram-positive cell walls have generally received little attention; many of the procedures used in preparing cell walls involve treatments which remove or degrade the cell-wall proteins (Rogers and Perkins, 1968).

An indication that the composition of cell walls may vary with growth phase was obtained by Young (1965) who reported differences in the galactosamine contents of cell walls from exponential- and stationary-phase cells of *Bacillus subtilis*. The galactosamine content was also dependent on the composition of the growth medium. Recent investigations have shown that the amino-acid composition of peptidoglycan of *Staphylococcus aureus* and *S. epidermidis* may be altered by changing the relative amounts of glycine and L-alanine in the growth medium (Schleifer *et al.*, 1969). The effect of limitation of different nutrients or ions on the composition of the cell wall may be studied using a chemostat. *Bacillus subtilis* var. *niger*, grown under conditions of magnesium, potassium, nitrogen or sulphur limitation, contains a wall teichoic acid (Ellwood, 1970a), whereas when grown under phosphate limitation the walls contain a teichuronic acid (a phosphorus-free polymer). Both polymers are anionic. The switch over from teichoic acid to teichuronic acid synthesis, or *vice versa*, proceeds very rapidly, much faster than can be accounted for by a diluting out of the appropriate polymer in the cell wall and involves turnover of both polymer and peptidoglycan (Ellwood and Tempest, 1969). The size of the cell, and hence the relative contribution of cell wall to cell mass, depends upon growth rate. Also, the faster the cells of *B. subtilis* are grown under magnesium-limitation, the more teichoic acid is present in the cell wall. A corresponding increase in teichuronic acid content occurs with phosphorus limitation (Ellwood, 1970a).

B. Morphology

The thickness of the cell walls is usually in the range 150–500 Å (Glauert and Thornley, 1969; Salton, 1964) but values of up to 800 Å have been reported for *Lactobacillus acidophilus* (Glauert, 1962). The thickness of the cell walls of *Clostridium welchii* is greater in stationary-phase cells (260 Å) than in exponential-phase cells (155 Å). When viewed in thin section in the electron microscope, the cell walls often

appear homogeneously electron dense (Salton, 1964; Glauert and Thornley, 1969) or sometimes as a less densely stained layer sandwiched between two thinner more densely stained layers (Higgins and Shockman, 1970b; Hughes *et al.*, 1970). In a very detailed study of *B. megaterium*, Nermut (1967) showed that the cell walls appeared uniformly electron dense when fixed with glutaraldehyde and stained with uranyl acetate, ruthenium red or phosphotungstate. The walls appeared trilamellar when stained with lead citrate, osmium tetroxide or potassium permanganate. Extraction of teichoic acid with formamide decreased the thickness of the cell wall by about half leaving a 100 Å-thick peptidoglycan layer. Accordingly, Nermut (1967) proposed that the cell wall of *B. megaterium* consists of a teichoic acid layer, 120 Å thick, on the outside of the peptidoglycan. This division of components did not correspond to any of the layers revealed in thin section and the layered appearance may be an artifact and not reflect differences in chemical composition. Extraction of teichoic acid produced no change in the trilamellar appearance of the cell walls of *L. arabinosus* (Archibald *et al.*, 1961). Attempts to determine the precise position of teichoic acid in cell walls by immunological experiments (Burger, 1966) have provided confusing results. Thus, the bulk of the teichoic acid present in the cell walls of *B. subtilis* is not available to antibodies, suggesting that the teichoic acid is distributed throughout the cell wall; but the opposite result was obtained for *B. licheniformis*.

Inhibition of protein synthesis in *Streptococcus faecalis* by chloramphenicol treatment or by depriving cells of an essential amino acid results in wall thickening by a factor of 2–3 (Shockman, 1965). A similar phenomenon has been described for *Staphylococcus aureus* (Giesbrecht and Ruska, 1968), *B. cereus* (Chung, 1967) and *B. subtilis* (Hughes *et al.*, 1970). With *B. subtilis* during wall thickening, the cell-wall profile is changed from a three- to a five-layered structure by the appearance of a densely stained layer in the centre of the cell wall, although the cell wall synthesized during thickening appears to be similar in composition to normal cell wall. Cells were fixed with glutaraldehyde, post-fixed with osmium tetroxide and post-stained with uranyl acetate.

Regular surface patterns are common to the cell walls of bacilli and clostridia (Holt and Leadbetter, 1969; Glauert and Thornley, 1969; Betz and Zeikus, 1968). The surface of the cell walls of *B. polymyxa* has been examined by several investigators and consists of an outer layer of units approximately 70 Å in diameter in tetragonal array, with a centre-to-centre spacing of 100 Å. Excellent agreement has been found between the dimensions obtained by shadowing (Goundry *et al.*, 1967), negative staining (Nermut and Murray, 1967) and freeze-etching (Holt and Leadbetter, 1969). Analysis of electron micrographs

WALLS AND MEMBRANES IN BACTERIA

of negatively stained preparations by optical techniques (Finch *et al.*, 1967) indicate that each unit consists of four subunits 40–50 Å in diameter. The structured layer is composed of protein together with some neutral sugars (Goundry *et al.*, 1967) and may be removed by treatment of cell walls with guanidine hydrochloride, trypsin, cold formamide or sodium dodecyl sulphate. When fixed and embedded by the Ryter-Kellenberger procedure, the cell walls of *B. polymyxa* appear trilamellar. Both the outer densely-stained layer and the inner intermediate layer disappear when the structured layer is removed from the cell wall (Nermut and Murray, 1967). The structured layer is removed by guanidine hydrochloride from the cell wall in the form of intact sheets, giving an indication that hydrogen bonds may be involved in holding the layer to an inner cell-wall layer; this inner layer probably contains

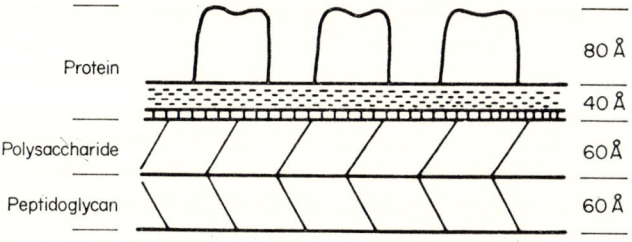

FIG. 6. Model of the cell wall of *Bacillus polymyxa*. Taken from Nermut and Murray (1967).

polysaccharide. The disruption of the structured layer by sodium dodecyl sulphate is indicative of the importance of hydrophobic bonds in maintaining structural integrity. There is also some evidence for the presence of salt linkages within the structured layer. A model for the cell wall of *B. polymyxa* is shown in Fig. 6.

The outer tetragonal layer of the wall of *B. brevis* has been isolated and purified (Brinton *et al.*, 1969). It consists of homogeneous protein subunits of molecular weight 140,000 which will, under suitable conditions, re-assemble to form a square network indistinguishable from the cell surface. Rather surprisingly, the re-assembled structure is cylindrical in shape with a radius equal to that of the organism. Arrays of particles are present on the surface of the cell walls of *B. cereus* (Holt and Leadbetter, 1969; Ellar and Lundgren, 1967) and *B. anthracis* (Holt and Leadbetter, 1969; Gerhardt, 1967) although they seem to be easily disrupted. A different type of surface is seen in freeze-etched replicas of *B. fastidiosus* (Holt and Leadbetter, 1969). Here the surface appears as a

34   D. A. REAVELEY AND R. E. BURGE

mesh of interwoven fibres. Similar fibres are seen on the surface of cell walls of some strains of *B. stearothermophilus* and *B. macroides*. In another strain of *B. macroides*, a globular layer is present on top of the mesh.

The visibility of the surface pattern depends to some extent on the growth phase of the organism (Nermut and Murray, 1967; Holt and Leadbetter, 1969). A globular layer not discernible in vegetative cells normally appears at the onset of sporulation on the surfaces of *B. sphaericus* and *B. psychrophilus*. However, some bacilli including *B. licheniformis*, *B. megaterium* and *B. subtilis*, lack structural details on the outer surface irrespective of stage of growth (Holt and Leadbetter, 1969). Extraction of lipid facilitates observations on the surface structure of *B. polymyxa* (Goundry *et al.*, 1967). A flocculent lipid layer external to the cell wall hinders observations on the delicate surface pattern of the cell wall of *Corynebacterium ovis* (Hard, 1969).

The composition and structure of the cell walls of the radiation-resistant coccus *Micrococcus radiodurans* have been investigated by Work and Griffiths (1968). The cell walls consist of an inner peptido-glycan layer about 240 Å thick (Work and Griffiths, 1968) penetrated by numerous holes with diameters of the order of 100 Å (Thornley *et al.*, 1965; Work and Griffiths, 1968). A small part (less than 5%) of this layer, composed of phosphorus, carbohydrate, and some amino acids, remains as a bag-shaped residue after enzymic removal of the components of peptidoglycan. The position and status of these components in the peptidoglycan layer are not clear. The outer wall is 160–300 Å thick (Work and Griffiths, 1968), and consists of at least two separate layers (Work and Griffiths, 1968; Glauert and Thornley, 1969). One is a network-like structure removable by trypsin digestion; another contains hexagonally-packed units and consists of carotenoids, lipids, protein and polysaccharide.

The surfaces of mycobacterial cell walls are covered with fibrils, often embedded in, or in old cells overlaid by, amorphous material (Imaeda *et al.*, 1968). The fibrils are best seen in cells from young cultures where they show an irregular arrangement sometimes twisting to form thick fibres; fibres are less frequent in cells from old cultures and are often parallel to each other. This outer fibrillar layer is of low electron density in thin section and consists of lipopolysaccharide. When removed by extraction with trichloroacetic acid, an underlying layer composed of densely packed fibrils is revealed. This middle layer of the cell wall is believed to consist of lipopolysaccharide, protein and lipid and is seen in thin section after permanganate fixation. The innermost layer is thought to consist of fibrils containing mycolic acid-polysaccharide embedded in a membranous arabinogalactan-peptidoglycan layer.

C. Peptidoglycans

After gently removing polysaccharide, protein or teichoic acid from the cell walls of Gram-positive bacteria, peptidoglycans remain as wall-shaped structures at least 100 Å thick (Ghuysen and Leyh-Bouille, 1969). Although detailed studies have so far been made on relatively few peptidoglycans, a general structure has evolved common to all (Ghuysen, 1968; Osborn, 1969; Ghuysen and Leyh-Bouille, 1969). This structure has three invariant features, namely:

(1) Glycan chains consisting of alternate N-acetylglucosamine and *N*-acetylmuramic acid residues; (2) Short peptide chains joined to the carboxyl group of muramic acid by an amide bond; (3) Cross bridges linking adjacent peptide chains.

(a) *Glycan chains*: The glycan chains of *S. aureus* strain Copenhagen (Tipper and Strominger, 1966; Tipper *et al.*, 1967) consist of alternate N-acetylglucosamine and N-acetylmuramic-acid residues in $\beta(1,4)$ linkage with N-acetylglycosamine residues at the reducing ends of the chains (Tipper, 1968). This basic structure is probably common to many organisms, including *M. lysodeikticus* (Leyh-Bouille *et al.*, 1966), *L. casei* (Hungerer *et al.*, 1969), *L. acidophilus* (Coyette and Ghuysen, 1970) and *S. epidermidis* (Tipper and Berman, 1969).

All the muramic acid residues of *S. aureus* strain Copenhagen (Ghuysen, 1968; Tipper and Berman, 1969), *S. epidermidis* Texas 26 (Tipper and Berman, 1969) and *L. acidophilus* (Coyette and Ghuysen, 1970) are

TABLE 3. Lengths of the Glycan Chains Present in the Peptidoglycans of some Bacteria

Organism	Number of hexosamine residues in glycan chains		
	Range	Average	Reference
Staphylococcus aureus Strain Copenhagen	12–90	24	Tipper *et al.* (1967)
Staphylococcus epidermidis Strain Texas 26	6–89	13	Tipper and Berman (1969)
Bacillus subtilis Marburg strain 168	—	10	Hughes (1970a)
Bacillus licheniformis N.C.T.C. 6346	—	10	Hughes (1970a)
Lactobacillus casei Strain R094	10–40	20	Hungerer *et al.* (1969)
Arthrobacter crystallopoietes Spherical form	14–63	33	Krulwich *et al.* (1967)
Rod form	114 and 135	124	Krulwich *et al.* (1967)

36   D. A. REAVELEY AND R. E. BURGE

substituted with peptide units. Approximately 50–70% of the muramic acid residues in *S. aureus* and *L. acidophilus* respectively are *O*-acetylated in the 6-position. The distribution of these residues in the glycan chains is not known; possibly some chains contain 6 *O*-acetylated muramic acid residues whilst others do not. Roughly 60% (Leyh-Bouille *et al.*, 1966; Munoz *et al.*, 1966) of the muramic acid residues of *M. lysodeikticus* possess free carboxyl groups and a few are not N-acetylated (Mirelman and Sharon, 1967). Muramic acid residues in the peptidoglycan of *Mycobacterium smegmatis* are probably *N*-glycolylated instead of N-acetylated (Petit *et al.*, 1969).

The lengths of the glycan chains vary with organism, and chains of widely differing lengths may be present in the same organism (Table 3). For instance, chain lengths of 12–90 hexosamine units have been reported for the peptidoglycan of *S. aureus* with an average value of 24 (Tipper *et al.*, 1967). Clearly the chains are not long enough to span the entire cell. The glycan chains of the spherical form of *Arthrobacter crystallopoietes* appear to be shorter than in the rod form of the organism (Krulwich *et al.*, 1967). An N-acetylmuramidase has been reported to be more active in the spherical form (Krulwich and Ensign, 1968) and the shorter chains may arise as a consequence of degradation during isolation of the peptidoglycan. An inspection of Table 3 reveals that there is no general correlation between shape of the organism and length of glycan chains.

(b) *Peptide chains*: These are normally peptides of the type: L-Ala-D-Glu-diamino acid-D-Ala (Table 4). The L-alanine residue is N-terminal and is bound in the peptidoglycan to the carboxyl group of muramic acid. Rather interestingly the γ-carboxyl group of D-glutamic acid is involved in the peptide bond to the diamino acid. The diamino acid is usually either L-lysine or *meso*-2,6-diaminopimelic acid (Table 4). *Meso*-2,6-Diaminopimelic acid is found in many rod-shaped organisms, such as bacilli, but is rarely present in spherical organisms. Where the *meso* acid is present, the L-asymmetric carbon atom is involved in peptide bonding (Dezélée and Bricas, 1970) thus preserving the general sequence of asymmetric carbon atoms involved in peptide bonding as L-D-L-D. The LL-isomer of 2,6-diaminopimelic acid is present in *Clostridium welchii* (Leyh-Bouille *et al.*, 1970) and some other organisms (Ghuysen, 1968); small amounts of the DD-isomer are present in the peptidoglycan of *B. megaterium*. A few other unusual diamino acids are present in the third position of some peptides (for review see Ghuysen, 1968). The peptides from several corynebacteria are unusual in that homoserine replaces the diamino acid and glycine replaces L-alanine (Perkins, 1965, 1967).

The α-carboxyl group of D-glutamic acid in the peptidoglycan of some organisms (Table 4; e.g. *S. aureus*; Munoz *et al.*, 1966) is amidated.

WALLS AND MEMBRANES IN BACTERIA

Amide residues are probably present on the free carboxyl groups of 2,6-diaminopimelic acid residues in *B. licheniformis* (Hughes, 1970a). Both the free carboxyl group of 2,6-diaminopimelic acid and D-glutamic acid in *Corynebacterium diphtheriae* are amidated (Kato *et al.*, 1968). In

TABLE 4. Peptide Chains of Peptidoglycans of Gram-positive Bacteria

Peptide chain[a]	Organism
$\overset{\displaystyle NH_2}{\underset{\alpha}{\mid}}\overset{}{\underset{\gamma}{}}$ Glycan-L-Ala-D-Glu-L-Lys-D-Ala	Staphylococci, streptococci, lactobacilli, some micrococci
Glycan-L-Ala-D-Glu-L-Lys-D-Ala Gly	Planococci, *Bacillus sphaericus*
$\overset{\displaystyle NH_2}{\underset{\alpha}{\mid}}\overset{}{\underset{\gamma}{}}$ Glycan-L-Ala-D-Glu-L-Lys-D-Ala	Some micrococci including *Micrococcus lysodeikticus*
(NH₂) Glycan-L-Ala-D-Glu-*meso* DAP-D-Ala	Bacilli
NH₂ Glycan-L-Ala-D-Glu-LL-DAP-D-Ala	*Clostridium welchii*, *Streptomyces* sp., some propionibacteria
NH₂ NH₂ Glycan-L-Ala-D-Glu-*meso* DAP-D-Ala	Mycobacteria, corynebacteria
NH₂ Glycan-L-Ala-D-Glu-L-Orn-D-Ala	*Micrococcus radiodurans*
Glycan-L-Ala-D-Glu-Homoserine-D-Ala	Some plant pathogenic corynebacteria
Glycan-L-Ser-D-Glu-L-Orn-D-Ala	*Butyribacterium rettgeri*

In some of the peptidoglycans large amounts of tripeptide lacking the terminal D-alanine residues are present. Data taken from Ghuysen (1968), Ghuysen and Leyh-Bouille (1969), Hughes (1970a), Kato *et al.* (1968), Leyh-Bouille *et al.* (1970), Schleifer and Kandler (1970), Schleifer *et al.* (1968), and Weitzerbin-Falszpan *et al.* (1970).

[a] NH₂ indicates the presence of an amide residue on a free carboxyl group.

Micrococcus lysodeikticus the α-carboxyl group of D-glutamic acid is normally substituted with glycine (Mirelman and Sharon, 1967) but under certain growth conditions glycine may be replaced by D-serine (Whitney and Grula, 1968).

A number of the peptide chains of the peptidoglycan of *S. aureus* possess an additional D-alanine residue at the *C*-terminal end (Veerkamp *et al.*, 1965; Tipper and Strominger, 1968). Approximately one third of the peptides of *S. epidermidis* possess this extra D-alanine (Schleifer *et al.*, 1969). Detailed studies of the biosynthesis of peptidoglycans (for review, see Osborn, 1969) have shown that new units fitted into the giant molecule have the peptide sequence:tetrapeptide-D-alanine. During the cross bridging of the peptides, this terminal D-alanine residue is lost; thus the additional D-alanine is indicative of an uncross-linked peptide. A tripeptide terminating in L-lysine is present in the peptidoglycan of *M. roseus* (Petit *et al.*, 1966) where it accounts for approximately 25% of the peptides and appears to be a genuine constituent of the peptidoglycan (Munoz *et al.*, 1966). The tripeptide is present in large amounts in some diaminopimelic acid-containing glycans including those from bacilli (Hughes *et al.*, 1968; Hungerer and Tipper, 1969; Grant and Wicken, 1970). Such peptides are thought to arise from the action of D-carboxy-peptidases (Kazuo and Strominger, 1968; Hungerer and Tipper, 1969).

(c) *Cross bridges*. The cross bridges linking peptide chains attached to different glycan chains appear to be species specific (see, for instance, Schleifer and Kandler, 1970) and subject to wide variation. Four different types of cross bridges have been recognised (Ghuysen, 1968). All involve the carboxyl group of the terminal D-alanine residue of a peptide chain.

(i) *Type I*. A direct peptide bond, between the free amino group of the diamino acid on one peptide chain and the carboxyl group of the terminal D-alanine residue of another, is found in the 2,6-diaminopimelic acid-containing glycans of bacilli (Hughes *et al.*, 1968). The amount of cross bridging in the peptidoglycan is low (Ghuysen, 1968; Warth and Strominger, 1968); only two thirds of the peptide chains of *B. subtilis* and *B. licheniformis* are joined together (Hughes, 1970a). Roughly 15% of the 2,6-diaminopimelic acid of *B. megaterium* is present as the DD isomer. Possibly this isomer is involved in linking different layers of the 100 Å-thick peptidoglycan (Ghuysen and Leyh-Bouille, 1969).

(ii) *Type II*. In these peptidoglycans, the peptide subunits are connected by bridges of one or more amino-acid residues. Some typical bridges are shown in Table 5. In the peptidoglycan of *S. aureus* strain Copenhagen, the terminal carboxyl group of a pentaglycine bridge is linked to the ε-amino group of the lysine residue in the peptide chain. About 75% of these units are linked through the N-terminal glycine residue to the C-terminal D-alanine residue on neighbouring peptide chains giving a tight structure (Fig. 7). Approximately 6% of the glycine bridges have the penultimate N-terminal glycine residue replaced by serine (Tipper and Berman, 1969). When *S. aureus* strain Copenhagen is grown in the presence of an unfavourable ratio of glycine to alanine (Schleifer, 1969)

only one third of the amino groups of lysine are substituted with the glycine pentapeptide, another third are free, whilst the remainder are substituted with L-alanine and not involved in cross linking. Although the amount of cross linking is considerably decreased the morphology of the cell remains unchanged.

TABLE 5. Some Type II Cross Bridges in Bacterial cell-wall Peptidoglycans

Cross bridge	Organism and Reference
Gly-Gly-Gly-Gly-Gly - - (L-Lys) (some glycine residues may be replaced by alanine or serine— see text)	Staphylococci (Ghuysen, 1968; Tipper, 1969; Schleifer et al., 1969)
L-Ala-L-Ala-L-Ala-L-Thr - - (L-Lys)	Micrococcus roseus Strain R27 (Petit et al., 1966)
L-Al-L-Ala-L-Ala - - (L-Lys)	Micrococci including some strains of Micrococcus roseus, streptococci (Schleifer and Kandler, 1970; Ghuysen, 1968)
D-Asparagine $\overset{\beta}{-}$ - (L-Lys)	Lactobacilli and streptococci of lactic group (Kandler, 1967; Ghuysen, 1968)
L-Glu$\overset{\gamma}{-}$Gly - - (L-Lys)	Micrococcus luteus (Schleifer and Kandler, 1970)
L-Glu$\overset{\gamma}{-}$L-Ala - - (L-Lys)	Micrococcus freudenreichii (Schleifer and Kandler, 1970)
Gly-Gly - - (L-Orn)	Micrococcus radiodurans (Schleifer and Kandler, 1970)
D-Glu$\overset{\gamma}{-}$- - (L-Lys)	Planococci (Schleifer and Kandler, 1970)
D-Glu$\overset{\gamma}{-}$Gly - - (L-Lys)	Sarcina ureae (Schleifer and Kandler, 1970)
Gly - - (LL-DAP)	Clostridium welchii and Streptomyces sp. (Leyh-Bouille et al., 1970); some propionobacteria (Schleifer et al., 1968)

N-Terminal amino acids are on the left. These amino acids are joined to C-terminal D-alanine residues of the peptide chains in the peptidoglycan. The C-terminal ends of the cross bridges are joined to the free amino groups of diamino acids (given in brackets) in other peptide chains.

The peptidoglycans of S. epidermidis are very similar to S. aureus although the cross bridges are more extensively substituted with serine. Four different bridges have been recognized (Tipper, 1969), namely: (i) pentaglycine, and replacement of glycine by serine in the (ii) third position (counting from the N-terminus) (iii) third and first position

40   D. A. REAVELEY AND R. E. BURGE

(iv) second position. The amount of cross linking is less than in *S. aureus*. When *S. epidermidis* strain 24 is grown in a glycine-deficient medium, the glycine content of the peptidoglycan drops by 40% and the L-alanine content increases. About 15% of the lysine residues are substituted on the ε-amino group with alanine residues which are not involved in cross linking; about 35% of the remaining lysine residues are unsubstituted. Addition of either serine or glycine to the growth medium leads to an increase in the glycine content of the peptidoglycan. With strain 66 an L-alanine residue substituted on the amino group of lysine forms part

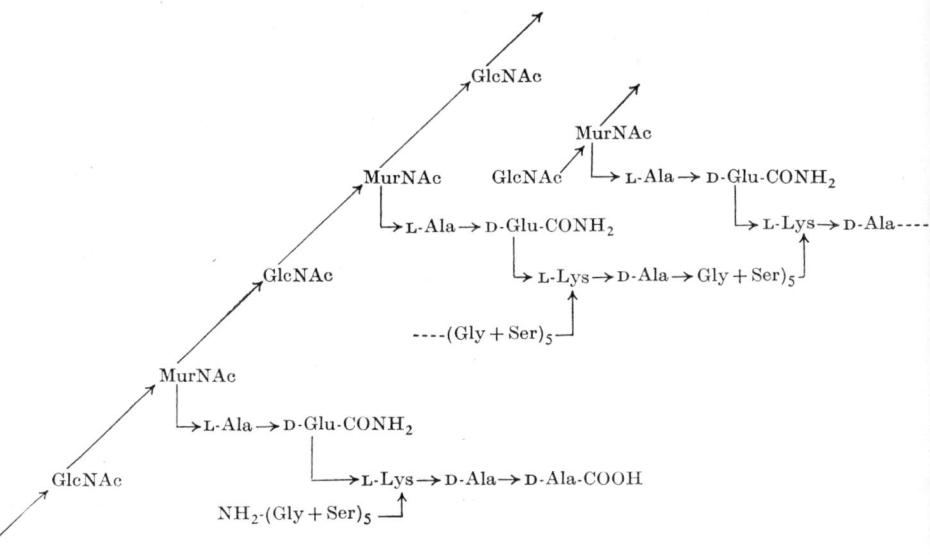

Fig. 7. Diagrammatic representation of the peptidoglycan of *Staphylococcus aureus* Copenhagen. Taken from Tipper and Berman (1969). GlcNAc indicates a residue of N-acetylglucosamine; MurNAc, of N-acetylmuramic acid; Ala, of alanine; Glu, of glutamate; Lys, of lysine; Ser, of serine.

of a cross bridge of sequence: Gly-Gly-Gly-Gly-L-Ala. This alanine residue cannot be replaced by glycine but some of the glycine residues may be replaced by serine (Schleifer, 1969).

The peptidoglycan of *B. sphaericus* differs from those of other bacilli in the presence of lysine instead of 2,6-diaminopimelic acid and, apart from the free carboxyl group of D-glutamic acid not being amidated, is identical with the peptidoglycan of *L. casei*. Both contain the peptide sequence: L-Ala-D-Glu-L-Lys-D-Ala cross-linked by an isoasparaginyl residue between lysine and D-alanine (Hungerer and Tipper, 1969; Hungerer *et al.*, 1969). Carboxypeptidases are particularly active in *B. sphaericus* since no un-crosslinked peptides containing terminal

WALLS AND MEMBRANES IN BACTERIA 41

D-alanine residues are found. The aspartic acid cross bridge is found in most lactobacilli (Kandler, 1967; Coyette and Ghuysen, 1970) and in *Streptococcus lactis* (Schleifer and Kandler, 1967a). The type of cross bridge is potentially very useful for taxonomic purposes. The motile halophilic cocci (planococci) may be distinguished from other micrococci and staphylococci by the presence of a D-glutamic acid cross bridge (Schleifer and Kandler, 1970). On the basis of physiological evidence, the enterococci have been designated *S. faecalis* or *S. faecium*. Kandler *et al.* (1968) showed that the species termed *faecium* possessed a cross bridge of D-aspartic acid whereas strains of *S. faecalis* possessed cross bridges of three L-alanine residues.

The peptide subunits of the rod forms of propionibacteria have the sequence: L-Ala-D-Glu-LL-DAP-D-Ala, and are cross linked by glycine through the amino group of 2,6-diaminopimelic acid to the C-terminal D-alanine residue of an adjacent tetrapeptide (Schleifer *et al.*, 1968). The peptide subunits from the cocci contain *meso*-2,6-diaminopimelic acid, alanine and glutamic acid; the nature of the cross bridge is not known.

(iii) *Type III.* In peptidoglycans of *M. lysodeikticus* (Ghuysen *et al.*, 1968a; Schleifer and Kandler, 1967b) and related organisms (Campbell *et al.*, 1969; Schleifer and Kandler, 1970) the cross bridges consist of complete peptide chains arranged head-to-tail. Two types of inter-peptide links can be distinguished in these peptidoglycans. The first, D-Ala-L-Ala, joins the peptide chains head-to-tail; the second, between a C-terminal D-alanine residue of the extended peptide chain and the ε-amino group of lysine in a single chain attached directly to muramic acid on another glycan chain, completes the cross bridging. The lengths of the cross bridges vary widely but they average about six peptide units. Since the constituents of the peptidoglycan of *M. lysodeikticus* are present in approximately equimolar amounts, long stretches of the glycan chain must be unsubstituted (see p. 36) with peptide, giving rise to a relatively open structure.

(iv) *Type IV.* In these peptidoglycans, the α-carboxyl group of glutamic acid in one peptide chain is connected *via* a bridging diamino acid to the C-terminal D-alanine residue of another subunit on a neighbouring glycan chain. Typical of this class is the peptidoglycan of *Butyribacterium rettgeri* (Guinand *et al.*, 1969) where the peptide chain L-Ser-D-Glu-L-Orn-D-Ala is cross linked to another chain by either D-ornithine or D-lysine bridges (ratio 1:2). Similar cross bridges are present in some corynebacteria (Perkins, 1967). A variation of the type IV peptidoglycan is found in *Microbacterium lacticum* and *M. liquefaciens* (Schleifer, 1970) where the cross bridges are of the type: glycine-diamino acid. The amino group of glycine is bound to the α-carboxyl group of glutamic acid (or *threo*-3-hydroxyglutamic acid) on one peptide chain

and an amino group of the diamino acid is linked to the carboxyl group of the terminal D-alanine residue on another peptide chain.

D. TEICHOIC AND TEICHURONIC ACIDS

1. *Teichoic acids*

Teichoic acids are found in many Gram-positive organisms where they may account for as much as 20–50% of the dry weight of the cell wall. In their simplest form teichoic acids are polymers of glycerol or ribitol phosphate substituted glycosidically with sugar residues. D-Alanine is often bound in an alkali-labile ester linkage to the polyol backbone. Phosphodiester bridges link the 1- and 3-positions of glycerol or the 1- and 5-positions of ribitol residues. Teichoic acids vary greatly in the extent, nature and regularity of carbohydrate substitution (Ghuysen *et al.*, 1968b; Archibald *et al.*, 1969). Thus, the ribitol teichoic acid of *B. subtilis* W23 (Chin *et al.*, 1966) is a mixture of completely unsubstituted and completely substituted polyol chains. The significance of this is not known. Nor is there any evidence as to the distribution of the two chains within the cell wall. The teichoic acid of *L. buchneri*, on the other hand, consists of partially glucosylated glycerol phosphate chains. Two teichoic acids are present in *Staphylococcus aureus* strain Copenhagen; one contains the sugar N-acetylglucosamine, in α-glycosidic linkage to ribitol, whereas the hexosamine is in β-linkage in the other (Tori *et al.*, 1964). Other strains may possess one or other of these teichoic acids. In the cell-wall teichoic acid of *B. stearothermophilus*, glucose is believed to be linked to the 1-position of glycerol and the phosphodiester linkage to involve positions 2 and 3 (Wicken, 1966).

A recently discovered variation is the presence of sugar residues within the glycerol backbone. Typical examples are the teichoic acids isolated from *B. licheniformis* ATCC 9945 (Burger and Glaser, 1966) where galactose or glucose is attached glycosidically to the 1-position of glycerol and phosphodiester bridges join the 3-position of glycerol to the 6-position of the sugar. Teichoic acids present in some strains of *L. plantarum* may have similar structures (Adams *et al.*, 1969b). A teichoic acid from strain I3 of *S. lactis* (Archibald *et al.*, 1968a) consists of alternating N-acetyl-glucosamine 1-phosphate and glycerol phosphate residues. The hexosamine is linked through phosphate to glycerol phosphate which is in turn bound through its phosphate residue to the 3-hydroxyl group of the amino sugar. In both this polymer and the polymer from *B. stearothermophilus*, D-alanine is ester bound to the 6-position of the carbohydrate. A related polymer consisting simply of N-acetylglucosamine 1-phosphate residues in 1-6 linkage is present in another strain of *S. lactis* (Archibald *et al.*, 1968b). The pneumococcal C-substance resembles a teichoic acid

in being a cell-wall polymer of ribitol phosphate which is extracted by cold trichloroacetic acid (Brundish and Baddiley, 1968). However analysis shows choline, phosphate, N-acetyl-D-galactosamine, D-glucose, N-acetyldiaminotrideoxyhexose and ribitol to be present in molar ratio of $1:2:1:1:1:1$. This polymer is far more complicated than any teichoic acid so far studied and its structure has yet to be determined.

Estimations of chain lengths have been made for teichoic acids extracted from cell walls and those synthesized *in vitro*. Results for ribitol teichoic acids agree well and give a chain length of 5-12 repeating units (Brooks and Baddiley, 1968). Glycerol teichoic acids synthesized *in vitro* are longer (30–32 repeating units) than glycerol teichoic acids (20 units) extracted from cell walls. The significance of this result is not known; possibly the trichloroacetic acid used for extraction slightly degrades glycerol teichoic acid.

2. *Teichuronic Acids*

A polysaccharide isolated from the cell walls of *M. lysodeikticus* consisted of equimolar amounts of D-glucose and 2-acetamido-2-deoxymannuronic acid (Perkins, 1963). Such uronic acid-containing polymers have been given the name 'teichuronic acids'. The teichuronic acid of *B. licheniformis* 6346 is probably composed of 3-*O*-glucuronosyl-galactosamine units linked from the reducing group of the galactosamine residue to the 4-position of glucuronic acid (Hughes and Thurman, 1970). The chain is composed of 35–40 disaccharide units (Hughes, 1970b). The occurrence of teichoic acids and teichuronic acids in the cell wall, depending on the limitation of essential metabolites, has already been described.

3. *Nature of Linkage to Peptidoglycan*

The nature of the method used to extract teichoic acid from cell walls (prolonged treatment with trichloroacetic acid in the cold) suggests that the process involves the breaking of covalent bonds. Enzymic lysis of the cell walls of *S. aureus* (Ghuysen *et al.*, 1965), *B. subtilis* (Young, 1966) and *B. licheniformis* (Hughes, 1965) produces cell-wall fragments containing components of teichoic acid and peptidoglycan. The results of Young *et al.* (1964) implicate the terminal phosphate of the teichoic acid in the linkage to peptidoglycan. Multiple linkages are ruled out (Rogers and Garrett, 1965). The linkage is labile to acids and alkalis by mechanisms which do not involve cyclic phosphates (Archibald and Baddiley, 1965) and stable towards ammonia and other amines at neutral pH values. Also, as teichoic acids were extracted from *B. subtilis*, *S. aureus* and *L. arabinosus* by hydrazines, Archibald and Baddiley (1965) suggest that the terminal phosphate residue of the teichoic-acid chain is linked by a phosphoramidate bond to the terminal reducing

44   D. A. REAVELEY AND R. E. BURGE

sugar of a glycan chain. The teichoic acid and the peptidoglycan of *B. stearothermophilus* are not separated by phenylhydrazine treatment, and Grant and Wicken (1968) suggest that, in this case, teichoic acid is joined to peptidoglycan by a phosphodiester bridge involving the terminal phosphate residue of teichoic acid and the 6-hydroxyl group of muramic acid.

Muramic acid 6-phosphate has been found in acid hydrolysates of the cell walls of many Gram-positive bacteria (Lui and Gotschlich, 1967; Knox and Holmwood, 1968; Heymann *et al.*, 1967; Montague and Moulds, 1967); for instance 3% of the muramic-acid residues are phosphorylated in the 6-position in *L. acidophilus* (Coyette and Ghuysen, 1970). Lui and Gotschlich (1967) suggest that the phosphate serves as the link between the cell-wall polysaccharides of Pneumococcus and group A streptococci and peptidoglycan. The possibility that muramic acid 6-phosphate is formed by phosphoryl migration of the 4-phosphate must not be overlooked (Button *et al.*, 1966). The results for *B. licheniformis* (Hughes, 1970b) and lactobacilli (Hall and Knox, 1965; Knox and Hall, 1965; Knox and Holmwood, 1968) are indicative of phosphodiester bridges between the reducing ends of cell-wall polysaccharides and muramic acid. In *L. acidophilus* there appears to be one bridge per polysaccharide unit (Coyette and Ghuysen, 1970).

Conclusive evidence for a phosphodiester bridge to muramic acid has been obtained in only one organism. Degradative studies on autolysis products of the cell walls of *S. lactis* strain I3 (Button *et al.*, 1966) clearly establish that the teichoic acid is linked to muramic acid through a phosphodiester bridge involving either the 4- or 6-hydroxyl group of muramic acid and the terminal glycerol residue of teichoic acid. It must be noted however (see p. 42) that this teichoic acid has an unusual structure. The cell walls of *Streptococcus pyogenes* Type 14 contain two polysaccharides. The G-polysaccharide is believed to be linked to muramic acid *via* a phosphodiester bridge. The other polysaccharide, the so-called C-polysaccharide, appears to be glycosidically linked to the C-4 position of N-acetylmuramic acid at the non-reducing end of the glycan chain (Munoz *et al.*, 1967).

4. *Function of Teichoic Acids*

The binding of cations, especially divalent cations, in the cell wall appears to be a function of teichoic acids (Cutinelli and Galdiero, 1967). Significantly, the greatest amount of teichoic acid is found in the cell walls of organisms grown under conditions in which the supply of magnesium is limited (Ellwood, 1970a). Heptinstall *et al.* (1970) demonstrated that the cell walls of *Staphylococcus aureus* bind more magnesium/ mole phosphate when sodium chloride (7·5%) is included in the growth

WALLS AND MEMBRANES IN BACTERIA
45

medium. The increased binding of magnesium ions is correlated with a decreased amount of ester-bound D-alanine in the cell-wall teichoic acid. Heptinstall *et al.* (1970) suggest that teichoic acids are responsible for maintaining a high concentration of divalent cations in the vicinity of the cell wall and cell membrane, and the requirement for divalent ions is efficiently controlled by varying the extent of alanine substitution of the teichoic acid. Magnesium ions are known to be necessary for membrane integrity and for the activity of many enzymes involved in cell-wall synthesis and oxidative phosphorylation. A similar role is suggested for the so-called intracellular teichoic acids. These teichoic acids appear to occupy a rather ill-defined region between the cell-wall and cytoplasmic membrane, or are possibly associated with mesosomes, and are always of the glycerol type even when the wall teichoic acid is of the ribitol variety. Significantly, intracellular teichoic acids are present under conditions of magnesium limitation (Ellwood and Tempest, 1968).

Teichoic acids are involved in phage adsorption. Phage-resistant mutants of *B. subtilis* 168 (Young, 1967) differ from the wild-type strain by the absence of glucose in the cell-wall teichoic acid. Experiments by Matthew and Rosenblum (1967) indicate that teichoic acid must be in combination with peptidoglycan for phage adsorption. Either the phage recognizes teichoic acid as part of the receptor site or possibly the correct spatial arrangement of polymers in the cell wall provided by a specific teichoic acid is necessary for phage adsorption. Bacteriophage-resistant mutants of *S. aureus* H have been separated into two classes depending on whether or not they adsorb phages (Chatterjee, 1969). Phage adsorption is associated with a normal teichoic acid in the cell wall. Inability to adsorb phages indicates loss of N-acetyl-glucosamine from the cell-wall teichoic acid. One mutant which does not adsorb phage has a greatly diminished content of teichoic acid in the cell wall and is believed to arise by a single mutation (Chatterjee, 1969; Chatterjee *et al.*, 1969). This teichoic acid-deficient mutant grows in clumps. Furthermore, although the mutant and wild strains autolyse equally well, the cell walls of the mutant strain autolyse more rapidly than those of the wild-type strain. There are several other recorded instances in which changes in teichoic acids are associated with changes in lytic enzymes. *Streptococcus zymogenes* (Davie and Brock, 1966) is not susceptible to lytic enzymes which the organism secretes into the medium until alanine residues are removed from the ribitol teichoic acid. Satisfying the choline requirement of pneumococci by ethanolamine drastically affects the activity of cell-wall lytic enzymes (Tomasz, 1968). This change in lytic activity is probably connected with the replacement of choline by ethanolamine in the C-substance. At the present time it is not clear whether these effects are simply due to the cation imbalance

being upset or whether teichoic acids have a specific secondary function in controlling autolytic enzymes in cell division (Heptinstall *et al.*, 1970).

E. OTHER CELL-WALL POLYMERS

With the advent of pure cell-wall preparations, it became clear that many important cell antigens were present in the cell walls. One such antigen is the type-specific M protein of group-A streptococci. The presence of this antigen is correlated with virulence and it functions by inhibiting phagocytosis. The manner in which the protein is linked to the cell wall is not known, but covalent bonds are likely to be involved since mild-acid hydrolysis or trypsin treatment is necessary for its removal (McCarty and Morse, 1964). Treatment of the cell walls with endohexosaminidase, a phage-associated enzyme muralytic for group-A streptococci, yields fractions in which the specific proteins are associated with polysaccharide and peptidoglycan components (Barkulis *et al.*, 1968). The elegant fluorescent studies of Hahn and Cole (1963) demonstrated that M protein was located in the outer layers of the cell wall and its appearance coincided with the synthesis of fresh cell-wall material. Nevertheless the site of synthesis of the M protein is the cell membrane (Karawaka *et al.*, 1965). M-Protein is a collection of proteins rather than a single polypeptide (Fox and Wittner, 1965; Pierce, 1964). The gel electrophoresis patterns of proteins from the same serotype are similar (Fox and Wittner, 1965). A series of purified M proteins has been isolated from Type 12 group-A streptococci (Lange *et al.*, 1969). All possess an N-terminal lysine residue and react with type-specific antisera; they appear to differ slightly in molecular weight. Possibly this might reflect different stages in peptide-chain synthesis. Acid-extractable M protein accounts for only about 25% of the protein of the group-A streptococci cell wall (Barkulis and Jones, 1957) and other protein antigens notably R and T antigens are present (McCarty and Morse, 1964). A fourth protein antigen has recently been described (Bray and Bass, 1968).

The cell walls of streptococci are the sites of most of the group-specific antigens (McCarty and Morse, 1964). The group-A polysaccharide is extracted from the cell wall of group-A streptococci by treatment with either cell-wall destroying enzymes or hot formamide. Both methods produce serologically active preparations but the former method yields material containing small amounts of peptidoglycan components (McCarty and Morse, 1964). An insight into the nature of antigenic determinants has come from studies of group-A polysaccharides and mutant A strains (A-variant strains). Both polysaccharides are comprised of rhamnose and N-acetylglycosamine, but the A-variant polysaccharide contains considerably less hexosamine, the ratio of the two being greater

than 20:1 compared with 2:1. Further progress was made using induced enzymes from soil bacteria capable of degrading carbohydrates. One of these enzymes, believed to be a β-N-acetylglucosaminidase, removes N-acetylglucosamine residues from the A polysaccharide (about 60–75% of the hexosamine is removed). The residual polysaccharide is then susceptible to an enzyme which destroys the A-variant polysaccharide and also is able to cross-react strongly with A-variant serum. McCarty and Morse (1964) believe that both A and A-variant polysaccharides consist of backbones containing only small amounts of glucosamine to which rhamnose-containing oligosaccharides are attached. The differences in specificities, it is suggested, are due to the presence of terminal β-linked N-acetylglucosamine residues on the rhamnose oligosaccharides in the A polysaccharide. Additional evidence for the importance of N-acetylglucosamine in antigenic specificity of group-A polysaccharides is that this sugar (and some of its derivatives) inhibits the precipitin reaction. The group-C polysaccharide differs primarily from the group-A polysaccharide in the replacement of the terminal N-acetylglucosamine residues by N-acetylgalactosamine (McCarty and Morse, 1964).

V. The Cell Envelopes of Gram-Negative Bacteria

A. Introduction

Many investigators using a wide range of methods, including bio-chemical, immunological, electron-microscope, X-ray diffraction and phage techniques, have studied the outer regions of the cell envelopes of Gram-negative bacteria (for reviews, see Salton, 1964, 1967a; Martin, 1966; Burge and Draper, 1967a, b, c; Rogers and Perkins, 1968; Glauert and Thornley, 1969). The results of these many investigations clearly establish that lipopolysaccharides, phospholipids and protein are present in this part of the cell envelope, and much of the work on Gram-negative cell envelopes has concentrated on the way these various components fit together in the cell envelope. It is generally accepted that these components, together with peptidoglycan, comprise the region of the cell envelope known as the cell wall, although in marine pseudomonads (Brown et al., 1962; Forsberg et al., 1970a) the peptidoglycan is closely associated with the cytoplasmic membrane.

A few workers have successfully separated cell walls and cytoplasmic membranes of Gram-negative bacteria, but it is too early to say whether the methods are of general applicability. From the supernatant fluid remaining after deposition of "cell walls" of *Pseudomonas aeruginosa*, Gray and Thurman (1967) obtained material composed principally of

protein (52%) and lipid (45%). This material was free of peptidoglycan and lipopolysaccharide and is believed to consist of fragments of the cytoplasmic membrane. Schnaitman (1970a) observed that disaggregation of *E. coli* in a French pressure cell, in the presence of high concentrations of salt, yielded cell-envelope fragments. If, however, the disintegration took place at low ionic strength in the presence of low concentrations of EDTA, a mixture of two kinds of fragments was obtained which could be separated by sucrose density-gradient centrifugation. The heavier fraction was composed of large fragments, having, in thin section, the characteristic appearance of cell walls (see p. 49). This material had a high content of lipopolysaccharide and a low content of cytochromes. Vesicles were concentrated in the less dense fraction. These vesicles were surrounded by a membrane which, in thin section, had the characteristic appearance of cytoplasmic membrane. The identification of these vesicles as cytoplasmic membrane is further indicated by the enrichment of cytochromes and succinate dehydrogenase in the less dense fraction. Both of these components are located in the cytoplasmic membranes of Gram-positive bacteria.

A method for separating the outer layers of the cells wall from the cytoplasmic membrane of *E. coli* has been developed by Miura and Mizushima (1968). Sphaeroplasts produced by the action of lysozyme and EDTA in a tris-HCl-sucrose solution were lysed in the presence of 5 mM-magnesium chloride and dialysed against 34 mM-EDTA. The material in the dialysate could be separated into dense and less-dense fractions by sucrose density-gradient centrifugation. The less dense fraction was composed of vesicles and its composition, both chemical and enzymic, was very reminiscent of the cytoplasmic membranes of Gram-positive bacteria. The dense fraction contained most of the envelope carbohydrate and some protein and lipid; components of the electron-transport system were essentially absent. In thin section in the electron microscope this fraction was shown to contain coiled membranous structures 60–100 Å thick. Birdsell and Cota-Robles (1967) have shown that, under the conditions used by Miura and Mizushima (1968) for sphaeroplast formation, the outer layers of the cell wall roll up into similar coiled structures leaving the cells surrounded principally by the cytoplasmic membrane. The identification of the coiled structures isolated by Miura and Mizushima (1968) as the outer regions of the cell walls is confirmed by the observation that only the fraction containing these structures will react with antibodies to whole cells (Miura and Mizushima, 1969).

B. Morphology

The appearance of Gram-negative bacteria in thin section in the electron microscope has formed part of an excellent review by Glauert and Thornley (1969). The appearance of the envelope depends on the conditions of fixation and the state of the culture (De Petris, 1967; Frank and Dekegel, 1967; Murray *et al.*, 1965; Watson and Remsen, 1970). The variable appearance of the envelope in normal fixed cells has led to the suggestion (De Petris, 1967) that the cell envelope in living cells may be in different physiological states. Nonetheless, a general picture of the Gram-negative cell in thin section has emerged, and consists of an inner triple-layered membrane (cytoplasmic membrane) and an outer triple-layered membrane (outer membrane) with a densely-staining layer between (Glauert and Thornley, 1969). This profile is not found in extremely halophilic bacteria, and these organisms will be discussed separately. The outer membrane is generally between 60 and 100 Å thick and in normal cells presents an undulating appearance (De Petris, 1967; Frank and Dekegel, 1967). The outer membrane of *E. coli* bulges out to produce extrusions which may in some cases separate as distinct vesicles (De Petris, 1967; Knox *et al.*, 1967). These vesicles are surrounded by a membrane identical in thickness and appearance with the outer membrane of the cell. The outer membrane is straight in isolated cell envelopes or in cells infected with phage (De Petris, 1967; Burge and Draper, 1967a). In some organisms, additional layers may be present outside the outer membrane (see p. 52).

The dense layer between the two triple-layered membranes has now been detected in a large number of Gram-negative organisms (Glauert and Thornley, 1969; Watson and Remsen, 1970; Wang *et al.*, 1970) although in some cases the layer is intimately associated with the outer membrane and is seen merely as a thickening of the densely staining inner layer of this membrane (Glauert and Thornley, 1969). The thickness of this layer is normally in the range 20–80 Å but is exceptionally thick in *Veillonella parvula* (100–500 Å; Bladen and Mergenhagen, 1964). It is susceptible to digestion by lysozyme (Murray *et al.*, 1965; De Petris, 1967; Frank and Dekegel, 1967) and is absent from sphaeroplasts. It is also missing from L forms of *Proteus mirabilis* (Hofschneider and Martin, 1968). This information has led to the belief that this densely staining inner layer is composed of peptidoglycan; the layer bears a striking resemblance both in thickness and staining properties to isolated intact peptidoglycans (De Petris, 1967; Frank and Dekegel, 1967; Wang *et al.*, 1970).

Treatment of *E. coli* (De Petris, 1967) with proteases detaches the outer membrane from the inner densely staining peptidoglycan. De

50   D. A. REAVELEY AND R. E. BURGE

Petris (1967) suggests that proteins are present in this region of the cell
envelope and some of these proteins serve to anchor the peptidoglycan
to the outer membrane. De Petris (1967) tentatively suggests, from
observations in thin sections, that some of the proteins may be present
in a globular form (see Section IV, C, p. 35).

In an extensive combined electron microscope and X-ray study of the
cell wall from *P. vulgaris*, Burge and Draper (1967a, b) found dimensions
of 92 Å for the width of the outer membrane and 52 Å for the peptido-
glycan/protein layer. The results of a synthesis of the electron-density

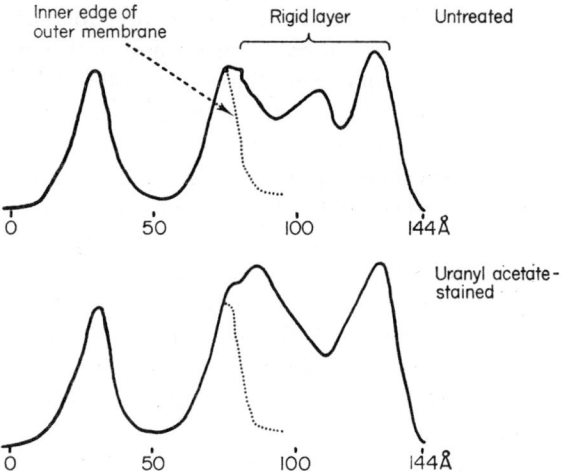

FIG. 8. Distribution of electron density across the cell wall of *Proteus vulgaris*.

distribution through the thickness of the *P. vulgaris* cell wall, without
staining and following treatment with uranyl acetate, are shown in Fig. 8.
The preferential staining of the cytoplasmic side of the outer membrane
is clearly observed and is consistent with the results found by electron
microscopy.

Ionic interactions, probably involving magnesium ions, are of prime
importance in the structural integrity of the cell envelopes of marine
pseudomonads (Forsberg *et al.*, 1970a). By carefully controlling the
ionic compositions of the solutions used to wash and resuspend marine
Pseudomonad B16, Forsberg *et al.* (1970b) were able to remove progressively layers of the envelope from the cell. They demonstrated that
the envelope was five-layered consisting of an outer loosely bound layer,
a triple-layered outer membrane, a layer situated between the triple-
layered membrane and peptidoglycan, peptidoglycan and a cytoplasmic

WALLS AND MEMBRANES IN BACTERIA 51

membrane. The outer three layers all contained carbohydrate, lipid and protein, although the outer membrane contained less carbohydrate than the other two layers.

As revealed by negative staining, the surfaces of many Gram-negative bacteria appear rough and wrinkled (Glauert and Thornley, 1969). When shadowed and examined in the electron microscope, the cell walls of most common Gram-negative bacteria appear smooth. The surface of *E. coli*, when examined by freeze-etching, appears smooth (Nanninga, 1970; Bayer and Remsen, 1970; Fiil and Branton, 1969) unless grown in the presence of glycerol (Nanninga, 1970). The wrinkled appearance observed using thin-section or negative staining techniques may be caused by cell shrinkage since measurements of thicknesses of cells after freeze-etching, negative staining and thin-sectioning demonstrate that, in the former case, the cell dimensions are greater by about 35% (Bayer and Remsen, 1970).

The multi-layered nature of the cell walls demonstrated by thin-section techniques is confirmed by freeze-etch studies. The existence of a globular layer within the cell wall has been demonstrated by Bayer and Remsen (1970) and Nanninga (1970). The globules observed by Nanninga are about 100 Å in diameter and may correspond to the globules seen on the surface of the peptidoglycan (De Petris, 1967). A hexagonal array of particles of diameter 104 Å has been seen in the cell walls of *E. coli* after heating at 90° for 10 min. and negative staining (Fischman and Weinbaum, 1966). The globules observed by Nanninga (1970) were not in hexagonal array; possibly the hexagonal pattern is produced by the heat treatment.

There seems general agreement on the appearance of the fracture faces revealed during freeze-etching of the cytoplasmic membranes of *E. coli*. A convex face exposed by freeze-etching contains a net-like arrangement of particles 20–100 Å in diameter, but a concave face so exposed contains fewer particles (Fiil and Branton, 1969; Nanninga, 1970). The distribution of particles over the concave surface is patchy. Smooth areas appear on the convex surface as a response to starvation of magnesium, calcium or carbon (Fiil and Branton, 1969). Magnesium starvation also produces an infolding of the membrane and the appearance of a para-crystalline array of particles. Surprisingly the magnesium content of the membrane does not change during magnesium starvation of the cells.

The surface layer of the cell wall of *Spirillum serpens* has been the subject of a detailed study using freeze-etch, negative staining and chemical techniques (Buckmire and Murray, 1970). The presence of a hexagonal array of subunits (90–110 Å in diameter) on the surface of the cell wall has been shown to be dependent on an adequate amount of

52   D. A. REAVELEY AND R. E. BURGE

calcium in the growth medium. The hexagonal layer consisting of protein
and a backing layer probably comprised of lipopolysaccharide may be
removed by heating the cells at 60° for 1 hr. The trilamellar appearance
of the outer membrane of the cell wall remains after such extraction. The
hexagonal layer can be separated from the backing layer by treatment
of the cell walls with a solution (1·6 M) of guanidine hydrochloride.

Three distinct surface patterns have been recognized in the marine
photosynthetic Thiorhodaceae using a combination of freeze-etch,
negative stain and thin-section techniques. The first, present in *Ecto-*
thiorhodospira mobilis, consists of spherical particles 50 Å in diameter
mounted 80 Å apart on the surface of the outer triple-layered membrane
(Remsen *et al.*, 1968). In *Chromatium buderi*, structures resembling wine
glasses 350 Å in diameter with stems 50 Å wide and 150 Å long rest on a
75 Å wide layer outside the outer membrane. A third type of structure
consisting of an hexagonal array of 50 Å diameter particles is present on
the surface of *C. gracile* (Remsen *et al.*, 1970).

Regular arrays of surface units are present on marine ammonia-
oxidizing organisms but not on their terrestrial counterparts nor on
nitrite-oxidizing bacteria (Watson and Remsen, 1970).

In *Nitrosocystis oceanus* the particles are of diameter 50 Å and in
hexagonal array. Since the layer, thought to be protein, is destroyed by
formic acid or EDTA, metal–oxygen bonds are believed to be involved in
maintaining the hexagonal array (Watson and Remsen, 1969). The
particles appear to be organized into three layers giving an overall
thickness of 150 Å (Watson and Remsen, 1970). The outer membrane
is situated below this layer and Watson and Remsen (1970) present
evidence that this membrane is constructed from a double layer of 40 Å
diameter globules (the presence of globules within a triple-layered
membrane is discussed on p. 30). Between the outer membrane and a
20 Å thick layer, believed to be peptidoglycan, another globular layer
(globules 100–125 Å in diameter) is situated. A further globular layer is
apparently present between the peptidoglycan and the cytoplasmic
membrane.

C. PEPTIDOGLYCANS OF GRAM-NEGATIVE BACTERIA

Gram-negative bacteria contain much less peptidoglycan than Gram-
positive organisms. Generally about 5–20% by weight of the "cell wall"
(Salton, 1964; Ghuysen, 1968; Clarke *et al.*, 1967a) is peptidoglycan;
possibly smaller amounts are present in marine pseudomonads (Brown,
1964) and the components of peptidoglycan have not been detected in the
cell envelopes of some extremely halophilic bacteria.

WALLS AND MEMBRANES IN BACTERIA 53

Treatment of the cell envelope of *Escherichia coli* with sodium dodecyl sulphate and phenol-water yields an insoluble cell-wall fraction (termed the rigid or R layer) which retains the original shape of the cell. Globular protein distributed over the surface of the R layer is easily removed by proteases (Martin and Frank, 1962; and see p. 55) without affecting the size or shape of the layer. The protein-free R layer consists principally of peptidoglycan components together with some glucose (21%) and lipid (8%). Both glucose, present as glycogen (Leutgeb and Weidel, 1963) and lipid (Leutgeb *et al.*, 1963) may be removed without disrupting the sac-like nature of the peptidoglycan. The intact peptidoglycan is often termed a sacculus (Weidel and Pelzer, 1964). Protein globules are also present on the surface of the sacculi of *P. mirabilis* (Martin, 1966) and *Shigella flexneri* (Frank and Dekegel, 1967). Protein is associated with the peptidoglycan of *Spirillum serpens* (Kolenbrander and Ensign, 1968) but not with the peptidoglycan of another *Spirillum* species (Martin, 1966).

The inability to obtain sacculi from some Gram-negative organisms containing relatively small amounts of peptidoglycan, for instance vegetative cells of *Myxococcus xanthus* (White *et al.*, 1968) and *Erwinia* sp. (Grula *et al.*, 1965), has led to the suggestion that, in these organisms, the peptidoglycan does not completely surround the cell but occurs in discrete patches separated from each other by non-peptidoglycan components. The inability to isolate intact sacculi is sometimes due to the action of autolytic enzymes during the preparation of the cell envelopes (Weidel *et al.*, 1963). Degradation of the peptidoglycans of *E. coli* (Weinbaum *et al.*, 1967) and *Chromobacterium violaceum* (Whiteside and Corpe, 1969) by a preparation of phospholipase C from *Clostridium welchii* is due to the presence in the enzyme preparation of a N-acetyl-glucosaminidase which breaks down the glycan chains of Gram-negative bacteria (Martin and Kemper, 1970). The enzyme is not active against the glycan chains of the peptidoglycans in the Gram-positive organisms *Micrococcus lysodeikticus* and *Sarcina lutea*.

Although few detailed analyses have been made of peptidoglycans from Gram-negative organisms, the data available at the present time indicate that muramic acid, glucosamine, alanine, 2,6-diaminopimelic acid and glutamic acid, in molar ratios around $1:1:2:1:1$, are invariant constituents (Mandelstam, 1962; Schocher *et al.*, 1962; Jušić *et al.*, 1964; Grula *et al.*, 1965; Martin, 1966; Kolenbrander and Ensign, 1968; Wang and Lundgren, 1968; Clarke *et al.*, 1967c; Verma and Martin, 1967). Apparently a diversity of amino-acid constituents is peculiar only to peptidoglycans from Gram-positive bacteria. As far as is known, 2,6-diaminopimelic acid is always present as the *meso* isomer (Ghuysen, 1968).

Detailed information about the peptidoglycans of Gram-negative bacteria is confined mainly to those of *E. coli*, studied by Weidel and his colleagues, and *P. mirabilis* which has been examined by Martin. Much of the work on the peptidoglycan of *E. coli* is summarized in the review of Weidel and Pelzer (1964). Digestion of the sacculus with lysozyme (a β1,4 N-acetylhexosaminidase) produces two major fragments: a disaccharide-tetrapeptide (GlcNAc-MurNAc-L-Ala-D-Glu-*meso*-2,6-DAP-D-Ala) designated C_6, and its dimer (C_3) in which the free amino group of *meso*-2,6-diaminopimelic acid of one C_6 unit is in peptide linkage with the terminal D-alanine residue of another C_6 unit. A third fragment, C_5, differing from C_6 by the absence of the terminal D-alanine residue, is believed to be formed from the C_6 unit by the action of a D-carboxy-peptidase during the isolation of the peptidoglycan (Weidel *et al.*, 1963). Dinitrophenylation of the 2,6-diaminopimelic acid residues in the sacculus shows that approximately 60% of the peptide chains are cross linked, indicating a monomer:dimer ratio of 1·33:1. This is in good agreement with the actual amounts of monomer and dimer produced on lysozyme digestion of the peptidoglycan (Takebe, 1965). As a result, the peptidoglycan of *E. coli* is thought to be built primarily from C_3 and C_6 units in random arrangement and joined by glycosidic linkages involving the 1-position of N-acetylmuramic-acid residues and the 4-position of N-acetylglucosamine residues. The cross bridges in the C_3 units serve to join adjacent glycan chains. The carboxyl groups of D-glutamic acid and *meso*-2,6-diaminopimelic acid which are not involved in peptide bonding are not amidated.

In general detail, therefore, the peptidoglycan of *E. coli* is very similar to those of the bacilli. As in *Bacillus megaterium* the amino group adjacent to the L-asymmetric carbon atom of *meso*-2,6-diaminopimelic acid is linked to the γ-carboxyl group of D-glutamic acid and the C-terminal D-alanine residue is joined to the carboxyl group of *meso*-2,6-diaminopimelic acid adjacent to the L-asymmetric carbon atom (Dezélée and Bricas, 1970).

In all of the peptidoglycans of Gram-negative bacteria so far studied, the cross bridges have involved the amino group of 2,6-diaminopimelic acid on one peptide chain and the terminal D-alanine residue on another chain (Ghuysen, 1968). The proportion of 2,6-diaminopimelic acid residues involved in cross bridging is roughly the same in *P. mirabilis* (33%; Katz and Martin, 1970) as in *E. coli* (30%; Takebe, 1965). However there are significant differences between the two peptido-glycans. Lysozyme breaks down essentially completely the peptidoglycan of *E. coli* into the small disaccharide fragments, whereas about 60% of the reducing end groups of the muramic acid residues of the peptido-glycan of *P. mirabilis* are not available for reduction after lysozyme

WALLS AND MEMBRANES IN BACTERIA

digestion of the sacculus (Katz and Martin, 1970). Complete degradation of this peptidoglycan into disaccharide fragments may be achieved using a N-acetylmuramidase from a Limax amoeba (Katz and Martin, 1970). The proportion of 2,6-diaminopimelic acid residues involved in cross linking is greater in *Spirillum serpens* (54%; Kolenbrander and Ensign, 1968) than in the other two organisms studied. The average length of the glycan chains in the peptidoglycan of *Sp. serpens* has been estimated as 99 hexosamine residues (Kolenbrander and Ensign, 1968).

D. THE LIPOPROTEIN OF THE RIGID LAYER OF THE CELL ENVELOPE OF *Escherichia coli*

The protein globules present on the surface of the peptidoglycan sacculus of *E. coli* B can be removed by the action of proteolytic enzymes (Martin and Frank, 1962). Such treatment of cell envelopes leads to a detachment of the peptidoglycan from the outer membrane (De Petris, 1967; Braun and Rehn, 1969) indicating that the protein globules may be instrumental in knitting the cell envelope together. The most effective way of removing the protein from the rigid layer involves brief treatment with trypsin (Braun and Rehn, 1969) which leaves lysine residues linked through their α-amino groups (Braun and Sieglin, 1970) to carboxyl groups of *meso*-2,6-diaminopimelic acid in the peptidoglycan. Analysis of the peptidoglycan shows one "foreign" lysine residue for every 10 disaccharide repeating units of the glycan chain (Braun and Rehn, 1969). Assuming an even distribution of the protein over the surface of the peptidoglycan and making several assumptions about the lengths and separation of the glycan chains, it is possible to calculate (Braun and Rehn, 1969) that there should be about 10^5 molecules of the protein per cell. This figure agrees well with the experimental results and confirms that the protein is distributed evenly over the surface of the peptido-glycan as was suggested by electron microscope observations. The protein which is covalently bound to lipid has a molecular weight of approxi-mately 7000 and an unusual amino-acid composition. Polar amino acids comprise 65% of the lipoprotein and glycine, cysteine, proline, phenyl-alanine and histidine are absent. The lipoprotein accounts for 40% of the rigid layer by weight. After brief trypsin treatment of the rigid layer, arginine is the N-terminal acid of the lipoprotein (Braun and Sieglin, 1970) and, after pronase digestion of the rigid layer, lysine and arginine in equimolar amounts remain attached to the peptidoglycan. The peptides depicted in Table 4 (p. 37) were isolated from partial acid hydrolysates of the pronase-treated rigid layer. Significantly, no alanine is linked directly to *meso*-2,6-diaminopimelic acid. In a subsequent

3

56 D. A. REAVELEY AND R. E. BURGE

paper (Braun and Wolff, 1970) the linkage between lysine and *meso*-2,6-diaminopimelic acid was shown to involve the carboxyl group of diamino-pimelic acid adjacent to the L-asymmetric carbon atom. This carbon atom, it will be recalled, is normally linked to the amino group of

TABLE 6. Some Peptides Present in a Partial Acid Hydrolysate of the Pronase-Treated Rigid Layer of *Escherichia coli* B (Braun and Sieglin, 1970)

1. Diaminopimelyl-lysyl-arginine
2. Alanyl-glutamyl-diaminopimelyl-lysyl-arginine
3. Glucosaminyl-muramyl-alanyl-glutamyl-diaminopimelyl-lysyl-arginine

D-alanine in the peptide chain of the peptidoglycan. Thus, where lipo-proteins are covalently linked to peptidoglycan, the terminal D-alanine residue of the peptidoglycan peptide chain is replaced by the N-terminal lysine residue of the lipoprotein.

E. LIPOPOLYSACCHARIDES

The complex lipopolysaccharides present in the outer regions of the cell envelopes of *Salmonella* spp. have been intensively studied by Westphal, Lüderitz and their collaborators (for recent reviews on the biological activity, structure, biosynthesis and genetic determination of structure, see Lüderitz, 1970; Osborn, 1969). The five sugars present in all of the lipopolysaccharides, together with ethanolamine and phosphate, constitute the core of the lipopolysaccharide. This core region is common to all *Salmonella* lipopolysaccharides and is probably very similar in related organisms (Lüderitz, 1970; Osborn, 1969). Two other regions are present in the lipopolysaccharides; these are the O-specific side chains which are responsible for serological specificity and are lacking in lipopolysaccharide from the so-called R-mutant strains, and a covalently bound lipid, termed lipid A.

The O-antigenic side chains consist of oligosaccharide repeating units. The nature of the sugars present in the repeating unit, together with the type of linkages involved, determine the serological specificity of lipo-polysaccharides (Lüderitz, 1970).

The O-antigenic side chains are synthesized separately from the rest of the lipopolysaccharide and may only be transferred to a completed core. Biosynthesis of the core takes place by the sequential addition of the basal sugars from their nucleoside diphosphate precursors onto the non-reducing end group of the growing core (Osborn, 1969). Depletion of an enzyme involved in this process or in the formation of the precursors leads to a partially completed core. Thus a whole series of R mutants

WALLS AND MEMBRANES IN BACTERIA

are possible and these are classified by the composition of their core structures (Lüderitz, 1970). R_a Mutants contain the completed core but are deficient in the ability to synthesize or attach the antigenic side chains to the core. The R_e mutants possess the simplest lipopolysaccharide consisting only of KDO (2-keto-3-deoxyoctonic acid) and lipid A. From a detailed study of the lipopolysaccharides of R mutants,

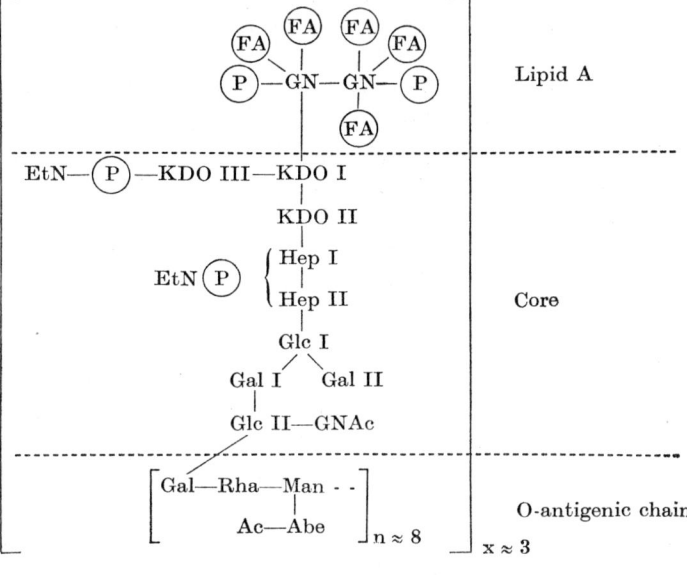

FIG. 9. Diagrammatic representation of the structure of the lipopolysaccharide of *Salmonella typhimurium*. Data taken from Luderitz (1970), Gmeiner *et al*. (1969), Dröge *et al*. (1970) and Romeo *et al*. (1970). Rha, indicates a rhamnose residue; Gal, galactose; Abe, abequose; Man, mannose; Glc, glucose; Hep, heptose; KDO, 2-keto-3-deoxyoctonic acid; GN, glucosamine; FA, fatty acid; EtN, ethanolamine; Ac, acetyl; P, phosphate. Molecular weight estimations suggest that x has a value of three. The subunits are linked through phosphodiester bridges involving lipid A or heptose.

the structure of the core of Salmonella lipopolysaccharides has been worked out (Lüderitz, 1970; Dröge *et al*., 1970). SR Mutants lack an enzyme which is involved in polymerizing the repeating units of the O-antigenic side chains. The lipopolysaccharides from such mutants contain only one oligosaccharide repeating unit joined to the core and have proved useful in determining the nature of the linkage between O-antigenic side chains and the core (Nikaido, 1970; Yuasha *et al*., 1970; Hammerling *et al*., 1970). The structure of the lipopolysaccharide of *S. typhimurium* is depicted diagrammatically in Fig. 9.

2-Keto-3-deoxyoctonic acid (KDO), a characteristic component of lipopolysaccharides, has been detected in the cell envelopes of a large number of species from many orders of Gram-negative bacteria (Ellwood, 1970b) and lipopolysaccharides have been isolated and their sugar composition examined (for a review, see Lüderitz *et al.*, 1968). Apart from *Salmonella* spp. detailed studies of R mutants and a comparison with the parent organism have not been made. In general, knowledge of the structure of other lipopolysaccharides is either fragmentary or non-existent. The general structure of core and O-antigenic side chains is common to *E. coli* (for a review, see Osborn, 1969) and *Shigella flexneri* (Johnston *et al.*, 1967) but the core structures for these organisms have characteristic compositions. Fensom and Meadow (1970) produced evidence for a core and side-chain regions in a lipopolysaccharide isolated from *Pseudomonas aeruginosa*. Glucose, heptose, galactosamine, KDO, phosphorus and alanine are present in the core region of the lipopoly-saccharide. Alanine has been reported to be present in the purified lipopolysaccharide of *Ps. alcaligenes* (Key *et al.*, 1970a). The poly-saccharide regions of both lipopolysaccharides are very rich in phosphorus (Fensom and Gray, 1969; Key *et al.*, 1970a).

Heptose, a characteristic component of the core region of lipopoly-saccharides of *Salmonella* spp. (Lüderitz *et al.*, 1968), is not present in lipopolysaccharides from *Xanthomonas* species (Volk, 1968). All 17 species examined contained uronic acid, phosphate and mannose in molar ratios of roughly $1:2:1$. 2-Keto-3-deoxyoctonic acid (KDO) and some other sugars were always present. Both KDO (another characteristic component of the core region of enterobacterial lipopolysaccharides) and heptose are absent from lipopolysaccharides of two species of *Moraxella* (Adams *et al.*, 1970) and *Neisseria catarrhalis* (Adams *et al.*, 1969a). Obviously the structure of these lipopolysaccharides is different from those of the Enterobacteriaceae.

Mild acid hydrolysis splits the bond between polysaccharide and lipid A, liberating lipid A as a water-insoluble precipitate. In KDO-containing lipopolysaccharides, lipid A and polysaccharide are thought to be joined by a ketosidic linkage involving the reducing end group of a KDO residue in the polysaccharide (Gmeiner *et al.*, 1969). In the lipo-polysaccharides of Moraxella, the linkage is believed to involve a repeating glucose unit in the polysaccharide. Preparations of lipid A from many Gram-negative organisms appear to be very similar and consist of glucosamine, fatty acids, phosphate and frequently small amounts of ethanolamine (Lüderitz *et al.*, 1968; Fensom and Gray, 1969; Key *et al.*, 1970a; Adams and Singh, 1970). The replacement of some of the glucosamine by galactosamine has been reported for some organisms (Adams *et al.*, 1969a; 1970). The basic unit of lipid A consists of two

WALLS AND MEMBRANES IN BACTERIA 59

glucosamine residues (Burton and Carter, 1964) esterified with phosphate and fatty acids. A hydroxy acid, often β-hydroxymyristic acid, accounts for at least half of the fatty acids present and N-acylates the glucosamine residues (Lüderitz et al., 1968). Of the remaining fatty acids, palmitic and lauric acids usually predominate (Burton and Carter, 1964; Fensom and Gray, 1969; Adams and Singh, 1970).

On the basis of the products isolated after hydrazinolysis and partial acid hydrolysis of the lipopolysaccharide of a R_e mutant of *Salmonella minnesota*, Gmeiner et al. (1969) suggest that the glucosamine residues are linked β-1,6. A 1,6 linkage, probably a β-linkage, joins the glucosamine residues in the lipid A of *Serratia marcescens* (Adams and Singh, 1970). The glucosamine units of lipid A are heterogeneously substituted in the 1- and probably 4-position with phosphate. Possibly disaccharide units are joined together through phosphodiester bonds (Gmeiner et al., 1969).

The structures presented for lipopolysaccharides leave open the possibility of cross bridges involving phosphate groups linking adjacent polysaccharide chains or adjacent lipid A molecules. Molecular weight estimations on a enzymically deacylated lipopolysaccharide of *Salmonella london* (Malchow et al., 1969) and a galactose-deficient lipopolysaccharide of *S. typhimurium* (Romeo et al., 1970) indicate that around three units form a lipopolysaccharide molecule.

Lipopolysaccharides contain both hydrophilic and hydrophobic regions and form micellar structures with particle weights of several millions in aqueous solution. In the electron microscope, isolated lipopolysaccharides may appear as hollow spheres (Frank and Dekegel, 1967; Rothfield and Horne, 1967), discs (Bladen and Mergenhagen, 1964) or stacks of membranes (De Petris, 1967). A branched ribbon-like network is also frequently observed (Wang et al., 1970; Lopes and Inniss, 1970). A transition from spheres to ribbons may be induced by EDTA (Frank and Dekegel, 1967). All of the structures when viewed in thin section appear as unit membranes (De Petris, 1967; Frank and Dekegel, 1967; Wang et al., 1970; Burge and Draper, 1967c). X-Ray diffraction patterns show that the lipid hydrocarbon chains are orientated normal to the plane of the lipopolysaccharide (Burge and Draper, 1967c). Rothfield and Horne (1967) suggest that the lipid moieties of the lipopolysaccharides are opposed with the polysaccharide chains pointing outwards and occupying the outer surface of the structure (Fig. 9, p. 57). The presence of O-antigenic chains on the surfaces of these structures has been indicated by the use of ferritin-labelled antibodies (Shands et al., 1967). Other evidence for the surface location of the polysaccharide chains has come from negative staining (Lopes and Inniss, 1970). The ribbons of a lipopolysaccharide from a smooth strain (i.e. possessing O-antigenic chains) of *E. coli* appear in negative-stained preparations

60   D. A. REAVELEY AND R. E. BURGE

to consist of a central unstained core enclosed within two parallel lines. Another unstained zone was present on either side of the central zone. This outer zone, the postulated site of the O-antigenic chains, was lacking

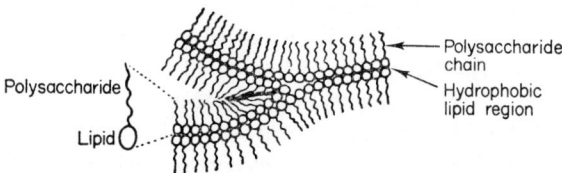

Assembly of lipopolysaccharide units to form ribbon shaped structure

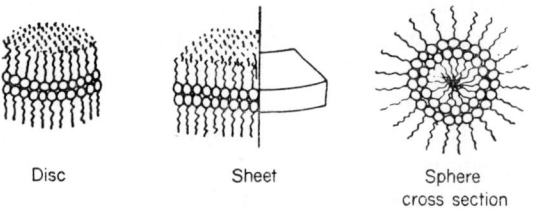

FIG. 10. Proposed assembly of lipopolysaccharide units in Gram-negative bacteria to form various structures previously observed. Taken from Lopes and Inniss (1970).

in preparations from rough mutant strains of *E. coli* and *S. typhimurium*. The possible structures of the various forms which lipopolysaccharides take are depicted in Fig. 10.

F. LIPIDS

As with Gram-positive bacteria, the lipids extracted from the cell with chloroform-methanol mixtures are concentrated in the cell membranes (Salton, 1964; White and Cox, 1967; Rizza et al., 1970). The lipids are located in the inner (cytoplasmic membrane) and outer regions of the cell envelope but, in some organisms, may also be present in complex internal membranes (Cho and Salton, 1966). Preliminary experiments (Hancock and Meadow, 1967; Gray and Thurman, 1967) indicate that the lipid compositions of cytoplasmic membrane and outer membrane of *Ps. aeruginosa* are similar. However, the fatty-acid

WALLS AND MEMBRANES IN BACTERIA 61

composition of the internal membranes and the cell envelope of *Rhodospirillum rubrum* differ markedly (Cho and Salton, 1966).

The extractable lipids of the cell envelopes of Gram-negative bacteria consist principally of phospholipids and free fatty acids (Gray and Wilkinson, 1965; Bobo and Eagon, 1968). At least half of the phospholipid of *E. coli* is phosphatidylethanolamine (Ames, 1968; Kanemasa *et al.*, 1967). Phosphatidylglycerol, diphosphatidylglycerol and small amounts of other phospholipids are also present (Ames, 1968; Cronan, 1968). Magnesium is associated with diphosphatidylglycerol extracted from the cell envelopes of *Ps. aeruginosa* and a marine pseudomonad (Gordon and MacLeod, 1966). The phospholipid compositions of whole cells of several Gram-negative bacteria are given in Table 7 (Randle *et al.*, 1969). The presence of phosphatidylcholine in *Ps. aeruginosa* has recently been challenged (Hancock and Meadow, 1969). Traces of phosphatidylcholine or choline in the growth medium could give rise to small amounts of phosphatidylcholine present in some bacterial phospholipids (Bobo and Eagon, 1968). Some workers have reported the presence of phosphatidylserine in *E. coli* (Ames, 1968) and *Ps. aeruginosa* (Clarke *et al.*, 1967b). On the basis of their survey, Randle *et al.* (1969) divided Gram-negative bacteria into two groups depending on whether or not N-methyl derivatives of phosphatidylethanolamine were present (Table 7). A third group was postulated to cover Hyphomicrobiales and some *Rhodopseudomonas* species which differ from the first group in containing ornithine esters of phosphatidylglycerol and by the variable occurrence of phosphatidylcholine.

Generally, the cellular concentrations of phosphatidylethanolamine, phosphatidylglycerol, phosphatidyl-N-monomethylethanolamine and phosphatidylcholine attain a maximum in the late logarithmic or early stationary phases of growth and then decrease (Hancock and Meadow, 1969; Randle *et al.*, 1969; Shively and Benson, 1967). The concentration of diphosphatidylglycerol, on the other hand, reached a maximum in the stationary phase. Such changes are independent of the composition of the growth medium. A conversion of phosphatidylglycerol into diphosphatidylglycerol is apparent in *E. coli* and *Thiobacillus thiooxidans* (Shively and Benson, 1967; Cronan, 1968). Cells of *S. typhimurium*, grown in a medium with a low concentration of phosphate, contained a diminished amount of phospholipid although relative amounts were unchanged. Growth rate was not appreciably affected (Ames, 1968).

The fatty acids present in the loosely bound lipids of the cell envelopes of Gram-negative bacteria have been examined (Cho and Salton, 1966). The fatty acids present in these lipids are markedly different from the fatty acids of lipid A (Nesbitt and Lennarz, 1965; Hancock and Meadow,

TABLE 7. The Phospholipid Composition[a] of some Gram-negative Bacteria (taken from Randle et al., 1969). The Cultures were Harvested in the Stationary Phase

Bacteria	Culture age (hr.)	Distribution of the total lipid phosphorus (%)							
		PC	PMME	PE	PG	PA	diPG	PGP	Other
Serratia marcescens	48	0	0	66.1	14.4	2.3	17.0	0	0
Enterobacter aerogenes	24	0	0	74.3	20.8	1.0	3.2	0.4	0
Escherichia coli	48	0	0	52.1	18.3	0.9	28.1	0.5	0
Proteus vulgaris	48	0	12.1	63.0	5.8	0.9	14.8	0	3.3[b]
Chromobacterium violaceum	24	0	0	76.6	17.9	0.9	4.6	0	0
Azotobacter tumefaciens	46	27.7	16.4	17.5	13.0	4.1	18.5	2.6	0
Azotobacter agilis	48	2.4	4.9	56.8[c]	13.3	Trace	22.6	0	0
Pseudomonas aeruginosa	32	1.4	0	69.4	14.9	1.7	9.4	2.1	0

[a] PC, indicates phosphatidylcholine; PMME, phosphatidylmonomethylethanolamine; PE, phosphatidylethanolamine; PG, phosphatidylglycerol; PA, phosphatidic acid; diPG, diphosphatidylglycerol; PGP, phosphatidylglycerol phosphate.
[b] Peak chromatographed in anion-exchange resin in position of glycerophosphorylserine but confirmatory evidence on the basis of acid hydrolysis products was not obtained.
[c] Includes phosphatidyldimethylethanolamine as established by chromatography of acid hydrolysis products.

WALLS AND MEMBRANES IN BACTERIA 63

1967). Generally, $C_{16:0}$, $C_{16:1}$, $C_{18:1}$ and cyclopropane acids are predominant (Cho and Salton, 1966; Hancock and Meadow, 1969; Thiele et al., 1969; Lewis et al., 1968). Branch-chain acids are only rarely present (Rizza et al., 1960). Cyclopropane acids are formed from unsaturated acids by the addition of a methylene group from S-adenosylmethionine across the double bond (Law et al., 1963) and in the stationary phase, the amount of cyclopropane acid increases whereas the amount of unsaturated acid decreases (Law et al., 1963; Cronan, 1968). Most of the unsaturated and cyclopropane acids are present in the 1 position of the phospholipid. Cyclopropane acids may serve to protect phospholipids from degradation (Cronan, 1968). Strains of E. coli resistant to tetracycline and strains of Klebsiella pneumoniae resistant to polymyxin contain higher concentrations of unsaturated fatty acids and lower concentrations of cyclopropane acids than sensitive organisms (Dunnick and O'Leary, 1970).

Changing the growth temperature of cultures of E. coli from 37° to 10° results in an increase in the relative amounts of unsaturated fatty acid (Shaw and Ingraham, 1965; Okuyama, 1969) whilst the concentration of other fatty acids in the cell remained constant (Okuyama, 1969). Such a change may be explained by the cell requiring, in the cold environment, fatty acids of lower melting points to prevent increased rigidity of the cell membranes. During the transition from high to lower temperature, degradation and synthesis of phospholipids, including phosphatidylethanolamine, occurs resulting in an overall 20% increase in phospholipid content. The fatty acids liberated during the degradation are recycled into fresh phospholipid (Okuyama, 1969). A mutant strain of E. coli has been described (Silbert et al., 1968) which requires unsaturated or cyclopropane acids for growth. This mutant may prove valuable in structural studies since it is possible to vary the fatty-acid composition of the envelope phospholipids by changing the fatty-acid growth supplement.

G. PROTEINS

The only protein of the Gram-negative cell envelope studied in any detail is the lipoprotein covalently attached to peptidoglycan remarkably little attention has been paid to other envelope proteins. Protein is fractionated into the phenol phase during aqueous phenol extraction of "cell walls" (Wardlaw, 1963; Clarke et al., 1967b). A fraction consisting almost exclusively of protein was isolated from the phenol phase after extraction of the cell envelopes of Ps. aeruginosa (Clarke et al., 1967b). The fraction accounted for 22–28% of the cell envelope and was rich in aspartic acid and glutamic acid residues; however the presence of large amounts of ammonia in acid hydrolysates of the fraction suggests

that at least some of these residues are amidated. Similar fractions have been isolated from *E. coli* (Wardlaw, 1963; Weinbaum *et al.*, 1970). A protein excreted as part of a lipopolysaccharide-phospholipid-protein complex during lysine deprivation of a lysine-requiring mutant of *E. coli* is also rich in aspartic acid, glutamic acid and glycine residues and low in cystine (Knox *et al.*, 1967). The protein fraction of *Ps. aeruginosa* contained traces of glucosamine; glucosamine-containing proteins are present in the cell envelope of *E. coli* (Weinbaum *et al.*, 1970). As with the glycoproteins of animal cell membranes, aspartic acid (or asparagine) is believed to be involved in binding the amino sugar to the protein.

Schnaitman (1970b) has examined the proteins of the cell envelope of *E. coli* by quantitative gel electrophoresis. Approximately 20–30 bands were observed in the gels; one protein, of molecular weight 44,000, accounted for as much as 40% of the total envelope protein. Preliminary experiments (Schnaitman, 1970c) suggest that similar proteins are present in high concentration in the cell envelopes of *Shigella sonnei*, *Xanthomonas sinensis*, *Proteus vulgaris*, *Pseudomonas aeruginosa* and *Neisseria catarrhalis*. The major protein in the envelope of *E. coli* is concentrated in the cell wall and accounts for 70% of the cell-wall protein; it is not associated with peptidoglycan. Studies with crude preparations of cell walls and cytoplasmic membranes (Schnaitman, 1970a) show that there are fewer proteins associated with the cell wall (probably around six) compared with the cytoplasmic membrane (27 bands were concentrated in this fraction). Although growth of *E. coli* under various conditions produced large changes in the concentrations of certain membrane-bound enzymes, these changes were not reflected in major differences in the protein compositions of the cell envelopes (Schnaitman, 1970b). A major membrane protein, of molecular weight 20,000, is missing from the cell envelope of a chlorate-resistant mutant of *E. coli*. The protein is not identical with either of the membrane-bound enzymes nitrate reductase or formate dehydrogenase which are absent from the mutant (Schnaitman, 1969).

H. Effect of EDTA on Gram-Negative Bacteria

The combination of EDTA and lysozyme in tris-HCl buffer lyses many Gram-negative organisms (Repaske, 1956, 1958). There is much evidence that tris is more than an inert buffer and in some way potentiates the action of EDTA (Asbell and Eagon, 1966; Neu *et al.*, 1967; Voss, 1967; Neu, 1969; Tucker and White, 1970a). Tris may be replaced (Voss, 1967) by other aliphatic amines or quaternary ammonium compounds, the most effective having C_{12}–C_{16} alkyl chains.

Both *Ps. aeruginosa* (Gray and Wilkinson, 1965) and *Ps. alcaligenes* (Key *et al.*, 1970b) are lysed by EDTA without the specific requirement for tris buffer and lysozyme. The sensitivity of pseudomonads to lysis by EDTA is related to the ability of EDTA to solubilize material from the cell wall (Wilkinson, 1967). The presence of divalent cations in the cell wall of *Ps. aeruginosa* has been demonstrated (Eagon *et al.*, 1965). The rod-shaped osmotically fragile forms of *Ps. aeruginosa* produced by incubation of cells in EDTA/sucrose/tris-HCl buffer may be restored to osmotic stability by adding multivalent cations to the solution preferably in the same proportion as present in the cell wall (Asbell and Eagon, 1966). The restored cells respired and, like normal cells, formed induced permeases to citrate. Asbell and Eagon (1966) suggest that EDTA acts by chelating cations involved in cross linking lipopolysaccharide in the cell wall. Degradation of the peptidoglycan sacculus by autolytic enzymes has been suggested as the reason for the non-requirement of lysozyme for the lysis of these pseudomonads (Leive, 1968). The possibility that some of the metal ions are involved in holding the peptidoglycan sacculus together should, however, not be overlooked. Buckmire and MacLeod (1965) suggest such a role for divalent cations in a marine pseudomonad. An interesting development was the observation by Brown and Melling (1968) that *Ps. aeruginosa* grown in a magnesium-limited medium has a greatly diminished sensitivity to EDTA.

Brief treatment of *E. coli* with EDTA in tris-HCl buffer at 0° renders the organism permeable to a wide range of small, charged molecules (Leive, 1968). Adding back magnesium ions is ineffective in restoring the permeability barrier, but the barrier is restored if the cells are allowed to grow (Leive, 1968); synthesis of protein, ribonucleic acid or peptidoglycan is not required. EDTA does not affect active transport but releases 30–50% of the surface lipopolysaccharide of *E. coli*. Both lipopolysaccharide release and permeability change are rapid and independent of temperature. However, the two effects are, at least partially, dissociated, since, although lipopolysaccharide release is independent of temperature in the pH range 6–9, permeability is altered radically (Lieve *et al.*, 1968).

The amount of lipopolysaccharide lost from a given strain appears to be constant and cannot be increased by variation in the extraction procedure (Levy and Leive, 1968). Newly synthesized lipopolysaccharide is not initially released but rapidly enters an equilibrium between retained and released material. The excreted lipopolysaccharide which contains small amounts of protein (5–10%) and phospholipid (5%) may be separated into two fractions by ultracentrifugation (Leive *et al.*, 1968). One lipopolysaccharide fraction appears to contain more side chains, whereas the other fraction contains most of the extracted protein and lipid. Since the side chains contain the antigenic determinants of whole

cells, Leive *et al.* (1968) suggest that this lipopolysaccharide fraction is present in the outermost layers of the cell envelope. On the basis of ^{32}P-labelling experiments, Tucker and White (1970a) suggest that EDTA/tris-HCl progressively strips phospholipid from the cell surface of *Haemophilus parainfluenzae*. After labelling for only a short time prior to EDTA treatment, the lipids initially released possessed a lower specific activity than the phospholipids remaining in the cell envelope. The extracted phospholipids were richer in diphosphatidylglycerol and phosphatidylglycerol and differed in fatty-acid composition when compared with the lipids in the residue (White and Tucker, 1970a, b). As the incubation with EDTA/tris-HCl continued, both the specific activity and fatty-acid composition of the extracted lipids approached the composition of the lipids in the rest of the cell envelope.

I. Osmotic Shock

If *E. coli* is treated with EDTA/tris-HCl buffer in the presence of 0·5 *M*-sucrose and the cells then exposed to cold aqueous magnesium chloride, a group of hydrolytic enzymes involved in phosphate, nucleotide and sugar degradation (Heppel, 1967; Neu and Chou, 1967) are released from the cells into the surrounding medium together with factors involved in active transport (Pardee, 1968; Piperno and Oxender, 1966). In the process, about 10% of the total cellular protein (Neu and Heppel, 1965; Nossal and Heppel, 1966) and all the acid-soluble nucleotides are lost from the cells, yet they remain viable. The enzymes are also lost when sphaeroplasts are formed by the enzyme/EDTA/tris-HCl treatment in a hypertonic sucrose solution but not when sphaeroplasts are formed by the action of penicillin (Heppel, 1967).

A survey (Neu and Chou, 1967) has shown that the enzymes, cyclic phosphodiesterase (which also functions as a 3′-nucleotidase; Neu, 1968a), 5′-nucleotidase (which in many cases is identical with uridine diphosphate sugar hydrolase; Heppel, 1967; Glaser *et al.*, 1967; Neu, 1968b), acid hexose phosphatase and acid phenylphosphatase, are released by osmotic shock from strains of *Shigella*, *Enterobacter*, *Citrobacter*, *Serratia* and some strains of *Salmonella*. Members of the Proteus and Providencia groups do not release the enzymes but they do release acid-soluble nucleotides. Lipopolysaccharide may be lost from cells at either the EDTA or the shocking stage. Winshell and Neu (1970) observed that an efficient release of 5′-nucleotidase occurred with organisms releasing the greatest amount of lipopolysaccharide at the shocking stage. *Proteus vulgaris* released only small amounts of lipopolysaccharide at both stages.

The cellular location of the proteins released on osmotic shock has been

investigated by several workers. It is well known that the activities of the enzymes liberated by osmotic shock can be measured in intact cells but, since the activities are in general lower than those present in an equivalent cell extract (Heppel, 1967), it has been implied that the substrates had to overcome a partial permeability barrier. This barrier is believed to be the cell wall, and the enzymes are thought to be located in the region between the cell wall and the cytoplasmic membrane (termed the periplasmic space). Shocking is known to affect the permeability of the cell wall to large molecules such as lysozyme, ribonuclease and deoxyribonuclease.

Cytochemical techniques have been used to attempt to confirm the location of these proteins in the cell envelope. Generally the results obtained have been inconclusive. Thus, alkaline phosphatase has been indicated to be present in the periplasmic region (Done *et al.*, 1965), whereas 5'-nucleotidase and 3'-nucleotidase have been suggested to be located on the cell surface (Nisonson *et al.*, 1969). Recent cytochemical experiments on *E. coli* (Wetzel *et al.*, 1970) placed alkaline phosphatase, hexose monophosphatase and cyclic phosphodiesterase both on the cell surface and in the periplasmic space. The location of the enzymes apparently depended on the technique used to detect the enzymes. Wetzel *et al.* (1970) reported high concentrations of the three enzymes in an enlarged periplasmic region at the poles of the cells. This observation was confirmed by studying "mini" cells which bud from the poles of rod-shaped cells during the growth cycle. These mini cells contained significantly larger concentrations of alkaline phosphatase, cyclic phosphodiesterase, acid hexose monophosphatase, 5'-nucleotidase, and ribonuclease I than the rod cells (Dvorak *et al.*, 1970).

J. Structure of the Outer Membrane of the Gram-Negative Cell Envelope

Treatment of cell envelopes with phenol, a procedure which removes lipopolysaccharide, lipid and protein from the envelopes (Martin and Frank, 1962; Clarke *et al.*, 1967b), removes the outer membrane. The presence of the O-antigenic chains of lipopolysaccharides on the surface of the outer membrane has been demonstrated by ferritin-labelled antibody experiments (Shands, 1966); significantly, areas of close contact between cells are only observed with smooth strains (De Petris, 1967). Phage studies have shown that both lipopolysaccharide and lipoprotein are present on the surface of *E. coli* (Frank and Dekegel, 1967). That lipopolysaccharide and phosphatidylethanolamine are associated together in the Gram-negative cell envelope is shown by the work of Rothfield and his coworkers. Phosphatidylethanolamine is required for two of the steps in the synthesis of the core of lipopolysaccharide (Weiser

68   D. A. REAVELEY AND R. E. BURGE

and Rothfield, 1968). An active complex, formed by heating and slowly cooling a mixture of lipopolysaccharide and phospholipid, will combine with soluble enzyme and act as acceptor for sugar-transfer reactions with a similar activity to the cell envelope. Electron microscope monitoring shows that, on cooling a mixture of phosphatidylethanolamine and lipopolysaccharide, there is continuity of structure between the lipopolysaccharide spheres and the phosphatidylethanolamine lamellae (Rothfield and Horne, 1967). A penetration of lipopolysaccharide into a monolayer of phosphatidylethanolamine has been revealed by surface pressure measurements; studies indicate that both components probably lie with their fatty-acyl chains directed upwards (Romeo et al., 1970). The inference that the outer membrane of the Gram-negative bacterium contains a bilayer formed from lipopolysaccharide and phospholipid has yet to be proven. The nature of the association between lipopolysaccharide and protein is not known but is amenable to further investigation. Lysine limitation of a lysine-requiring mutant of E. coli leads to excretion into the medium of a complex consisting of lipopolysaccharide (60%), protein (11%) and phospholipid (mainly phosphatidylethanolamine, 26%). The excretion is believed to be caused by a shedding of the outer layers of the cell wall due to unbalanced growth. Similar material is excreted in large amounts from Salmonella typhimurium or E. coli by inhibiting protein synthesis (Rothfield and Pearlman-Kothencz, 1969). Isopycnic centrifugation and immune co-precipitation showed lipopolysaccharide protein and lipid to be in a single complex. The protein was present largely as a single species (see Section V, G, p. 63).

K. The Cell Envelope of Halophilic Bacteria

Extremely halophilic bacteria, such as *Halobacterium halobium*, *H. cutirubrum* and *H. salinarium*, require salt concentrations in the range 3–5 M for growth and preservation of ultrastructure. In thin sections the isolated cell envelopes appear multilayered consisting of an outer "cell wall" and an inner cytoplasmic membrane (Cho et al., 1967; Stoeckenius and Rowen, 1967; Steensland and Larsen, 1969). In appearance and thickness, the cytoplasmic membrane is similar to other bacterial cytoplasmic membranes. The thickness of the cell wall has been quoted as 60–150 Å in *H. salinarium* and *Halobacterium* sp. (Steensland and Larsen, 1969), and 75–90 Å (Cho et al., 1967) and 130–150 Å (Stoeckenius and Rowen, 1967) in *H. halobium*. In thin section the surface of the cell wall appears dented (Steensland and Larsen, 1969), beaded (Stoeckenius and Rowen, 1967) or spikey (Cho et al., 1967). This appearance may be related to the hexagonal arrangement of spherical particles 120–150 Å apart demonstrated by shadowing to be present on the

WALLS AND MEMBRANES IN BACTERIA

surface of the cell envelopes of halophilic bacteria (Larsen, 1967). The outer region of the cell wall stains more densely than the inner region suggesting a two-layered structure. Cell envelopes isolated by mechanical disintegration are in the form of closed vesicles (Stoeckenius and Rowen, 1967; Steensland and Larsen, 1969).

On lowering the sodium chloride concentration of a suspension of envelopes, the first observable effect is the loss of the hexagonal surface pattern of the cell wall, followed by a loss of the cell wall itself. The sodium chloride concentration necessary is dependent upon the organism, but loss of cell wall occurs at $2 \cdot 2$ M for H. $salinarium$ and $1 \cdot 0$–$1 \cdot 4$ M for H. $halobium$. On further dilution, the cytoplasmic membrane breaks up into fragments.

The components of peptidoglycan are absent from the cell envelope of halophilic bacteria (Steensland and Larsen, 1969; Stoeckenius and Rowen, 1967), the cell wall probably consists of protein and amino sugar, whereas the cytoplasmic membrane consists of protein and lipid (Steensland and Larsen, 1969; Stoeckenius and Rowen, 1967). A characteristic feature of the envelope proteins of halophilic bacteria is the much greater excess of acidic amino acids compared with basic amino acids than in envelope proteins of other Gram-negative bacteria (Steensland and Larsen, 1969).

VI. Structure Arrangements of Polymeric Components Within Bacterial Cell Walls and Cytoplasmic Membranes

The detailed connection between the bulk chemical constitution of bacterial wall components and their appearance in the electron microscope has yet to be established. This may be accomplished by: (1) model building procedures (Kelemen and Rogers, 1969), e.g. from the detailed knowledge now available of the chemical structure of the peptidoglycan layer in Gram-positive cells; and (2) from X-ray diffraction methods which can both provide direct structural information and establish the validity of molecular models produced from chemical evidence alone.

To establish the relationship between chemical structure and polymeric arrangement as seen in the electron microscope, it is essential to evaluate the effects of the usual fixation, embedding and staining procedures; efforts in this direction were made by Burge and Draper (1967a, b, c) for the wall of $Proteus$ $vulgaris$ and also an extracted lipopolysaccharide component. Similar X-ray studies to those for P. $vulgaris$ cell walls have been carried out for $Escherichia$ $coli$ cell walls (R. E. Burge, D. A. Reaveley and J. Draper, unpublished results) and the diffraction patterns for the two organisms show the presence of very similar layering both for the outer membrane and the complete cell

70   D. A. REAVELEY AND R. E. BURGE

wall. These similarities are expected on the basis of the similar chemical compositions of the walls.

X-Ray diffraction studies on isolated Gram-positive cell walls and peptidoglycan layers have been reported by Carito et al. (1967). These authors found several characteristic diffraction lines in the X-ray patterns of three Gram-positive cell walls (Bacillus subtilis, B. cereus and Clostridium sporogenes) the strongest being at spacings of 6·58 and 5·24 Å. These spacings remained when the corresponding peptidoglycan from one organism (B. subtilis) was examined. A study of the X-ray diffraction patterns from the cell walls and peptidoglycans isolated from a number of Gram-positive organisms has been undertaken (H. H. M. Balyuzi, R. E. Burge and D. A. Reaveley, unpublished results). The organisms include two strains of Staphylococcus aureus and strains of Bacillus licheniformis and Micrococcus lysodeikticus. Initially with some specimens of S. aureus walls and peptidoglycans, following very extensive washing in order to remove salt contamination, a pattern of sharp diffraction lines, of a different character from those observed by Carito et al. (1967), was observed superimposed on a background of (liquid-like) continuous X-ray scattering. In other specimens of S. aureus walls and peptidoglycans, as in all of the specimens of walls and peptidoglycans from the other organisms, no sharp diffraction lines were found. Subsequently, the sharp lines have been attributed to contamination with lead salts not removable by normal washing procedures and arising from either the glass beads used to break the cells or new glassware.

Bearing in mind the similarities between the organisms examined by Carito et al. (1967) and by Balyuzi, Burge and Reaveley (unpublished results), it seems likely that all of these cell walls and peptidoglycans give an amorphous X-ray pattern, and the crystalline X-ray reflexions observed by Carito et al. (1967) in all of their preparations arise due to salt contamination.

The lack of crystalline X-ray reflexions from the whole wall and the isolated peptidoglycans means that the molecular ordering as seen from any point within the structures is of limited extent and governed by local stereochemical effects. This type of structure can be examined (Balyuzi and Burge, 1971) by the methods of X-ray examination originally developed for elucidating the structure of liquids. It is anticipated that the different lengths of glycan chains will contribute in different cell walls to regions of varying extent over which a reasonable degree of molecular ordering is maintained. Similarly, any co-operative contribution to the X-ray scattering from the peptide chain bridges will depend on the stereochemical constraints to which the chains are subject both within the chains themselves and from their local environments. These amorphous patterns, which may be interpreted with the aid of molecular

models up to a maximum size of (say) 20 Å, should be distinguished from the X-ray results corresponding to the layering (on the 100 Å scale) observed with walls and membranes. The amorphous patterns deal with molecular ordering, the layered patterns with macromolecular sizes and associations. The amorphous patterns cannot be used, as for example can the highly crystalline and extensive diffraction found in the crystalline proteins, to determine a unique structure for a wall or a wall component. However, the amorphous patterns can be compared with calculated X-ray results from molecular models built on the basis of chemical evidence; this is the approach being applied by Balyuzi, Burge and Reaveley (unpublished results).

So far as X-ray diffraction patterns of the bacterial cytoplasmic membrane are concerned, an important need is to isolate pure stable membrane fractions of both Gram-positive and Gram-negative organisms and establish, first, the degree of structural similarity corresponding to similarity of their chemical compositions and, second, the relationship between their structure and those of mammalian membranes. Preliminary X-ray studies of the (Gram-positive) cytoplasmic membranes from *B. licheniformis* (D. A. Reaveley and R. E. Burge, unpublished results) show a close similarity in gross structural features between this cytoplasmic membrane and erythrocyte membranes (Finean *et al.*, 1968). This result is interesting in relation to the work on optical rotatory dispersion and circular dichroism (e.g. Lenard and Singer, 1966) which shows that the protein components of erythrocyte membranes and cytoplasmic membranes from *B. subtilis* are very similar in molecular conformation.

VII. Acknowledgements

We thank Dr. A. H. Fensom, Dr. G. R. Millward, and Dr. S. G. Wilkinson for critical reading and discussion of the manuscript. One of us (D.A.R.) acknowledges the receipt of a Medical Research Council Fellowship.

REFERENCES

Abram, D. (1965). *J. Bact.* **89**, 855.
Abrams, A. and Baron, C. (1968). *Biochemistry, N.Y.* **7**, 501.
Abrams, A., Nielsen, L. and Thaemert, J. (1964). *Biochim. biophys. Acta* **80**, 325.
Adams, G. A. and Singh, P. P. (1970). *Can. J. Biochem.* **48**, 55.
Adams, G. A., Tornabene, T. G. and Yaguchi, M. (1969a). *Can. J. Microbiol.* **15**, 365.

Adams, G. A., Quadling, C., Yaguchi, M. and Tornabene, T. G. (1970). *Can. J. Microbiol.* **16**, 1.

Adams, J. B., Archibald, A. R., Baddiley, J., Coapes, H. E. and Davison, A. L. (1969b). *Biochem. J.* **113**, 191.

Ames, G. F. (1968). *J. Bact.* **95**, 833.

Archibald, A. R. and Baddiley, J. (1965). *Biochem. J.* **95**, 19 c.

Archibald, A. R., Armstrong, J. J., Baddiley, J. and Hay, J. B. (1961). *Nature, Lond.* **191**, 570.

Archibald, A. R., Baddiley, J. and Button D. (1968a). *Biochem. J.* **110**, 543.

Archibald, A. R., Baddiley, J., Button, D., Heptinstall, S. and Stafford, G. H. (1968b). *Nature, Lond.* **219**, 855.

Archibald, A. R., Baddiley, J. and Heptinstall, S. (1969). *Biochem. J.* **111**, 245.

Aronson, A. (1966). *J. molec. Biol.* **15**, 505.

Asbell, M. A. and Eagon, R. G. (1966). *J. Bact.* **92**, 380.

Balyuzi, H. H. M. and Burge, R. E. (1971). *Biopolymers* **10**, 777–788.

Barkulis, S. S. and Jones, M. F. (1957). *J. Bact.* **74**, 207.

Barkulis, S. S., Loh, V., Boltralik, J. J. and Smith, C. E. (1968). *Bact. Proc.* 110.

Bayer, M. E. (1968). *J. gen. Microbiol.* **53**, 395.

Bayer, M. E. and Remsen, C. C. (1970). *J. Bact.* **101**, 304.

Beebe, J. L. (1970). *Bact. Proc.* 50.

Bertsch, L. L., Bonson, P. P. M. and Kornberg, A. (1969). *J. Bact.* **98**, 75.

Betz, J. V. and Zeikus, J. G. (1968). *Bact. Proc.* 21.

Birdsell, D. C. and Cota-Robles, E. H. (1967). *J. Bact.* **93**, 427.

Bishop, D. J., Rutberg, L. and Samuelsson, B. (1967a). *Eur. J. Biochem.* **2**, 448.

Bishop, D. J., Rutberg, L, and Samuelsson, B. (1967b). *Eur. J. Biochem.* **2**, 454.

Bladen, H. A. and Mergenhagen, S. E. (1964). *J. Bact.* **88**, 1482.

Bobo, R. A. and Eagon, R. G (1968) *Can. J. Microbiol.* **14**, 503.

Bodman, H. and Welker, N. E. (1969). *J. Bact.* **97**, 924.

Bond, E. C. and Landman, O. E. (1970). *Bact. Proc.* 36.

Boylen, C. W. and Ensign, J. C. (1968). *J. Bact.* **96**, 421.

Braun, V. and Rehn, K. (1969). *Eur. J. Biochem.* **10**, 426.

Braun, V. and Sieglin, U. (1970). *Eur. J. Biochem.* **13**, 336.

Braun, V. and Wolff, H. (1970). *Eur. J. Biochem.* **14**, 387.

Bray, J. P. and Bass, J. A. (1968). *Bact. Proc.* 110.

Brenner, S., Dark, F. A., Gerhardt, P., Jeynes, M. H., Kandler, O., Kellenberger, E., Klieneberger-Nobel, E., McQuillen, K., Rubio-Hurios, M., Salton, M. R. J., Strange, R. E., Tomcsik, J. and Weibull, C. (1958). *Nature, Lond.* **181**, 1713.

Brinton, C. C., McNary, J. C. and Carnahan, J. (1969). *Bact. Proc.* 48.

Brooks, D. and Baddiley, J. (1968). *Biochem. J.* **113**, 635.

Brown, A. D. (1964). *Bact. Rev.* **28**, 926.

Brown, A. D., Drummond, D. G. and North, R. J. (1962). *Biochim. biophys. Acta* **58**, 514.

Brown, M. R. W. and Melling, J. (1968). *J. gen. Microbiol.* **54**, 439.

Brundish, D. E. and Baddiley, J. (1968). *Biochem. J.* **110**, 573.

Buckmire, F. L. A. and MacLeod, R. A. (1965). *Can. J. Microbiol.* **11**, 677.

Buckmire, F. L. A. and Murray, R. G. E. (1970). *Can. J. Microbiol.* **16**, 1011.

Burdett, I. D. J. and Rogers, H. J. (1970). *J. Ultrastruct. Res.* **30**, 354.

Burge, R. E. and Draper, J. (1967a). *J. molec. Biol.* **28**, 173.

Burge, R. E. and Draper, J. (1967b). *J. molec. Biol.* **28**, 189.

Burge, R. E. and Draper, J. (1967c). *J. molec. Biol.* **28**, 205.

Burger, M. M. (1966). *Proc. natn. Acad. Sci. U.S.A.* **56**, 910.

WALLS AND MEMBRANES IN BACTERIA

Burger, M. M. and Glaser, L. (1966). *J. biol. Chem.* **241**, 494.
Burton, A. J. and Carter, H. E. (1964). *Biochemistry, N.Y.* **3**, 411.
Butler, T. F., Smith, G. L. and Grula, E. A. (1967) *Can. J. Microbiol.* **13**, 1471.
Button, D., Archibald, A. R. and Baddiley, J. (1966). *Biochem. J.* **99**, 11 c.
Campbell, J. N., Leyh-Bouille, M. and Ghuysen, J. M. (1969). *Biochemistry, N.Y.* **8**, 193.
Card, G. L. (1969). *Bact. Proc.* 120.
Card, G. L., Georgi, C. E. and Militzer, W. E. (1969). *J. Bact.* **97**, 186.
Carito, S. L., Bazil, S. L. and DiGiacomo, G. (1967). *J. Bact.* **93**, 122.
Chatterjee, A. N. (1969). *J. Bact.* **98**, 519.
Chatterjee, A. N., Ward, J. B. and Perkins, H. R. (1967). *Nature, Lond.* **214**, 1311.
Chatterjee, A. N., Mirelman, D., Singer, H. J. and Park, J. T. (1969). *J. Bact.* **100**, 846.
Chin, T., Burger, M. M. and Glaser, L. (1966). *Archs Biochem. Biophys.* **110**, 358.
Cho, K. Y. and Salton, M. R. J. (1966). *Biochim. biophys. Acta* **116**, 73.
Cho, K. Y., Doy, C. H. and Mercer, E. H. (1967). *J. Bact.* **94**, 196.
Chung, K. L. (1967). *Can. J. Microbiol.* **13**, 341.
Clarke, K., Gray, G. W. and Reaveley, D. A. (1967a). *Biochem. J.* **105**, 749.
Clarke, K., Gray, G. W. and Reaveley, D. A. (1967b). *Biochem. J.* **105**, 755.
Clarke, K., Gray, G. W. and Reaveley, D. A. (1967c). *Biochem. J.* **105**, 759.
Clive, D. and Landman, O. E. (1970). *J. gen. Microbiol.* **61**, 233.
Cohn, F. (1872). *Beitr. Biol. Pfl.* **1**, Heft 2, 126.
Cole, R. M., Popkin, T. J., Boylan, R. J. and Mendelson, N. H. (1970). *J. Bact.* **103**, 793.
Coleman, G. (1969). *Biochem. J.* **112**, 533.
Conti, S. F., Jacobs, N. J. and Gray, C. T. (1968). *J. Bact.* **96**, 554.
Corner, T. R. and Marquis, R. E. (1969). *Biochim. biophys. Acta* **183**, 544.
Costerton, J. W., Forsberg, C., Matula, T. I., Buckmire, F. L. A. and MacLeod, R. A. (1967). *J. Bact.* **94**, 1764.
Cota-Robles, E. H. (1966). *J. Ultrastruct. Res.* **16**. 626.
Coyette, J. and Ghuysen, J. M. (1970). *Biochemistry, N.Y.* **9**, 2935.
Crane, F. L. and Hall, J. D. (1969). *Biochem. biophys. Res. Commun.* **36**, 174.
Cronan, J. E. (1968). *J. Bact.* **95**, 2054.
Cutinelli, C. and Galdiero, F. (1967). *J. Bact.* **93**, 2022.
Danielli, J. F. and Davson, H. (1935). *J. cell. comp. Physiol.* **5**, 495.
Daron, H. H. (1970). *J. Bact.* **101**, 145.
Da Silva, P. P. and Branton, D. (1970). *J. Cell Biol.* **45**, 598.
Davie, J. M. and Brock, T. D. (1966). *J. Bact.* **92**, 1623.
Davies, S. L. and Whittenbury, R. (1970). *J. gen. Microbiol.* **61**, 227.
De Petris, S. (1967). *J. Ultrastruct. Res.* **19**, 45.
De Voe, I. W., Thompson, J., Costerton, J. W. and MacLeod, R. A. (1970). *J. Bact.* **101**, 1014.
Dezélée, P. and Bricas, E. (1970). *Biochemistry, N.Y.* **9**, 823.
Done, J., Shorey, C. D., Loke, J. P. and Pollak, J. K. (1965). *Biochem. J.* **96**, 27 c.
Dröge, W., Lehmann, J., Lüderitz, O. and Westphal, O. (1970). *Eur. J. Biochem.* **14**, 175.
Dunnick, K. K. and O'Leary, W. M. (1970). *J. Bact.* **101**, 892.
Dvorak, H. F., Wetzel, B. K. and Heppel, L. A. (1970). *J. Bact.* **104**, 543.
Eagon, R. G., Simmons, G. P. and Carson, K. J. (1965). *Can. J. Microbiol.* **11**, 1041.
Edwards, M. R. and Stevens, R. W. (1963). *J. Bact.* **86**, 414.
Eisenberg, R. C., Yu, L. and Wolin, M. J. (1970a). *J. Bact.* **102**, 161.

Eisenberg, R. C., Yu, L. and Wolin, M. J. (1970b). *J. Bact.* **102**, 172.

Ellar, D. J. (1969). *J. gen. Microbiol.* **57**, vi.

Ellar, D. J. (1970). *Symp. Soc. gen. Microbiol.* **20**, 167.

Ellar, D. J. and Freer, J. H. (1969). *J. gen. Microbiol.* **58**, vii.

Ellar, D. J. and Lundgren, D. G. (1967). *J. Bact.* **94**, 1778.

Ellar, D. J., Lundgren, D. G. and Slepecky, R. A. (1967). *J. Bact.* **94**, 1189.

Ellwood, D. C. (1970a). *Biochem. J.* **118**, 367.

Ellwood, D. C. (1970b). *J. gen. Microbiol.* **60**, 373.

Ellwood, D. C. and Tempest, D. W. (1968). *Biochem. J.* **108**, 40P.

Ellwood, D. C. and Tempest, D. W. (1969). *Biochem. J.* **111**, 1.

Engelman, D. M. (1970). *J. molec. Biol.* **47**, 115.

Engelman, D. M., Terry, T. M. and Morowitz, H. J. (1967). *Biochim. biophys. Acta* 135, 381.

Fensom, A. H. and Gray, G. W. (1969). *Biochem. J.* **114**, 185.

Fensom, A. H. and Meadow, P. M. (1970). *FEBS Lett.* **9**, 81.

Fernández-Moran, H., Oda, T., Blair, P. V. and Green, D. E. (1964). *J. Cell Biol.* **22**, 63.

Ferrandes, B., Chaix, P. and Ryter, A. (1966). *C. r. hebd. Séanc. Acad. Sci., Paris* **263**, 1632.

Fiil, A. and Branton, D. (1969). *J. Bact.* **98**, 1320.

Finch, J. T., Klug, A. and Nermut, M. V. (1967). *J. Cell Sci.* **2**, 587.

Finean, J. B., Coleman, R., Knutton, S., Limbrick, A. R. and Thompson, J. E. (1968). *J. gen. Physiol.* **51**, 19s.

Fischer, A. (1891). *Ber. Verhandl. Kgl. sächs. Gesellsch. Wissensch. math-physik. Cl.* (*Leipzig*) **43**, 52.

Fischman, D. A. and Weinbaum, G. (1966). *Science, N.Y.* **155**, 472.

Fitz-James, P. C. (1960). *J. biophys. biochem. Cytol.* **8**, 507.

Fitz-James, P. C. (1964). *J. Bact.* **87**, 1483.

Fitz-James, P. C. (1965). *Bact. Rev.* **29**, 293.

Fitz-James, P. C. (1967). *In* "Protides of the Biological Fluids", (H. Peeters, ed.), Vol. 15, pp. 289–301. Elsevier Pub. Co. Amsterdam.

Fitz-James, P. C. (1968). *In* "Microbial Protoplasts, Spheroplasts and L-Forms", (L. B. Guze, ed.), pp. 124–143. Williams and Wilkins, New York.

Fleischer, S., Fleischer, B. and Stoeckenius, W. (1966). *J. Cell Biol.* **32**, 193.

Forsberg, C. W., Costerton, J. W. and MacLeod, R. A. (1970a). *J. Bact.* **104**, 1338.

Forsberg, C. W., Costerton, J. W. and MacLeod, R. A. (1970b). *J. Bact.* **104**, 1354.

Fox, E. N. and Wittner, M. K. (1965). *Proc. natn. Acad. Sci. U.S.A.* **54**, 1118.

Frank, H. and Dekegel, D. (1967). *Folia Microbiol., Praha*, **12**, 227.

Frerman, F. E. and White, D. C. (1967). *J. Bact.* **94**, 1868.

Gel'man, N. S., Lukoyanova, M. A. and Ostrovskii, D. N. (1967). "Respiration and Phosphorylation of Bacteria" (English trans.), Plenum Press, New York.

Gerhardt, P. (1967). *Fedn. Proc. Fedn. Am. Socs exp. Biol.* **26**, 1504.

Gerhardt, P. and Judge, J. A. (1964). *J. Bact.* **87**, 945.

Ghosh, B. K. and Carroll, K. K. (1968). *J. Bact.* **95**, 688.

Ghosh, B. K. and Murray, R. G. E. (1967). *J. Bact.* **93**, 411.

Ghosh, B. K. and Murray, R. G. E. (1969). *J. Bact.* **97**, 426.

Ghosh, B. K., Sargent, M. G. and Lampen, J. O. (1968). *J. Bact.* **96**, 1314.

Ghuysen, J. M. (1968). *Bact. Rev.* **32**, 425.

Ghuysen, J. M. and Leyh-Bouille, M. (1969). *J. gen. Microbiol.* **57**, i.

Ghuysen, J. M., Tipper, D. J. and Strominger, J. L. (1965). *Biochemistry, N.Y.* **4**, 474.

Ghuysen, J. M., Bricas, E., Lache, M. and Leyh-Bouille, M. (1968a). *Biochemistry, N.Y.* **7**, 1450.

Ghuysen, J. M., Strominger, J. L. and Tipper, D. J. (1968b). *In* "Comprehensive Biochemistry", (M. Florkin and E. H. Stotz, eds.), Vol. 26A, pp. 53–104. American Elsevier, New York.

Giesbrecht, P. and Ruska, C. H. (1968). *Klin. Wschr.* **46**, 575.

Glaser, L., Melo, A. and Paul, R. (1967). *J. biol. Chem.* **242**, 1944.

Glaser, M., Simpkins, H., Singer, S. J., Sheetz, M, and Chan, S. I. (1970). *Proc. natn. Acad. Sci. U.S.A.* **65**, 721.

Glauert, A. M. (1962). *Br. med. Bull.* **18**, 245.

Glauert, A. M. and Thornley, M. J. (1969). *A. Rev. Microbiol.* **23**, 159.

Gmeiner, J., Lüderitz, O. and Westphal, O. (1969). *Eur. J. Biochem.* **7**, 370.

Gooder, H. (1968). *In* "Microbial Protoplasts, Spheroplasts and L-Forms", (L. B. Guze, ed.), pp. 40–51. Williams and Wilkins, New York.

Gordon, R. C. and MacLeod, R. A. (1966). *Biochem. biophys. Res. Commun.* **24**, 684.

Goundry, J., Davison, A. L., Archibald, A. R. and Baddiley, J. (1967). *Biochem. J.* **104**, 1c.

Grant, W. D. and Wicken, A. J. (1968). *Biochem. biophys. Res. Commun.* **32**, 122.

Grant, W. D. and Wicken, A. J. (1970). *Biochem. J.* **118**, 859.

Gray, G. W. and Thurman, P. F. (1967). *Biochim. biophys. Acta* **135**, 947.

Gray, G. W. and Wilkinson, S. G. (1965). *J. gen. Microbiol.* **39**, 385.

Green, D. E. and Perdue, J. F. (1966). *Proc. natn. Acad. Sci. U.S.A.* **55**, 1295.

Grula, E. A., Smith, G. L. and Grula, M. M. (1965). *Can. J. Microbiol.* **11**, 605.

Grula, E. A., Butler, T. F., King, R. D. and Smith, G. L. (1967). *Can. J. Microbiol.* **13**, 1499.

Guinand, M., Ghuysen, J. M., Schleifer, K. H. and Kandler, O. (1969). *Biochemistry, N.Y.* **8**, 200.

Hahn, J. J. and Cole, R. M. (1963). *J. exp. Med.* **118**, 659.

Hall, E. A. and Knox, K. W. (1965). *Biochem. J.* **96**, 310.

Hammerling, G., Lüderitz, O. and Westphal, O. (1970). *Eur. J. Biochem.* **15**, 48.

Hancock, I. C. and Meadow, P. M. (1967). *J. gen. Microbiol.* **46**, x.

Hancock, I. C. and Meadow, P. M. (1969). *Biochim. biophys. Acta* **187**, 366.

Hard, G. C. (1969). *J. Bact.* **97**, 1480.

Heppel, L. A. (1967). *Science, N.Y.* **156**, 1451.

Heptinstall, S., Archibald, A. R. and Baddiley, J. (1970). *Nature, Lond.* **225**, 519.

Heymann, H., Manniello, J. M. and Barkulis, S. S. (1967). *Biochem. biophys. Res. Commun.* **26**, 486.

Higgins, M. L. and Shockman, G. D. (1970a). *J. Bact.* **101**, 643.

Higgins, M. L. and Shockman, G. D. (1970b). *J. Bact.* **103**, 244.

Highton, P. J. (1969). *J. Ultrastruct. Res.* **26**, 130.

Highton, P. J. (1970a). *J. Ultrastruct. Res.* **31**, 247.

Highton, P. J. (1970b). *J. Ultrastruct. Res.* **31**, 260.

Hirota, Y., Ryter, A. and Jacob, F. (1968). *Cold Spring Harb. Symp. quant. Biol.* **33**, 677.

Hofschneider, P. H. and Martin, H. H. (1968). *J. gen. Microbiol.* **51**, 23.

Holt, S. C. and Leadbetter, E. R. (1969). *Bact. Rev.* **33**, 346.

Houtsmuller, U. M. T. and Van Deenen, L. L. M. (1964). *Biochim. biophys. Acta* **84**, 96.

Houtsmuller, U. M. T. and Van Deenen, L. L. M. (1965). *Biochim. biophys. Acta* **106**, 564.

Hughes, R. C. (1965). *Biochem. J.* **96**, 700.

76   D. A. REAVELEY AND R. E. BURGE

Hughes, R. C. (1970a). *Biochem. J.* **119**, 849.

Hughes, R. C. (1970b). *Biochem. J.* **117**, 431.

Hughes, R. C. and Thurman, P. F. (1970). *Biochem. J.* **117**, 441.

Hughes, R. C., Rogers, H. J., Pavlik, J. G. and Tanner, P. J. (1968). *Nature, Lond.* **219**, 642.

Hughes, R. C., Tanner, P. J. and Stokes, E. (1970). *Biochem. J.* **120**, 159.

Hungerer, K. D. and Tipper, D. J. (1969). *Biochemistry, N.Y.* **8**, 3577.

Hungerer, K. D., Fleck, J. and Tipper, D. J. (1969). *Biochemistry, N.Y.* **8**, 3567.

Hurst, A. and Stubbs, J. M. (1969). *J. Bact.* **97**, 1466.

Ibbott, F. A. and Abrams, A. (1964). *Biochemistry, N.Y.* **3**, 2008.

Imaeda, T., Kanetsuna, F. and Galindo, B. (1968). *J. Ultrastruct. Res.* **25**, 46.

Ishida, M. and Mizushima, S. (1969a). *J. Biochem., Tokyo* **66**, 33.

Ishida, M. and Mizushima, S. (1969b). *J. Biochem., Tokyo* **66**, 133.

Johnston, J. H., Johnston, R. J. and Simmons, D. A. R. (1967). *Biochem. J.* **105**, 79.

Jušić, D., Roy, C. and Watson, R. W. (1964). *Can. J. Biochem.* **42**, 1553.

Kakefuda, T., Holden, J. T. and Utech, N. M. (1967). *J. Bact.* **93**, 472.

Kandler, O. (1967). *Zentlb. Bakt. ParasitKde (Abt.* 1.) **205**, 197.

Kandler, O., Schleifer, K. H. and Dandl, R. (1968). *J. Bact.* **96**, 1935.

Kaneda, T. (1967). *J. Bact.* **93**, 894.

Kanemasa, Y. Y., Akamatsu, Y. and Nojima, S. (1967). *Biochim. biophys. Acta* **144**, 382.

Karakawa, W. W., Rotta, J. and Krause, R. M. (1965). *Proc. Soc. exp. Biol. Med.* **118**, 198.

Kato, K., Strominger, J. L. and Kotani, S. (1968). *Biochemistry, N.Y.* **7**, 2762.

Katz, W. and Martin, H. H. (1970). *Biochem. biophys. Res. Commun.* **39**, 744.

Kazuo, I. and Strominger, J. L. (1968). *J. biol. Chem.* **243**, 3193.

Kelemen, M. V. and Rogers, H. J. (1969). *J. gen. Microbiol.* **57**, xiii.

Kellenberger, E. and Ryter, A. (1958). *J. biophys. biochem. Cytol.* **4**, 323.

Key, B. A., Gray, G. W. and Wilkinson, S. G. (1970a). *Biochem. J.* **120**, 559.

Key, B. A., Gray, G. W. and Wilkinson, S. G. (1970b). *Biochem. J.* **117**, 721.

King, J. R. and Gooder, H. (1970). *J. Bact.* **130**, 686.

Knox, K. W. and Hall, E. A. (1965). *Biochem. J.* **96**, 302.

Knox, K. W. and Holmwood, K. J. (1968). *Biochem. J.* **108**, 363.

Knox, K. W., Cullen, J. and Work, E. (1967). *Biochem. J.* **103**, 192.

Kocun, F. J. (1970). *Biochim. biophys. Acta* **202**, 277.

Kohn, A. (1960). *J. Bact.* **79**, 697.

Kolenbrander, P. E. and Ensign, J. C. (1968). *J. Bact.* **95**, 201.

Korn, E. D. (1969). *A. Rev. Biochem.* **38**, 263.

Krulwich, T. A. and Ensign, J. C. (1968). *J. Bact.* **96**, 857.

Krulwich, T. A., Ensign, J. C., Tipper, D. J. and Strominger, J. L. (1967). *J. Bact.* **94**, 734.

Lampen, J. O. (1965). *Sym. Soc. gen. Microbiol.* **15**, 115.

Landman, O. E. (1968). *In* "Microbial Protoplasts, Spheroplasts and L-Forms". (L. B. Guze, ed.), pp. 319–332. Williams and Wilkins, New York.

Landman, O. E. and Forman, A. (1969). *J. Bact.* **99**, 576.

Landman, O. E., Ryter, A. and Fréhel, C. (1968). *J. Bact.* **96**, 2154.

Lang, D. R. and Lundgren, D. G. (1970). *J. Bact.* **101**, 483.

Lange, C. F., Lee, R. and Merdinger, E. (1969). *J. Bact.* **100**, 1277.

Larsen, H. (1967). *Adv. microbial Physiol.* **1**, 97.

Law, J., Zalkin, H. and Kaneshiro, T. (1963). *Biochim. biophys. Acta* **70**, 143.

WALLS AND MEMBRANES IN BACTERIA

Leadbetter, E. R. and Holt, S. C. (1968). *J. gen. Microbiol.* **52**, 299.
Leene, W. and Van Iterson, W. (1965). *J. Cell Biol.* **27**, 241.
Leive, L. (1968). *J. biol. Chem.* **243**, 2373.
Leive, L., Shovlin, V. K. and Mergenhagen, S. E. (1968). *J. biol. Chem.* **243**, 6384.
Lenard, J. and Singer, S. J. (1966). *Proc. natn. Acad. Sci. U.S.A.* **56**, 1828.
Leutgeb, W. and Weidel, W. (1963). *Z. Naturf.* **18B**, 1060–1062.
Leutgeb, W., Maass, D. and Weidel, W. (1963). *Z. Naturf.* **18B**, 1062.
Levy, S. B. and Leive, L. (1968). *Proc. natn. Acad. Sci. U.S.A.* **61**, 1435.
Lewis, V. J., Weaver, R. E. and Hollis, D. G. (1968). *J. Bact.* **96**, 1.
Leyh-Bouille, M., Ghuysen, J. M., Tipper, D. J. and Strominger, J. L. (1966). *Biochemistry, N.Y.* **5**, 3079.
Leyh-Bouille, M., Bonaly, R., Ghuysen, J. M., Tinelli, R. and Tipper, D. J. (1970). *Biochemistry, N.Y.* **9**, 2944.
Lopes, J. and Inniss, W. E. (1970). *J. Bact.* **103**, 238.
Lowry, O. H., Rosebrough, N. J., Farr, L. and Randall, R. J. (1951). *J. biol. Chem.* **193**, 265.
Lüderitz, O. (1970). *Angew. Chemie. Internat. Ed.* **9**, 649.
Lüderitz, O., Jann, K. and Wheat, R. (1968). *In* "Comprehensive Biochemistry", (M. Florkin and E. H. Stotz, eds.), Vol. 26A, pp. 105–228. American Elsevier, New York.
Lui, T. Y. and Gotschlich, E. C. (1967). *J. biol. Chem.* **243**, 471.
Malchow, D., Lüderitz, O., Kickhöfen, B., Westphal, O. and Gerisch, G. (1969). *Eur. J. Biochem.* **7**, 239.
Mandelstam, J. (1962). *Biochem. J.* **84**, 294.
Mandelstam, J. (1969). *Symp. Soc. gen. Microbiol.* **19**, 377.
Martin, H. H. (1966). *A. Rev. Biochem.* **35**, 457.
Martin, H. H. and Frank, H. (1962). *Z. Naturf.* **17B**, 190.
Martin, H. H. and Kemper, S. (1970). *J. Bact.* **102**, 347.
Matthew, D. D. and Rosenblum, E. D. (1967). *Bact. Proc.* 144.
Mauck, J. and Glaser, L. (1970). *Biochem. biophys. Res. Commun.* **39**, 699.
McCarty, M. and Morse, S. I. (1964). *Adv. Immun.* **4**, 249.
McQuillen, K. (1960). *In* "The Bacteria", (I. C. Gunsalus and R. Y. Stanier, eds.), Vol. 1, pp. 249–359. Academic Press, New York.
Miller, I. L., Zsigray, R. M. and Landman, O. E. (1967). *J. gen. Microbiol.* **49**, 513.
Mirelman, D. and Sharon, N. (1967). *J. biol. Chem.* **242**, 3414.
Mirsky, R. (1969). *Biochemistry, N.Y.* **8**, 1164.
Mitchell, P. and Moyle, J. (1956). *Symp. Soc. gen. Microbiol.* **6**, 150.
Miura, T. and Mizushima, S. (1968). *Biochim. biophys. Acta* **150**, 159.
Miura, T. and Mizushima, S. (1969). *Biochim. biophys. Acta* **193**, 268.
Mizushima, S. (1968). *J. Biochem., Tokyo* **63**, 317.
Mizushima, S., Ishida, M. and Miura, T. (1966). *J. Biochem., Tokyo* **60**, 256
Montague, M. D. and Moulds, J. D. (1967). *Biochim. biophys. Acta* **135**, 565.
Moss, C. W., Dowell, V. R., Farshtchi, D., Raines, L. T. and Cherry, W. B. (1969). *J. Bact.* **97**, 561.
Munoz, E., Ghuysen, J. M., Leyh-Bouille, M., Petit, J. F., Heymann, H., Bricas, E. and Lefrancier, P. (1966). *Biochemistry, N.Y.* **5**, 3748.
Munoz, E., Ghuysen, J. M. and Heymann, H. (1967). *Biochemistry, N.Y.* **6**, 3659.
Munoz, E., Nachbar, M. S., Schor, M. T. and Salton, M. R. J. (1968a). *Biochem. biophys. Res. Commun.* **32**, 539.
Munoz, E., Freer, J. H., Ellar, D. J. and Salton, M. R. J. (1968b). *Biochim. biophys. Acta* **150**, 531.

Munoz, E., Salton, M. R. J., Ng, M. H. and Schor, M. T. (1969). *Eur. J. Biochem.* **7**, 490.
Murray, R. G. E., Steed, P. and Elson, H. E. (1965). *Can. J. Microbiol.* **11**, 547.
Myerholtz, L. E. and Hartsell, S. E. (1952). *Bact. Proc.* 34.
Nanninga, N. (1968). *J. Cell Biol.* **39**, 251.
Nanninga, N. (1970). *J. Bact.* **101**, 297.
Neale, E. K. and Chapman, G. B. (1970). *J. Bact.* **104**, 518.
Nermut, M. V. (1967). *J. gen. Microbiol.* **49**, 503.
Nermut, M. V. and Murray, R. G. E. (1967). *J. Bact.* **93**, 1949.
Nesbitt, J. A. and Lennarz, W. J. (1965). *J. Bact.* **89**, 1020.
Neu, H. C. (1968a). *Biochemistry, N.Y.* **7**, 3774.
Neu, H. C. (1968b). *Biochemistry, N.Y.* **7**, 3766.
Neu, H. C. (1969). *J. gen. Microbiol.* **57**, 215.
Neu, H. C. and Chou, J. (1967). *J. Bact.* **94**, 1934.
Neu, H. C. and Heppel, L. A. (1965). *J. biol. Chem.* **240**, 3685.
Neu, H. C., Ashman, D. F. and Price, T. D. (1967). *J. Bact.* **93**, 1360.
Nikaido, H. (1970). *Eur. J. Biochem.* **15**, 57.
Nisonson, I., Tannenbaum, M. and Neu, H. C. (1969). *J. Bact.* **100**, 1083.
Nossal, N. G. and Heppel, L. A. (1966). *J. biol. Chem.* **241**, 3055.
Okuyama, H. (1969). *Biochim. biophys. Acta* **176**, 1125.
Op Den Kamp, J. A. F., Van Iterson, W. and Van Deenen, L. L. M. (1967). *Biochim. biophys. Acta* **135**, 862.
Op Den Kamp, J. A. F., Redai, I. and Van Deenen, L. L. M. (1969). *J. Bact.* **99**, 298.
Osborn, M. J. (1969). *A. Rev. Biochem.* **38**, 501.
Ou, Li-Tse and Marquis, R. E. (1970). *J. Bact.* **101**, 92.
Owen, P. and Freer, J. H. (1970). *J. gen. Microbiol.* **60**, xi.
Pardee, A. B. (1968). *Science, N.Y.* **162**, 632.
Patterson. D., Weinstein, M., Nixon, R. and Gillespie, D. (1970). *J. Bact.* **101**, 584.
Patterson, P. H. and Lennarz, W. J. (1970). *Biochem. biophys. Res. Commun.* **40**, 408.
Perkins, H. R. (1963). *Biochem. J.* **86**, 475.
Perkins, H. R. (1965). *Biochem. J.* **97**, 3 c.
Perkins, H. R. (1967). *Biochem. J.* **102**, 29 c.
Petit, J. F., Munoz, E. and Ghuysen, J. M. (1966). *Biochemistry, N.Y.* **5**, 2764.
Petit, J. F., Adam, A., Wietzerbin-Falszpan, J., Lederer, E. and Ghuysen, J. M. (1969). *Biochem. biophys. Res. Commun.* **35**, 468.
Pfister, R. M. and Lundgren, D. G. (1967). *J. Bact.* **88**, 1119.
Pierce, W. A. (1964). *J. Bact.* **88**, 912.
Piperno, J. R. and Oxender, D. L. (1966). *J. biol. Chem.* **241**, 5732.
Pontefract, R. D., Bergeron, G. and Thatcher, F. S. (1969). *J. Bact.* **97**, 367.
Popkin, T. J., Theodore, T. S. and Cole, R. M. (1970). *Bact. Proc.* **24**.
Randle, C. L., Albro, P. W. and Dittmer, J. C. (1969). *Biochim. biophys. Acta* **187**, 214.
Razin, S. (1969). *A. Rev. Microbiol.* **23**, 317.
Reaveley, D. A. (1968). *Biochem. biophys. Res. Commun.* **30**, 649.
Reaveley, D. A. and Rogers, H. J. (1969). *Biochem. J.* **113**, 67.
Redwood, W. R., Müldner, H. and Thompson, T. E. (1969). *Proc. natn. Acad. Sci. U.S.A.* **64**, 989.
Reinert, J. C. and Steim, J. M. (1970). *Science, N.Y.* **168**, 1580.
Remsen, C. C. (1968). *Arch. Mikrobiol.* **61**, 40.
Remsen, C. C., Valois, F. W. and Watson, S. W. (1967). *J. Bact.* **94**, 422.

WALLS AND MEMBRANES IN BACTERIA

Remsen, C. C., Watson, S. W., Waterbury, J. B. and Trüper, H. G. (1968). *J. Bact.* **95**, 2374.

Remsen, C. C., Watson, S. W. and Trüper, H. G. (1970). *J. Bact.* **103**, 245.

Repaske, R. (1956). *Biochim. biophys. Acta* **22**, 189.

Repaske, R. (1958). *Biochim. biophys. Acta* **30**, 225.

Rieber, M., Imaeda, T. and Cesari, I. M. (1969). *J. gen. Microbiol.* **55**, 155.

Rizza, V., Tucker, A. N. and White, D. C. (1970). *J. Bact.* **101**, 84.

Robinow, C. F. (1962). *Circulation* **26**, 1092.

Rodwell, A. W., Razin, S., Rottem, S. and Argaman, M. (1967). *Archs Biochem. Biophys.* **122**, 621.

Rogers, H. J. (1970). *Bact. Rev.* **34**, 194.

Rogers, H. J. and Garrett, A. J. (1965). *Biochem. J.* **96**, 231.

Rogers, H. J. and Perkins, H. R. (1968). *In* "Cell Walls and Membranes". E. F. and N. Spon, Ltd. London.

Rogers, H. J., Reaveley, D. A. and Burdett, I. D. J. (1967). *In* "Protides of the Biological Fluids", (H. Peeters, ed.), Vol. 15, pp. 303–313. Elsevier Pub. Co. Amsterdam.

Rogers, H. J., McConnell, M. and Burdett, I. D. J. (1970). *J. gen. Microbiol.* **61**, 155.

Romeo, D., Girard, A. and Rothfield, L. (1970). *J. molec. Biol.* **53**, 475.

Rothfield, L. and Horne, R. W. (1967). *J. Bact.* **93**, 1705.

Rothfield, L. and Pearlman-Kothencz, M. (1969). *J. molec. Biol.* **44**, 477.

Rottem, S., Stein, O. and Razin, S. (1968). *Archs Biochem. Biophys.* **125**, 46.

Ryter, A. (1968). *Bact. Rev.* **32**, 39.

Ryter, A. and Jacob, F. (1964a). *Annls Inst. Pasteur, Paris* **107**, 384.

Ryter, A. and Jacob, F. (1964b). *Annls Inst. Pasteur, Paris* **110**, 801.

Ryter, A., Fréhel, C. and Ferrandes, B. (1967). *C. r. hedb. Séanc. Acad. Sci., Paris* **265**, 1259.

Salton, M. R. J. (1964). *In* "The Bacterial Cell Wall". Elsevier Pub. Co. Amsterdam.

Salton, M. R. J. (1967a). *A. Rev. Microbiol.* **21**, 417.

Salton, M. R. J. (1967b). *In* "Protides of the Biological Fluids", (H. Peeters, ed.), Vol. 15, pp. 270–288. Elsevier Pub. Co. Amsterdam.

Salton, M. R. J. and Freer, J. H. (1965). *Biochim. biophys. Acta* **107**, 531.

Salton, M. R. J. and Netschey, A. (1965). *Biochim. biophys. Acta* **107**, 539.

Salton, M. R. J. and Schmitt, M. D. (1967). *Biochem. biophys. Res. Commun.* **27**, 529.

Salton, M. R. J., Schmitt, M. D. and Trefts, P. E. (1967). *Biochem. biophys. Res. Commun.* **29**, 728.

Salton, M. R. J., Freer, J. H. and Ellar, D. J. (1968). *Biochem. biophys. Res. Commun.* **33**, 909.

Sargent, M. G., Ghosh, B. K. and Lampen, J. O. (1968). *J. Bact.* **96**, 1329.

Sargent, M. G., Ghosh, B. K. and Lampen, J. O. (1969). *J. Bact.* **97**, 820.

Schleifer, K. H. (1969). *J. gen. Microbiol.* **57**, xiv.

Schleifer, K. H. (1970). *Arch. Mikrobiol.* **71**, 271.

Schleifer, K. H. and Kandler, O. (1967a). *Arch. Mikrobiol.* **57**, 365.

Schleifer, K. H. and Kandler, O. (1967b). *Biochem. biophys. Res. Commun.* **28**, 965.

Schleifer, K. H. and Kandler, O. (1970). *J. Bact.* **103**, 387.

Schleifer, K. H., Plapp, R. and Kandler, O. (1968). *FEBS Lett.* **1**, 287.

Schleifer, K. H., Huss, L. and Kandler, O. (1969). *Arch. Mikrobiol.* **68**, 387.

Schnaitman, C. A. (1969). *Biochem. biophys. Res. Commun.* **37**, 1.

Schnaitman, C. A. (1970a). *J. Bact.* **104**, 890.
Schnaitman, C. A. (1970b). *J. Bact.* **104**, 882.
Schnaitman, C. A. (1970c). *J. Bact.* **104**, 1404.
Schocher, A. J., Bayley, S. T. and Watson, R. W. (1962). *Can. J. Microbiol.* **8**, 89.
Sedar, A. and Burde, R. (1965). *J. Cell Biol.* **27**, 53.
Shands, J. W. (1966). *Ann. N.Y. Acad. Sci.* **133**, 292.
Shands, J. W., Graham, J. A. and Nath, K. (1967). *J. molec. Biol.* **25**, 15.
Shaw, M. K. and Ingraham, J. L. (1965). *J. Bact.* **90**, 141.
Shaw, N. and Baddiley, J. (1968). *Nature, Lond.* **217**, 142.
Shen, P. Y., Coles, E., Foote, J. L. and Stenesh, J. (1970). *J. Bact.* **103**, 479.
Shively, J. M. and Benson, A. A. (1967). *J. Bact.* **94**, 1679.
Shockman, G. D. (1965). *Bact. Rev.* **29**, 345.
Shockman, G. D., Kolb, J. J., Bakay, B., Conover, M. J. and Toennies, G. (1963). *J. Bact.* **85**, 168.
Silbert, D. F., Ruch, F. and Vagelos, P. R. (1968). *J. Bact.* **95**, 1658.
Silva, M. T. (1967). *Expl Cell Res.* **46**, 245.
Steensland, H. and Larsen, H. (1969). *J. gen. Microbiol.* **55**, 325.
Steim, J. M., Tourtellotte, M. E., Reinert, J. C., McElhaney, R. N. and Rader, R. L. (1969). *Proc. natn. Acad. Sci. U.S.A.* **63**, 104.
Stoeckenius, W. and Rowen, R. (1967). *J. Cell Biol.* **34**, 365.
Suganuma, A. (1966). *J. Cell Biol.* **30**, 208.
Takebe, I. (1965). *Biochim. biophys. Acta* **101**, 124.
Theodore, T. S., Popkin, T. J. and Cole, R. M. (1970). *Bact. Proc.* 52.
Thiele, O. W., Lacave, C. and Asselineau, J. (1969). *Eur. J. Biochem.* **7**, 393.
Thornley, M. J., Horne, R. W. and Glauert, A. M. (1965). *Arch. Mikrobiol.* **51**, 267.
Tipper, D. J. (1968). *Bact. Proc.* 48.
Tipper, D. J. (1969). *Biochemistry, N.Y.* **8**, 2192.
Tipper, D. J. and Berman, M. F. (1969). *Biochemistry, N.Y.* **8**, 2183.
Tipper, D. J. and Strominger, J. L. (1966). *Biochem. biophys. Res. Commun.* **22**, 48.
Tipper, D. J. and Strominger, J. L. (1968). *J. biol. Chem.* **243**, 3169.
Tipper, D. J., Strominger, J. L. and Ensign, J. C. (1967). *Biochemistry, N.Y.* **6**, 906.
Tomasz, A. (1968). *Proc. natn. Acad. Sci. U.S.A.* **59**, 86.
Tori, M., Kabat, E. A. and Bezer, A. (1964). *J. exp. Med.* **120**, 13.
Tornabene, T. G., Bennett, E. O. and Oro, J. (1967). *J. Bact.* **94**, 344.
Trembley, G. Y., Daniels, M. J. and Schaechter, M. (1969). *J. molec. Biol.* **40**, 65.
Tucker, A. N. and White, D. C. (1970a). *J. Bact.* **102**, 498.
Tucker, A. N. and White, D. C. (1970b). *J. Bact.* **102**, 508.
Vanderkooi, G. and Green, D. E. (1970). *Proc. natn. Acad. Sci. U.S.A.* **66**, 615.
Vanderkooi, G. and Sundaralingam, M. (1970). *Proc. natn. Acad. Sci. U.S.A.* **67**, 233.
Van Dijk-Salkinoja, M. S., Stoof, T. J. and Planta, R. J. (1970). *Eur. J. Biochem.* **12**, 474.
Van Iterson, W. (1965). *Bact. Rev.* **29**, 299.
Van Iterson, W. and Op Den Kamp, J. A. F. (1969). *J. Bact.* **99**, 304.
Veerkamp, J. H. (1970). *Archs Biochem. Biophys.* **201**, 267.
Veerkamp, J. H., Lambert, R. and Saito, Y. (1965). *Archs Biochem. Biophys.* **112**, 120.
Verma, J. P. and Martin, H. H. (1967). *Folio Microbiol., Praha* **12**, 248.
Volk, W. A. (1968). *J. Bact.* **95**, 980.
Vorbeck, M. L. and Marinetti, G. V. (1965). *Biochemistry, N.Y.* **4**, 296.
Voss, J. G. (1967). *J. gen. Microbiol.* **48**, 391.
Wang, W. S. and Lundgren, D. G. (1968). *J. Bact.* **95**, 1851.

WALLS AND MEMBRANES IN BACTERIA

Wang, W. S., Korczynski, M. S. and Lundgren, D. G. (1970). *J. Bact.* **104**, 556.
Ward, J. B. and Perkins, H. R. (1968). *Biochem. J.* **106**, 391.
Wardlaw, A. C. (1963). *Can. J. Microbiol.* **9**, 41.
Warth, A. D. and Strominger, J. L. (1968). *Bact. Proc.* 64.
Watson, S. W. and Remsen, C. C. (1969). *Science, N.Y.* **163**, 685.
Watson, S. W. and Remsen, C. C. (1970). *J. Ultrastruct. Res.* **33**, 148.
Weibull, C. (1953). *J. Bact.* **66**, 688.
Weibull, C. (1956). *Symp. Soc. gen. Microbiol.* **6**, 111.
Weibull, C. (1965). *J. Bact.* **89**, 1151.
Weibull, C. (1968). *In* "Microbial Protoplasts, Spheroplasts and L-Forms", (L. B. Guze, ed.), pp. 62–73. Williams and Wilkins, New York.
Weidel, W. and Pelzer, H. (1964). *Adv. Enzymol.* **26**, 193.
Weidel, W., Frank, H. and Leutgeb, W. (1963). *J. gen. Microbiol.* **30**, 127.
Weigand, R. A., Shively, J. M. and Greenawalt, J. W. (1970). *J. Bact.* **102**, 240.
Weinbaum, G., Rich, R. and Fischman, D. A. (1967). *J. Bact.* **93**, 1693.
Weinbaum, G., Fischman, D. A. and Okuda, S. (1970). *J. Cell Biol.* **45**, 493.
Weiser, M. and Rothfield, L. (1968). *J. biol. Chem.* **423**, 1320.
Wetzel, B. K., Spicer, S. S., Dvorak, H. F. and Heppel, L. A. (1970). *J. Bact.* **104**, 529.
White, D. C. and Cox, R. H. (1967). *J. Bact.* **93**, 1079.
White, D., Dworkin, M. and Tipper, D. J. (1968). *J. Bact.* **95**, 2186.
Whiteside, T. L. and Corpe, W. A. (1969). *J. Bact.* **97**, 1449.
Whitney, J. G. and Grula, E. A. (1968). *Biochim. biophys. Acta* **158**, 124.
Wicken, A. J. (1966). *Biochem. J.* **99**, 108.
Weitzerbin-Falszpan, J., Das, B. C., Azuma, I., Adam, A., Petit, J. F. and Lederer, E. (1970). *Biochem. biophys. Res. Commun.* **40**, 57.
Wilkinson, S. G. (1967). *J. gen. Microbiol.* **47**, 67.
Winshell, E. B. and Neu, H. C. (1970). *J. Bact.* **102**, 537.
Work, E. and Griffiths, H. (1968). *J. Bact.* **95**, 641.
Yamaguchi, T., Tamura, G. and Arima, K. (1967). *J. Bact.* **93**, 483.
Young, F. E. (1965). *Nature, Lond.* **207**, 104.
Young, F. E. (1966). *J. Bact.* **92**, 839.
Young, F. E. (1967). *Proc. natn. Acad. Sci. U.S.A.* **58**, 2377.
Young, F. E., Tipper, D. J. and Strominger, J. L. (1964). *J. biol. Chem.* **239**, PC 3600.
Yu, L. and Wolin, M. J. (1970). *J. Bact.* **103**, 467.
Yuasha, R., Nakane, K. and Nikaido, H. (1970). *Eur. J. Biochem.* **15**, 63.
Yudkin, M. D. (1966). *Biochem. J.* **98**, 923.
Yudkin, M. D. and Davis, B. (1965). *J. molec. Biol.* **12**, 193.

Effects of Environment on Bacterial Wall Content and Composition

D. C. ELLWOOD AND D. W. TEMPEST

*Microbiological Research Establishment, Porton, Nr. Salisbury,
Wiltshire, England*

I. Introduction		83
II. Bacterial Wall Content		84
III. Teichoic Acids and Teichuronic Acids		86
A. Effects of Specific Nutrient Limitations . . .		88
B. Effects of Growth Rate		95
C. Effects of Ionic Environment		95
D. Effects of Medium pH Value		98
E. Control of Synthesis of Teichoic Acid and Teichuronic Acid	.	100
IV. Intracellular Teichoic Acids		102
V. Mucopeptide		102
A. Comparative Effects of Lysozyme on *Bacillus subtilis*	.	105
B. Effect of Growth Condition		106
VI. Lipopolysaccharide		110
VII. Protein		113
VIII. Lipids		113
IX. Concluding Remarks		114
References		115

I. Introduction

A striking feature of the bacterial cell is its capacity to undergo substantial changes in structure and functioning in response to changes in the growth environment (Herbert, 1961a; Neidhardt, 1963). Since, in Nature, bacteria inhabit environments that generally fluctuate markedly (and often rapidly) this characteristic must be, ecologically, most advantageous to the organism and of considerable evolutionary significance. However, we do not wish to consider here the broader aspects of phenotypic variation but to review, in this context, some recently obtained information regarding one particular organelle—the bacterial wall.

The walls of bacteria are complex structures containing several heteropolymers (mucopeptide, teichoic acid or lipopolysaccharide, lipoprotein and/or protein), the compositions of which have been much studied. But

84   D. C. ELLWOOD AND D. W. TEMPEST

almost without exception these studies have been carried out on materials obtained from organisms grown in uncontrolled environments, and therefore must be viewed with some caution since it is now clear that the composition of wall polymers possibly may vary markedly with the growth condition (Ellwood, 1970). At the heart of the problem lies the fact that, traditionally, bacteria are grown in batch-type cultures. These cultures are contained in a "closed system" (Herbert, 1961b) and because bacteria interact chemically with their environment (they take up nutrients and excrete end products of metabolism) the environment changes continuously as growth proceeds. Thus, the properties of the organisms are caused to change progressively throughout the culture growth cycle making it difficult to obtain reliable data. This difficulty can only be avoided by growing bacteria in an "open system", that is in a continuous-flow culture system; then one can obtain truly steady-state conditions in which the environment, and the properties of the organisms in the culture, do not change substantially with time.

Of the many "open" culture systems, the use of a "chemostat" type of continuous culture (Novick and Szilard, 1950) allows one to obtain not only a range of controlled environments but a range of unique environments—environments that can never be obtained in batch-type cultures, or obtained only transiently. And, in these unique environments, organisms express properties that may never otherwise be expressed. Thus, the application of this type of continuous culture (with its capacity to provide both controlled environments and new environments) facilitates the acquisition of reliable data (and new data) on the manner in which the content and composition of bacterial walls may vary. Using this technique, we have now accumulated sufficient of the data to attempt an assessment of the extent to which bacterial walls may change phenotypically; it is the purpose of this article to assemble and review this evidence.

As suggested above, it is intended only to consider those changes in bacterial wall content and composition that clearly result from changes in genetic expression (that is, phenotype and not genotype). Moreover we have ignored the capsule and other loosely attached surface structures and concentrated on the basic wall components—namely, mucopeptide, teichoic acid and/or teichuronic acid, lipopolysaccharide, protein and lipid.

II. Bacterial Wall Content

The average size of bacteria in a growing culture has been found to vary progressively with the growth rate (Gould, 1958; Herbert, 1958, 1961), and with the chemical nature of the growth environment (Tempest et al., 1965; Tempest and Ellwood, 1969). Since the surface area : volume

ENVIRONMENTAL EFFECTS ON BACTERIAL WALLS 85

ratio of organisms must vary inversely with their size then, assuming the wall thickness to be constant, the bacterial wall content also must change with the rate of growth. Quantitative determinations of the wall content of *Bacillus subtilis* var. *niger* and *Aerobacter aerogenes* organisms, growing in a chemostat at several different rates, showed this prediction to be correct (Table 1). It might be mentioned here that determination of the cell-wall content of these bacteria was made using a gravimetric procedure, i.e. disrupting the organisms and quantitatively separating

TABLE 1. Influence of Growth Rate and Growth-Limiting Component of the Medium on the Average Wall Content of *Bacillus subtilis* var. *niger* and *Aerobacter aerogenes* Organisms, grown in a Chemostat Culture.

The organisms were grown in 0·5 l. chemostats (Herbert *et al.*, 1965) in simple salts media (Evans *et al.*, 1970). The temperature was maintained constant at 35° and the pH value at 7·0 (for the bacilli) or 6·5 (for *A. aerogenes*). Bulk samples were collected from the overflow line into an ice-cooled receiver; organisms were separated from the extracellular constituents by centrifugation and washed with de-ionized water. Bacterial pastes were disrupted in a Braun MSK homogenizer and the walls quantitatively separated from the cytoplasmic constituents by differential centrifugation.

Dilution rate (hr.$^{-1}$)	Bacterial wall content (%, w/w)					
	Bacillus subtilis var. *niger*[a]				*Aerobacter aerogenes*[b]	
	Mg^{2+}-limited		PO_4^{3-}-limited		Mg^{2+}-limited	C-limited
0·05	—		36·8	37·7	—	—
0·1	26·7	27·1	22·8	23·1	19·1 20·5	20·0 21·0
0·2	23·2	23·9	20·8	21·1	—	—
0·3	18·2	18·7	17·0	17·5	15·0	14·6
0·7	—		—		13·0	11·0

[a] Data of Ellwood (1970).
[b] Data of Tempest and Ellwood (1969).

the walls from the cytoplasmic constituents. Such a procedure can be criticized on the grounds that it is difficult to ensure that the isolated wall material is homogeneous and free from occluded cytoplasm. However, the method used in obtaining the data shown in Table 1 clearly gave results that were reproducible. Moreover, the use of different techniques for cell disruption gave results that were in close agreement with those reported here (see Tempest and Ellwood, 1969). Essentially similar results were obtained by Collins (1964) and by Sud and Schaechter (1964) using a different approach. These workers measured hexosamine and diaminopimelic acid as indicators of cell-wall content and assumed

(probably erroneously) that wall composition did not vary with growth rate.

Not all of the bacterial species that have been examined showed marked changes in wall content in response to changes in growth rate. Thus, the wall contents of Mg^{2+}-limited or PO_4^{3-}-limited *Micrococcus lysodeikticus* and *Staphylococcus aureus* changed only slightly when growth rate was progressively increased (Table 2). It may not be co-incidental that the

TABLE 2. Influence of Growth Rate on the Wall Content of *Micrococcus lysodeikticus* (NCTC 2665) and *Staphylococcus aureus* H Grown in a Chemostat Culture.

The organisms were grown in 0·5 l. chemostats (Herbert *et al.*, 1965) in complex media. The temperature was maintained constant at 35° and the pH value at 7·0. Other experimental details are as in Table 1.

| Dilution rate (hr.$^{-1}$) | Bacterial wall content (%, w/w) | | | |
| | *Micrococcus lysodeikticus* | | *Staphylococcus aureus* | |
	Mg^{2+}-limited	PO_4^{3-}-limited	Mg^{2+}-limited	PO_4^{3-}-limited
0·1	25·9 26·9	22·6 23·4	—	13·5 13·5
0·2	—	—	—	14·2 16·3
0·3	22·6 22·9	23·8 23·9	16·3 16·5	—
0·4	—	—	—	12·4 12·8
0·5	22·2 —	23·8 27·0	—	—

average size of bacillary organisms changed more obviously than did that of the coccal forms in response to changes in growth rate (Gould, 1958). However, the reason for this difference in response is not immediately obvious.

III. Teichoic Acids and Teichuronic Acids

These polymers are present in the walls of a wide variety of Gram-positive bacteria (including members of the genera *Bacillus*, *Lactobacillus* *Streptococcus*, *Staphylococcus* and *Micrococcus*), but seemingly are not synthesized by any Gram-negative bacteria. Teichoic acids have been found to occur both as an integral part of the wall structure ("wall teichoic acid") and attached to the cytoplasmic membrane ("membrane teichoic acid"). Though similar in structure (and presumably in their functioning) the former show a greater degree of diversity in structure (Baddiley, 1970) and have been the more extensively studied.

The wall-bound teichoic acids are polymers of either ribitol phosphate of glycerol phosphate to which different sugars (such as glucose and

ENVIRONMENTAL EFFECTS OF BACTERIAL WALLS

(1) Glycerol teichoic acid:

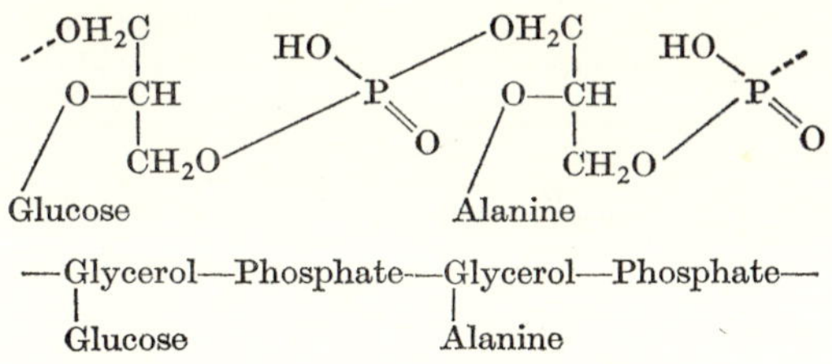

—Glycerol—Phosphate—Glycerol—Phosphate—
    | |
  Glucose   Alanine

(2) Ribitol teichoic acid:

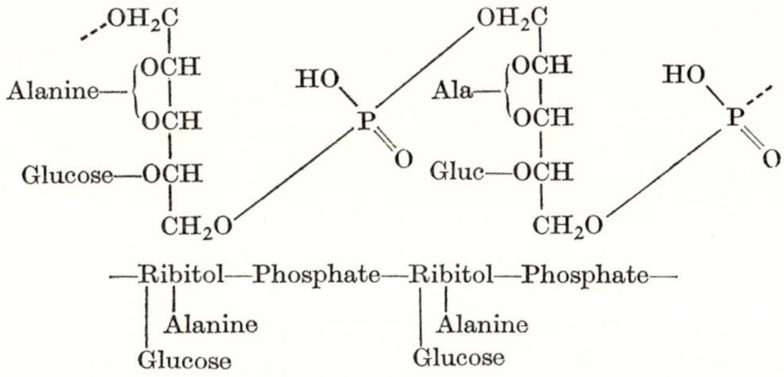

—Ribitol—Phosphate—Ribitol—Phosphate—
    | | | |
    | Alanine | Alanine
   Glucose   Glucose

(3) Burger-type polymer:

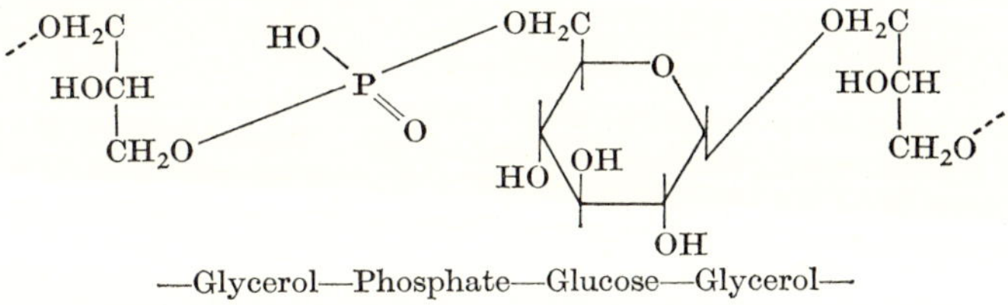

—Glycerol—Phosphate—Glucose—Glycerol—

(4) Polymer of *Staphylococcus lactis* I3:

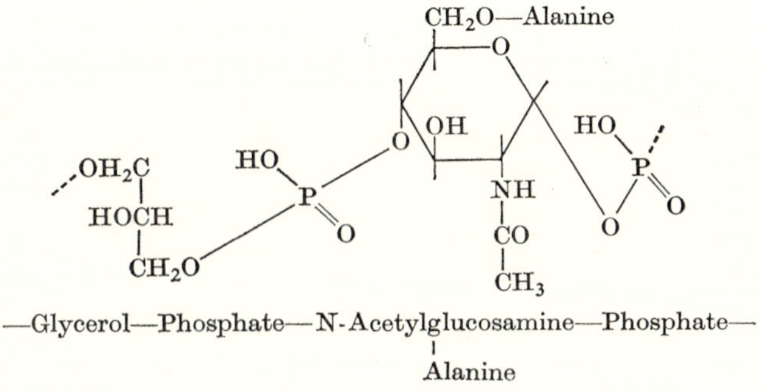

—Glycerol—Phosphate—N-Acetylglucosamine—Phosphate—
                                  |
                               Alanine

Fig. 1. Structures of several teichoic acids and a teichuronic acid.

(5) Teichuronic acid:

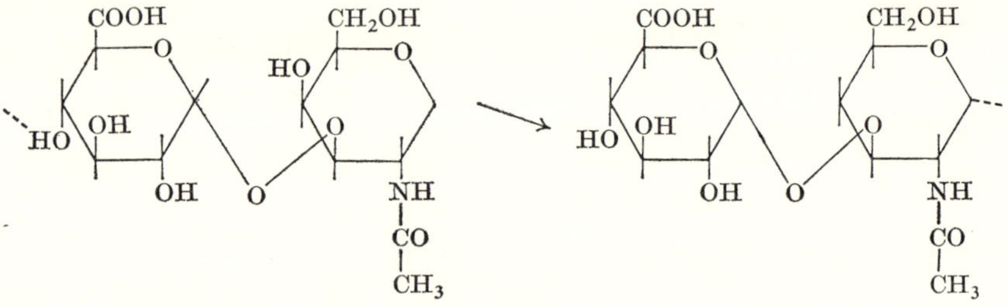

—Glucuronic acid—N-Acetylgalactosamine—Glucuronic acid—

Fig. 1.

galactose) and alanine are attached in amounts that vary widely. Several different basic teichoic acid structures have been recognised (these being shown in Fig. 1) and more than one type may be present in the walls of a particular organism at any time (Chin *et al.*, 1966). Teichuronic acids, like teichoic acids, are also anionic polymers but, in contrast to them, they lack phosphate and owe their anionic nature to the presence of acidic sugars in the molecule (Fig. 1).

Not only may bacterial walls contain more than one type of teichoic acid, they seemingly may contain both teichoic acid and teichuronic acid components (Janczura *et al.*, 1961; Hughes, 1966). But the types and amounts of these different polymers present in the wall of any particular bacterium depend very largely on the environmental conditions prevailing during growth. The interrelationships between environment and the anionic polymer content and composition of the walls of different bacteria have been much studied and are detailed below.

A. Effects of Specific Nutrient Limitations

In a chemostat culture, growth rate is controlled by restricting the supply of one essential nutrient whilst maintaining the concentrations of all others in excess of requirement. This "growth-limiting nutrient" may, of course, be any compound or element for which the organism has an absolute growth requirement but it is reasonable to assume that, when growth is so limited, the organisms will contain the minimum concentration of that substance necessary for them to grow and function in the prescribed environment at the imposed rate.

With cultures of *B. subtilis* var. *niger* the nature of the growth limitation profoundly affected the content and composition of wall anionic polymer. In particular, when growth occurred in the presence of adequate concentrations of phosphate, the walls invariably contained a teichoic acid; but when growth was limited by the availability of phosphate, wall

TABLE 3. Effect of Different Growth Limitations on the Cell-Wall Content and Composition of *Bacillus subtilis* var. *niger* Grown in a Chemostat Culture with the Dilution Rate Fixed at 0·3 hr.$^{-1}$, the Temperature at 35° and the pH Value Regulated at 7·0.

Growth of the organisms and isolation of the bacterial walls were carried out as described in Table 1. Teichoic acid was extracted from the purified wall preparations by treatment with 10%, w/v, trichloroacetic acid (37°, 24 hr.) followed by precipitation with ethanol.

Growth-limiting nutrient	Cell-wall content (% dry wt.)	Content (in g./100 g. dry weight isolated cell walls) of					
		Protein	Phosphorus	Teichoic acid[a]	Hexose	Glucuronic acid	Galactosamine
Glucose	12·0; 13·2	10·0	3·9	35–42	20	<3	<3
Ammonia	21·0; 21·0	11·5	4·8	45–52	23	<3	<3
Sulphate	21·0; 21·0	12·5	4·9	45–53	23	<3	<3
K$^+$	15·3; 15·5	10·5	4·3	39–46	20	<3	<3
Mg^{2+}	17·8; 18·0	12·5	6·9	62–74	32	<3	<3
Phosphate	17·0; 17·5	10·0	0·5	<3	<2	25	17

[a] Assuming the teichoic acid to be fully glucosylated but without alanine.

teichoic acid synthesis ceased and this polymer was totally replaced by teichuronic acid (Table 3). Thus it seemed that, when a constraint was applied to the supply of phosphate, this organism could switch off the synthesis of wall teichoic acid (but not the intracellular teichoic acid;

TABLE 4. Comparison of Phosphorus Contents of Walls from Different Gram-Positive Bacteria Grown in Mg^{2+}-Limited or PO$_4^{3-}$-Limited Chemostat Cultures (dilution rate of 0·2 hr.$^{-1}$; 35°; pH 7·0).

Organism	Grams phosphorus/100 g. dried walls	
	Mg^{2+}-limitation	PO$_4^{3-}$-limitation
Bacillus subtilis W23	4·1	0·1
Bacillus subtilis 168	4·2	0·1
Bacillus subtilis[a]	3·9	0·1
Bacillus subtilis var. *niger*	6·0	0·4
Bacillus licheniformis NCTC 6346	3·6	0·2
Bacillus megaterium KM	0·8	0·04
Micrococcus lysodeikticus NCTC 2665	3·1	0·1
Staphylococcus aureus H	3·0	0·5

[a] Strain of *Bacillus subtilis* obtained from Prof. J. Baddiley and used by Armstrong *et al.* (1961).

90   D. C. ELLWOOD AND D. W. TEMPEST

Ellwood & Tempest, 1968) and replace it, presumably functionally, with the non-phosphorus-containing anionic polymer, teichuronic acid.

The effect of phosphate limitation has now been examined in a wide variety of different Gram-positive bacteria whose walls were known to contain teichoic acid. As detailed below (and summarized in Tables 4

TABLE 5. Comparison of the Anionic Polymers Present in the Walls of Mg^{2+}-Limited and PO_4^{3-}-Limited Bacteria, Grown in a Chemostat at a Dilution Rate of 0·2 hr.$^{-1}$ (35°, pH 7).

Organism	Anionic polymers	
	Mg^{2+}-limited	PO_4^{3-}-limited
Bacillus subtilis W23	Ribitol teichoic acid	Teichuronic acid (*N*-Acetylgalactosamine and glucuronic acid)
Bacillus subtilis 168	Ribitol teichoic acid	Teichuronic acid
Bacillus subtilis var. niger	Glycerol teichoic acids	Teichuronic acid
Bacillus licheniformis (6346)	Glycerol teichoic acid	Teichuronic acid
Bacillus megaterium KM	Teichoic acid-type compound $\begin{bmatrix} \text{-GlucNH}_2\text{-Gluc-} \\ \text{Glycerol-phosphate} \end{bmatrix}$	$\begin{bmatrix} \text{-GlucNH}_2\text{-Gluc-} \\ N\text{-acetylamino-} \\ \text{glucuronic acid} \end{bmatrix}$
Micrococcus lysodeikticus	Glycerol teichoic acid	A polymer of glucose and *N*-acetylamino-mannuronic acid
Staphylococcus aureus H	Ribitol teichoic acid	An aminoglucuronic acid-containing polymer

and 5), in every case the wall teichoic acid was found to be displaced by a teichuronic acid-type polymer.

1. *Bacillus subtilis* var. *niger*.

The teichoic acids extracted from the walls of this organism (grown under all conditions of phosphate excess) generally had a hexose:phosphate molar ratio of 1·0. However, the polymer extracted from glucose-limited organisms had a decreased hexose content and a hexose:phosphate molar ratio of only 0·88. There were also differences in the extractability of the teichoic acids from the different bacterial walls. About 60% of the phosphorus was removed from the walls of organisms grown in

either Mg^{2+}-limited, SO_4^{2-}-limited or glucose-limited environments, but only 30% from the walls of organisms grown in a K^+-limited medium, when treated with 10%, w/v, trichloroacetic acid at 8° for 24 hr. (Ellwood 1970). With all of these organisms (grown at a dilution rate of 0·3 hr.$^{-1}$, 30°, pH 7) the ester-bound alanine to phosphorus ratio was less than 0·1.

The walls of *B. subtilis* var. *niger* contained a mixture of teichoic acids, the principal one being a polyglucosyl-glycerol phosphate polymer (Fig. 1) similar to that found in the walls of *B. licheniformis* (ATCC 9945) (Burger and Glaser, 1966). This type of teichoic acid accounted for between 65 and 75%, w/w, of the teichoic acid fraction isolated from the walls of either Mg^{2+}-limited, K^+-limited, SO_4^{2-}-limited or NH_3-limited organisms (grown at a dilution rate of 0·3 hr.$^{-1}$, 35°, pH 7). And this polymer, together with small amounts of a polygalactosyl-glycerol phosphate polymer (about 10% of the total teichoic acid content) accounted for over 90% of the extractable teichoic acid from glucose-limited organisms. The remaining teichoic acid extracted from the walls of Mg^{2+}-limited, K^+-limited, SO_4^{2-}-limited and NH_3-limited organisms was a $(1 \rightarrow 3)$ linked polyglycerol phosphate which was largely substituted in the 2-position of glycerol with D-glucose residues. The polymer extracted from the walls of glucose-limited organisms also contained small amounts of this type of teichoic acid which had little or no glucose substituents.

The different types of teichoic acid could not be separated by electrophoresis since they moved as diffuse bands, but, as judged by gel filtration and ultracentrifugation studies, they all had molecular weights of about 8,000 and, when examined in a polarimeter, gave a slight positive rotation.

The teichuronic acid isolated from the walls of PO_4^{3-}-limited *B. subtilis* var. *niger* and *B. subtilis* W23 had similar properties to that described for the polymer from *B. licheniformis* (Janczura *et al.*, 1961; Hughes and Thurman, 1970). The two preparations were chemically alike and had $[\alpha]_D$ values of $+35°$ and $+37°$, respectively. They moved similarly on electrophoresis and had similar infrared spectra (with an absorption band at 840 nm. indicative of α-linkages).

2. *Bacillus subtilis* W23.

The walls of this organism have been found to contain ribitol teichoic acids which are glucosylated to different extents depending on the growth condition (Chin *et al.*, 1966). When grown in a chemostat culture, with phosphate in excess of requirement (i.e. Mg^{2+}-, K^+-, SO_4^{2-}-, NH_3-, or glucose-limited), organisms had walls containing much teichoic acid; and this teichoic acid had a glucose:phosphorus molar ratio of 1·0. The glucose was apparently β-linked to the ribitol since it was readily removed

with β-glucosidase. Thus this teichoic acid resembled that described by Armstrong et al. (1961) except that it contained very little ester-bound alanine.

Recently Mauck and Glaser (1970) reported that, when B. subtilis W23 was grown in a minimal medium containing low concentrations of phosphate (5 mM), there was considerable turnover of wall material during the logarithmic phase of growth. In contrast, however, with chemostat cultures of this organism we found no evidence for wall turnover once the culture was in a steady state. And even during the transition period from Mg^{2+}-limited growth to PO_4^{3-}-limited growth we could find no evidence of turnover of the wall components (cf. turnover of wall components in cultures of B. subtilis var. niger; Ellwood and Tempest, 1969).

3. Bacillus subtilis 168.

This organism has been shown to contain in its walls a ribitol teichoic acid containing galactosamine (Young et al., 1964). This polymer was evident in the walls of organisms grown in a Mg^{2+}-limited chemostat culture but was totally replaced by a teichuronic acid when growth was limited by the availability of phosphate. When grown at a dilution rate of 0·3 hr.$^{-1}$, and at pH 7, the walls of Mg^{2+}-limited B. subtilis 168 contained no detectable ester-bound alanine.

4. Bacillus megaterium KM.

When grown in a batch culture, the walls of this organism contained a phosphomucopolysaccharide polymer composed of glucose, N-acetyl-glucosamine and glycerol phosphate in the molar proportions of $2:1·3:1$ (Ghuysen, 1964). The same material was present in the walls of this organism when grown in a Mg^{2+}-limited chemostat culture but, when grown in a phosphate-limited culture, it was totally replaced by a similar polymer of glucose and glucosamine with the glycerol phosphate moiety replaced by an aminohexuronic acid (probably N-acetylaminoglucuronic acid).

5. Staphylococcus aureus H.

This organism has many nutritional requirements and could only be grown in a complex nutrient medium. Nevertheless it was possible to devise such media which contained growth-limiting concentrations of Mg^{2+} or phosphate. When S. aureus was grown in a Mg^{2+}-limited chemostat culture (dilution rate of 0·3 hr.$^{-1}$, 35°, pH 7) and the walls analysed they were found to contain a ribitol teichoic acid similar to that described by Baddiley et al. (1962). This teichoic acid contained glucosamine in the

same molar proportion as phosphorus but had little ester-bound alanine. When grown in a PO_4^{3-}-limited medium, however, this wall teichoic acid ceased to be synthesized and was replaced by a non-phosphorus containing anionic polymer. This latter polymer was composed of N-acetylglucosamine and N-acetylaminoglucuronic acid residues. If the growth rate of this PO_4^{3-}-limited culture was increased to 0·4 hr.$^{-1}$ (i.e. near to the critical dilution rate) then the wall phosphorus content increased substantially (from 0·2 to 0·8%, w/w) and teichoic acid re-appeared in the walls. It should be stressed that, at near-maximum dilution rates, the extracellular phosphate concentration would be close to a non-limiting value.

Since the teichoic acid of *S. aureus* is known to be antigenic, these variations in wall-polymer composition possibly could be of importance in the preparation of high potency vaccines.

6. *Micrococcus lysodeikticus*.

Perkins (1963) showed that the walls of this organism contain a polymer of glucose and N-acetylmannuronic acid but no phosphorus; and this latter observation was confirmed by Prasad and Litwack (1965). But Salton (1953) found that the walls of this organism contained between 0·09 and 0·13%, w/w, phosphorus, and Ghuysen (1968) also reported the presence of about 0·08%, w/w, phosphorus in *M. lysodeikticus* cell walls. Clearly, though, there was little or no teichoic acid present in these walls.

Like *S. aureus*, *M. lysodeikticus* has complex nutritional requirements and will not grow readily in a simple salts medium. Nevertheless, this organism could be grown in a chemostat in a complex medium with the supply of either Mg^{2+} or PO_4^{3-} limiting growth.

When grown in a phosphate-limited medium (dilution rate of 0·2 hr.$^{-1}$, 35°, pH 7) the walls contained about 35–45%, w/w, of a polymer containing glucose and aminomannuronic acid. However, when grown in a Mg^{2+}-limited medium, the walls contained 3·1% phosphorus and extraction of these walls with 10%, w/v, trichloroacetic acid (8°, 24 hr.) yielded a teichoic acid composed of $(1 \rightarrow 3)$-linked polyglycerol phosphate residues substituted (on the 2-position of glycerol) with N-acetylglucosamine. The glucosamine:phosphorus ratio was only about 0·3 but the teichoic acid also contained much ester-bound alanine. The ester-bound alanine:phosphorus molar ratio was 0·5 at a dilution rate of 0·1 hr.$^{-1}$, but fell to a value of 0·1 when the organisms were grown at a dilution rate of 0·5 hr.$^{-1}$.

From the above data one is compelled to conclude that, at least so far as these anionic polymers are concerned, the wall composition of Gram-positive bacteria is by no means fixed and unvarying. But it is possible that these changes in wall composition could be the result of mutation

94   D. C. ELLWOOD AND D. W. TEMPEST

and selection (rather than changes of phenotype) and this is not un-reasonable to suppose since some mutant organisms would always be present in the culture and the chemostat is known to provide a fiercely selective environment (Novick and Szilard, 1950). This possibility can be easily checked (see Ellwood and Tempest, 1969). From the kinetics of growth in a chemostat it is easy to deduce the rate at which the con-centration of any cellular component should change in response to a change in the growth condition, assuming that its synthesis was closely

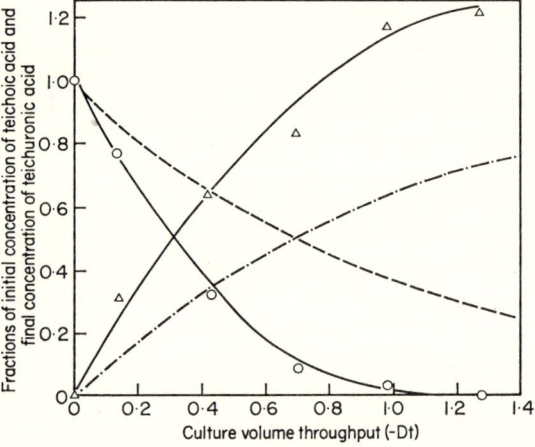

Fig. 2. Changes in the teichoic acid (○) and teichuronic acid (△) contents of cell walls of *Bacillus subtilis* var. *niger* following changeover from conditions of Mg^{2+}-limitation to phosphate-limitation in a chemostat culture. The regular broken line represents the theoretical washout rate assuming that teichoic acid synthesis ceased immediately the environment became depleted of inorganic phosphate. The irregular broken line represents the theoretical rate of increase in teichuronic acid, assuming that its synthesis started immediately the culture became phosphate-limited and continued at a rate proportional to the rate of synthesis of biomass.

coupled to growth. When a chemostat culture of *B. subtilis* var. *niger* was changed from being Mg^{2+}-limited to being limited in its growth by the availability of phosphate, the wall-teichoic acid content diminished at a rate that was much greater than the predicted rate (Fig. 2); con-versely, the wall teichuronic acid content increased substantially faster than the growth rate. Thus, not only was this change-over in wall composition most obviously phenotypic, but there was clearly turnover of the existing wall structures during the transition period. This latter conclusion was re-inforced by the finding of fragments of teichoic acid (and of the pentapeptide moiety of mucopeptide) in the culture extra-cellular fluids.

ENVIRONMENTAL EFFECTS OF BACTERIAL WALLS 95

Clearly, then, these Gram-positive bacteria possess mechanisms for drastically altering not only their wall content, but its composition. The change over from a wall containing largely teichoic acid (and no teichuronic acid) to one containing largely teichuronic acid may be an extreme example, but, as mentioned previously, less obvious changes also were found to have occurred when the nature of the growth limitation was changed (but phosphate was present in excess of requirement throughout). Thus, glucose was generally found to be the sole hexose component of all teichoic acid samples extracted from the walls of *B. subtilis* var. *niger* but, in preparations from glucose-limited and Mg^{2+}-limited organisms grown at a faster rate (0.3 hr.$^{-1}$), galactose was also found to be present in the purified polymer (see Ellwood, 1970).

The functional significance of these changes in teichoic-acid content and composition is not presently known although one fact is clear, and may be relevant; that is, bacterial walls invariably contain at least one anionic polymer and carry a substantial net negative charge. But whether this facilitates the uptake of electrolytes or serves some other purpose cannot be decided from the evidence currently available.

B. Effects of Growth Rate

Both Young (1965) and Chin *et al.* (1966) observed that the teichoic-acid content of bacterial walls varied quantitatively throughout the culture growth cycle, being greater in mid-log phase cultures than in cells taken from cultures in the late stationary phase.

Similarly, with cultures of *B. subtilis* var. *niger* that had been grown in a Mg^{2+}-limited chemostat culture, the teichoic-acid content of the walls varied with growth rate, increasing from about 50%, w/w, to about 75%, w/w, as the dilution rate was increased from 0.1 to 0.3 hr.$^{-1}$ (Ellwood, 1970). But this trend was not consistent and, with Mg^{2+}-limited cultures of either *B. subtilis* W23 (the organism studied by Chin *et al.*, 1966) or *M. lysodeikticus*, increasing the growth rate led to a progressive decrease in wall-teichoic acid content (from about 42%, w/w, at a dilution rate of 0.1 hr.$^{-1}$, to 35% at a dilution rate of 0.3 hr.$^{-1}$ in the former case, and from 29%, w/w, at a dilution rate of 0.1 hr.$^{-1}$ to 24% at a dilution rate of 0.5 hr.$^{-1}$ in the case of *M. lysodeikticus*).

C. Effect of Ionic Environment

The precise function(s) that teichoic acid and teichuronic acid serve in the growing bacterium is still unknown. In reviewing the various possibilities, Baddiley (1964, 1970) suggested that these polymers must create a densely charged environment in or beneath the wall which must,

96 D. C. ELLWOOD AND D. W. TEMPEST

in some way, influence the flux of electrolytes. With bacteria that are growing in a simple salts medium of moderate ionic strength, the movement of Mg^{2+}, K^+, H^+, PO_4^{3-} and SO_4^{2-} must account for the bulk of the total ion flux across the wall and plasma membrane. Therefore it is reasonable to suppose that, if teichoic acids and teichuronic acids are concerned with regulating ion flow, then some correlation should exist between wall anionic polymer content (and composition) and cation metabolism. In this connexion it might be significant that some Mg^{2+} is generally found loosely bound to the surface structures of bacteria

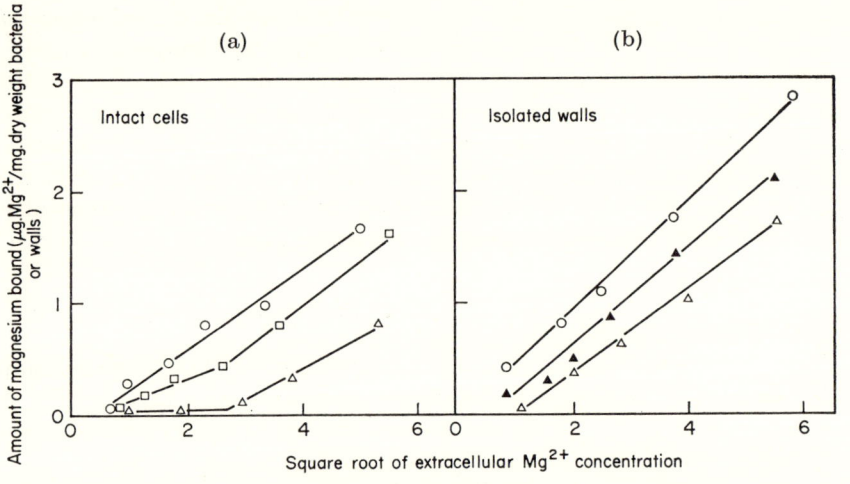

FIG. 3. Magnesium ion-binding affinity of intact cells and isolated walls of *Bacillus subtilis* var. *niger* grown in simple salts medium with the supply of Mg^{2+} (○), K^+ (□) or phosphate (△) limiting growth. The organisms were grown at a dilution rate of 0·3 hr.$^{-1}$, with the temperature controlled at 35° and the pH at 7·0, in the absence (open symbols) and presence (blocked symbols) of 4% (w/v) sodium chloride. Magnesium adsorption occurred from solutions containing graded amounts of magnesium chloride in 0·85% (w/v) sodium chloride.

(Strange and Shon, 1964) and that this adsorbed Mg^{2+} possibly serves an important functional role in the growing bacterium (Tempest and Strange, 1966). Heptinstall *et al.* (1970) have shown that, in *S. aureus* at least, teichoic acid is largely responsible for the Mg^{2+}-binding properties of the wall. It is not unreasonable to suppose, then, that by virtue of these ion-binding properties the wall anionic polymers concentrate cations at the cell surface as the first step in the assimilation process.

The binding of Mg^{2+} to bacterial cell walls is competitively inhibited by Na^+ (Strange and Shon, 1964). Therefore some measure of the strength of Mg^{2+} binding (i.e. binding "affinity" as opposed to ion-binding "capacity") could be obtained by quantitatively studying the

uptake of Mg^{2+} from solutions containing high concentrations of NaCl. A detailed comparison has been made of the effects of growth condition on Mg^{2+}-binding affinity of *B. subtilis* var. *niger* walls (Meers and Tempest, 1970) and these data have now been correlated with differences in wall polymer content and composition (Ellwood, 1971).

Clearly (Fig. 3a), Mg^{2+}-limited organisms bound Mg^{2+} much more avidly than did PO_4^{3-}-limited organisms, and this was demonstrably a property of the walls (Fig. 3b). It follows, then, that if adsorption of Mg^{2+} is a necessary prerequisite to assimilation, addition of high concentrations of NaCl to PO_4^{3-}-limited cultures should totally inhibit Mg^{2+} uptake and therefore growth. However, when increasing concentrations

TABLE 6. Effect of Medium NaCl Concentration on the Anionic Polymer Content and Composition of the Walls of PO_4^{3-}-Limited *Bacillus subtilis* var. *niger*.

The organisms were grown in a chemostat culture with the temperature regulated at 35°, the pH controlled at 7·0 and the dilution rate at 0·3 hr.$^{-1}$. Samples were collected and wall fractions obtained as described in Table 1.

Medium NaCl content (%, w/v)	Amount of wall (g./100 g.)	Amount in walls (g./100 g. dry weight)			
		Phosphorus	Hexose	Glucuronic acid	Galactos- amine
0	18·9	0·4	<2	24·0	17·0
2	22·3	0·7	10·4	19·7	17·0
4	19·1	3·3	16·7	6·1	3·0
6	16·1	4·3	18·8	<3	<3
		Teichoic acid		Teichuronic acid	

of NaCl were added to the medium feeding a PO_4^{3-}-limited culture of *B. subtilis* var. *niger*, the organisms responded by synthesizing walls that had a much increased affinity for Mg^{2+} (Fig. 3b). Analysis of these PO_4^{3-}-limited walls revealed that this increase in Mg^{2+}-binding affinity was accompanied by a marked change in wall composition (Table 6). As the medium NaCl concentration was increased progressively, from almost zero to 2, 4 and finally 6%, w/v, the wall teichuronic acid content progressively diminished and was replaced by teichoic acid; in the presence of 6%, w/v, NaCl, PO_4^{3-}-limited *B. subtilis* var. *niger* organisms had walls that totally lacked teichuronic acid, teichoic acid being the sole anionic polymer present (Table 6).

The above data strongly suggest that the wall teichoic acids are involved in the processes of cation assimilation. Under conditions of PO_4^{3-}-limitation, but where Mg^{2+} is present in excess of the growth

98   D. C. ELLWOOD AND D. W. TEMPEST

requirement, teichoic acids can be functionally replaced by teichuronic
acids. However, as soon as a constraint is applied to Mg^{2+} uptake, the
organisms respond by synthesizing a high-affinity Mg^{2+}-binding wall
component—namely teichoic acid.

D. EFFECTS OF MEDIUM pH VALUE

With batch cultures (and particularly those grown simply in Erlen-
meyer flasks) a variation in pH value generally occurs during the
exponential phase, and usually is accepted uncritically. But pH value is
an important environmental parameter and large shifts in culture H^+
concentration (due to assimilation of ionic nutrients and excretion of
acidic end products of metabolism) are frequently the cause of growth
ceasing (i.e. the onset of the stationary phase) in this type of culture.

Almost always bacterial growth leads to a change in culture pH value.
This can be minimized by adding substantial amounts of a suitable
buffer (e.g. phosphate) to the medium, but complete control can only be
effected by continuously monitoring the culture pH value and correcting
drift, as it occurs, by the addition of a suitable titrant (Callow and
Pirt, 1956).

By using such a system of automatic pH value regulation, incorporated
into a chemostat (see Herbert et al., 1965), it is possible not only to
control closely the culture pH value but also to vary it at will without
simultaneously affecting parameters such as growth rate. With an Mg^{2+}-
limited chemostat culture of B. subtilis W23, progressively lowering the
culture pH value from 7·5 to 5·5 caused a progressive decrease in wall
teichoic acid content and a detectable increase in wall-bound uronic acid
(Table 7). Also (and probably more significantly) there was a progressive
increase in the concentration of ester-bound alanine in the wall teichoic
acid. To check whether alanylation of the teichoic acid was a necessary
consequence of the lowering of the culture pH value, cultures of B.
subtilis var. niger were grown in a chemostat with growth limited by the
availability of ammonia (the sole utilizable nitrogen source); thus, it
could be supposed, only those nitrogen-containing compounds that were
essential for the organism to function in the prescribed environment
would be synthesized. Gradually lowering the pH value of this culture
again led to a progressive increase in the amount of ester-bound alanine
in the wall teichoic acid (Table 7) confirming that this alanine was indeed
necessary for the proper functioning of teichoic acid at low pH values.

The walls of PO_4^{3-}-limited organisms generally do not contain teichoic
acid (Tables 4 and 5, p. 89, 90) and it was therefore of importance to
determine whether chemical modifications of teichuronic acid, analogous
to those observed with teichoic acid, accompanied a lowering of the pH

of a PO_4^{3-}-limited culture. Progressively lowering the pH value of a PO_4^{3-}-limited *B. subtilis* var. *niger* culture from 8·0 to 5·0 and then back to 8·0 effected a marked, and reversible, decrease in wall teichuronic acid content and, more significantly, a very large increase in wall teichoic acid content (Fig. 4). And this teichoic acid contained about 0·2 moles of ester-bound alanine per mole of phosphorus. Thus it seemed likely that,

TABLE 7. Influence of the Medium pH Value on the Content and Composition of *Bacillus subtilis* var. *niger* and *Bacill ussubtilis* W23 Cell Walls

Organisms were grown in chemostat cultures with the temperature regulated at 35° and the dilution rate at 0·3 hr.$^{-1}$. Bulk samples were collected from the overflow line into an ice-cooled receiver; organisms were separated from the extracellular constituents by centrifugation and washed with de-ionized water. Bacterial pastes were disrupted in a Braun MSK homogenizer and the walls quantitatively separated from the cytoplasmic constituents by differential centrifugation. Teichoic acid was extracted from the purified wall preparations by treatment with trichloroacetic acid (24 hr. at 37°) followed by precipitation with ethanol.

Medium pH value	Bacterial wall content (%, w/w)		Content (g./100 g.) in purified wall of			Molar ratio: alanine/ phosphorus in teichoic acid
			Phosphorus	Teichoic acid	Uronic acid	
Bacillus subtilis W23[a] (Mg^{2+}-limited)						
7·5	24·2		3·3	30	Nil	0·1
6·5	24·8	26·4	2·8	25	3·8	0·2
5·5	21·7	33·4	2·4	22	8·5	0·3
Bacillus subtilis var. *niger*[b] (NH$_3$-limited)						
8·0	17·2		4·1	35	Nil	Nil
7·0	19·1	21·0	6·2	52	Nil	0·06
6·0	21·1		4·7	39	Nil	0·08
5·5	20·4		4·0	35	Nil	0·13
5·0	18·9		3·0	27	Nil	0·20

[a] Ribitol teichoic acid.
[b] Glycerol teichoic acid.

whatever function these anionic polymers serve in growing bacteria, only teichoic acid is effective at low environmental pH values and even then only when it is alanylated.

As stated previously, it seems probable that teichoic acids are involved in the processes of cation assimilation by Gram-positive bacteria. In support of this hypothesis, Heptinstall *et al.* (1970) showed that the Mg^{2+}-binding properties of staphylococcal cell walls was due largely to their teichoic acid content. However these workers also found that the

presence of ester-bound alanine markedly decreased the avidity with which teichoic acid bound Mg^{2+} at neutral pH values. In view of the finding (Table 7) that esterification of the wall teichoic acids only occurs at low pH values it would be of considerable value to determine whether the affinity with which Mg^{2+} is bound at acid pH values is similarly decreased, or enhanced, by the presence of ester-bound alanine. In this

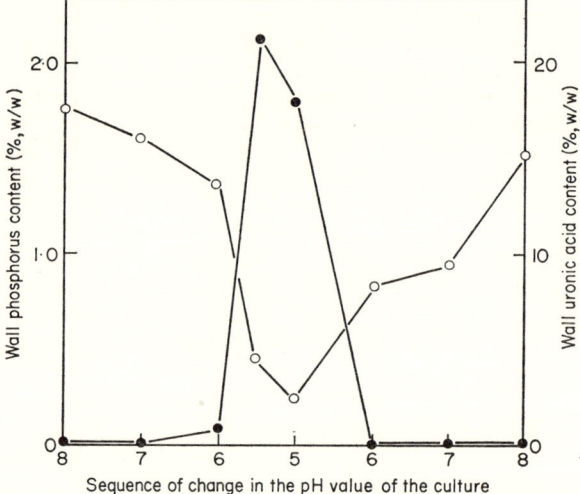

FIG. 4. Influence of medium pH value on the teichuronic acid and teichoic acid contents of walls of phosphate-limited *Bacillus subtilis* var. *niger*. Organisms were grown in a chemostat with the temperature regulated at 35° and the dilution rate set at 0·3 hr.$^{-1}$. The medium pH value was varied progressively from 8·0 to 5·0, and *vice-versa*, by adjusting the controls on the automatic pH indicator-controller. The open symbols (o) show the contents of teichuronic acid in wall preparations and the blocked symbols (●) the contents of teichoic acid.

connexion, it is interesting to note that there has been found to be an accumulation of Zwitterionic (amino-acyl) phospholipid components in the membranes of Gram-positive bacteria when they were cultured at low pH values (Houtsmuller and van Deenen, 1964, 1965; Op den Kamp *et al.*, 1965).

E. CONTROL OF SYNTHESIS OF TEICHOIC ACID AND TEICHURONIC ACID

If, in chemostat cultures of organisms, the change-over from teichoic acid-containing walls to teichuronic acid-containing walls involved solely the upgrowth of variant organisms in the culture then necessarily

ENVIRONMENTAL EFFECTS ON BACTERIAL WALLS 101

it would be a slow process; clearly (Fig. 2) it is a rapid process indicating that it is purely phenotypic. But the nature of the control process is unknown at present. The fact that teichuronic acid is found only in the walls of PO_4^{3-}-limited organisms, whereas teichoic acid is found in the walls of all organisms that have been grown in chemostat cultures containing ample phosphate (Tables 4 and 5), suggests that phosphate, or some phosphorus-containing metabolite, is a key regulatory substance. Thus, presumably, teichuronic acid synthesis is repressed by some phosphorus-containing intermediate(s) involved in teichoic acid synthesis, and only under conditions of phosphate depletion is this intermediate(s) present intracellularly at a sufficiently low concentration to

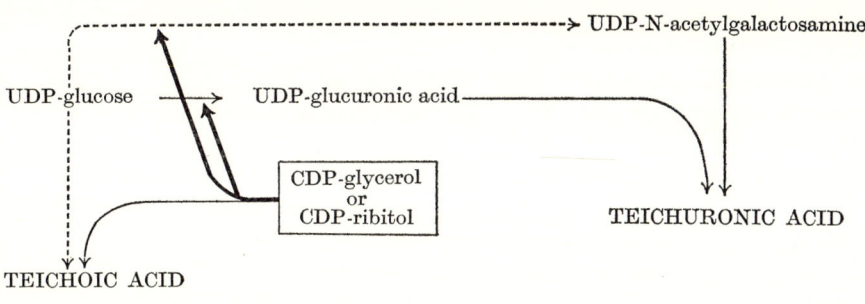

FIG. 5. Possible scheme for regulation of teichoic acid and teichuronic acid syntheses in Gram-positive bacteria. It is assumed that a threshold level of CDP-glycerol (or CDP-ribitol) is necessary for teichoic acid synthesis to proceed. At this threshold concentration (and at all greater concentrations) the nucleotide-sugar inhibits one or more reactions leading to teichuronic acid synthesis. The heavy arrows represent repressor activity.

allow teichuronic acid synthesis to proceed. And presumably teichoic acid synthesis could not occur under these conditions. In this connexion, CDP-glycerol (or CDP-ribitol) could be the repressor and, say, UDP-glucose dehydrogenase the sensitive enzyme (Fig. 5); thus, in the presence of ample phosphate, CDP-glycerol (or CDP-ribitol) would be synthesised but not UDP-glucuronic acid, and hence not teichuronic acid.

The walls of exponential-phase batch-grown *B. licheniformis* (NCTC 6346) contain both teichoic acid and teichuronic acid (Hughes, 1966), and so too have the walls of *B. subtilis* W23 (when grown in a chemostat at low pH values; Table 7), and the walls of *B. subtilis* var. *niger* (when grown in a PO_4^{3-}-limited culture in the presence of NaCl; Table 6). These observations are compatible with the scheme shown in Fig. 5 if one assumes that the primary effect of changing each parameter is to alter the intracellular concentration of pool nucleotide sugars. This

102   D. C. ELLWOOD AND D. W. TEMPEST

remains to be investigated experimentally but it has been found recently that growth rate, environmental pH value and, most particularly, medium salinity each affected markedly the intracellular concentrations of pool free-amino acids (Tempest *et al.*, 1970).

IV. Intracellular Teichoic Acids

Baddiley (1970) reported that all the Gram-positive bacteria that had been examined in his laboratory contained intracellular (membrane-bound) teichoic acids as well as the wall-bound polymers. Invariably these membrane teichoic acids were glycerol phosphate polymers (as opposed to the ribitol phosphate polymers frequently found in the walls) and were structurally similar to the wall-bound glycerol teichoic acids. Intracellular teichoic acids were present in all organisms that synthesized wall teichoic acids, irrespective of the nature of the growth limitation. Thus, phosphate-limited organisms which lacked totally a wall-bound teichoic acid (Table 4, p. 89) nevertheless still synthesized the "intracellular" polymer (Ellwood and Tempest, 1968). It must be concluded, therefore, that, whereas the wall-bound teichoic acids can be functionally replaced by other anionic polymers (teichuronic acids), the presence in the cell of membrane-bound teichoic acid is essential for their proper functioning.

We do not have any evidence of phenotypic variation in the content or composition of these intracellular teichoic acids but clearly their synthesis is controlled differently from that of the wall-bound acids. This is particularly evident when, as with *B. subtilis* var. *niger* organisms, both the wall and membrane-bound teichoic acids are glycerol phosphate polymers.

V. Mucopeptide

The mucopeptide component of bacterial walls (otherwise termed "murein", "peptidoglycan" or "glycosaminopeptide") is the polymer which acts as the main supporting material of the wall and confers on the cell its shape. Mucopeptides are built up from a close network of identical chains and form large "bag-shaped" macromolecules that totally enclose the plasma membrane. These polymers are mechanically very strong and can withstand the high internal hydrostatic pressures that generally are generated within the bacterial cell; thus they protect the plasma membrane and hold the bacterial cell intact.

The mucopeptides of several different bacteria have been analysed in detail and the results obtained have been authoritatively reviewed by Ghuysen (1968). With almost all of the bacterial walls so far examined, mucopeptides were found to contain a backbone of alternating β-1:4

ENVIRONMENTAL EFFECTS OF BACTERIAL WALLS 103

linked N-acetylglucosamine and N-acetylmuramic acid residues (Fig. 6); but in one case (*Staphylococcus aureus*, Copenhagen strain) the presence of O-acetyl substituents on the C-6 of 60% of the N-acetylmuramic acid residues was reported (Tipper and Strominger, 1966) and, in another case (*Mycobacterium smegmatis*), the muramic acid was found to be substituted with glycolic acid on the amino nitrogen (Petit *et al.*, 1969).

The composition of the peptide moiety of mucopeptides is seemingly much more variable than is the polysaccharide backbone. On the basis of an extensive survey, Ghuysen (1968) proposed that there were five basic types of peptide subunit in the mucopeptides of bacteria (Fig. 7) but that a degree of diversity also occurred in the composition of the "bridge" structures that linked the peptide moieties (and hence polysaccharide chains) together. Ghuysen (1968) again grouped these

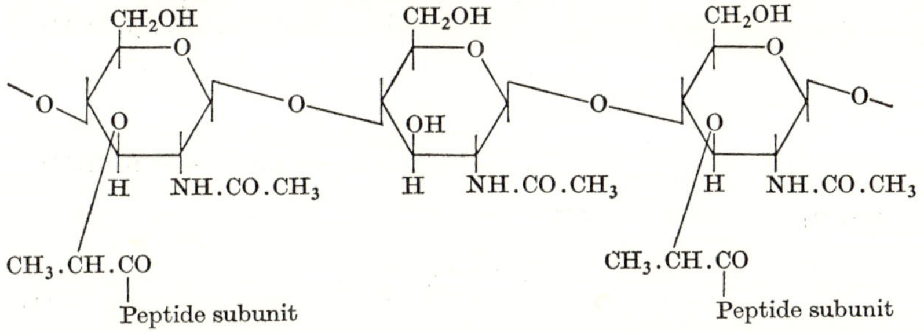

FIG. 6. "Backbone" structure of a typical mucopeptide molecule showing the attachment of pentapeptide moieties to the muramic acid residues.

"bridge" structures into five basic types, with a number of minor variations too large to illustrate here.

As a result of the detailed knowledge now available, it has been proposed to classify bacteria according to their mucopeptide fine structure (Ghuysen, 1968). But such a classification might be premature since, in spite of the extensive knowledge regarding the detailed structure of mucopeptides from different bacteria, little work has yet been done to determine whether the structure of these polymers can vary phenotypically. Young (1965) grew *Bacillus subtilis* 168 in several different media, in batch cultures, and reported that the mucopeptide composition did not vary significantly. But more recently Schleifer *et al.* (1969) (also see p. 36 of this volume) showed that, with *Staphylococcus aureus* strain Copenhagen and *S. epidermidis* strain 66, substantial changes in the composition of the "bridge" unit could occur in response to changes in the amino-acid composition of the growth medium.

Some preliminary work on the influence of growth condition on

104 D. C. ELLWOOD AND D. W. TEMPEST

mucopeptide composition has been carried out in our laboratory using the chemostat to provide a wide range of controlled environments. Our work, which we have not published previously, has been largely confined to two strains of *B. subtilis*, namely *B. subtilis* var. *niger* and *B. subtilis* W23.

A detailed investigation of the structure of the mucopeptide of Bacillus organisms was previously carried out by Drs. R. C. Hughes and

Type A:

GLYCAN
|
L-Ala $\longrightarrow$ D-Glu
$\gamma\llcorner\longrightarrow$ L $\longrightarrow$ D-Ala
|
D

Type B:

GLYCAN
|
L-Ala $\longrightarrow$ D-Glu $\overset{\alpha}{\longrightarrow}$ NH$_2$
$\gamma\llcorner\longrightarrow$ L-Lys $\longrightarrow$ D-Ala

Type C:

GLYCAN
|
L-Ala $\longrightarrow$ D-Glu $\overset{\alpha}{\longrightarrow}$ Gly
$\gamma\llcorner\longrightarrow$ L-Lys $\longrightarrow$ D-Ala

Type D:

GLYCAN
|
Gly $\longrightarrow$ D-Glu
$\gamma\llcorner\longrightarrow$ HomoSer $\longrightarrow$ D-Ala

Type E:

GLYCAN
|
L-Ser $\longrightarrow$ D-Glu
$\gamma\llcorner\longrightarrow$ L-Orn $\longrightarrow$ D-Ala

FIG. 7. Five types of peptide subunit that are found attached to the glycan portions of different mucopeptides. *Type A*: the peptide subunit of mucopeptides in *Escherichia coli*, *Bacillus megaterium* and *B. subtilis*. In *B. subtilis* the carboxyl groups of the D-carbon of *meso*-DAP of some peptide subunits are amidated. The vertical bar separating "L" and "D" represents diaminopimelic acid. *Type B*: the peptide subunits of mucopeptides in *Staphylococcus aureus* and *Streptococcus pyogenes*. *Type C*: the peptide subunits of mucopeptides in *Micrococcus lysodeikticus* and *Sarcina lutea*. *Type D*: the peptide subunits of mucopeptides in *Corynebacterium poinsettiae*. *Type E*: the peptide subunits of *Butyribacterium rettgeri* mucopeptide.

H. J. Rogers. These workers showed that walls of batch-grown *B. licheniformis* (NCTC 6346) organisms were composed of four different polymers: (i) a protein which accounted for about 10% of the wall weight; (ii) a teichoic acid, initially thought to be a ribitol-containing polymer but now known to be a glycerol teichoic acid; (iii) a teichuronic acid; and (iv) mucopeptide, which was of Type 1 in the classification of Ghuysen (1968). This mucopeptide was found to be composed of N-acetyl-

muramic acid, N-acetylglucosamine, L-alanine, D-glutamic acid, *meso*-diaminopimelic acid and D-alanine in the molar proportions of 0·7:0·8: 1·0:1·0:1·0:0·6 respectively. The values observed for the two amino sugars were always less than unity (0·7:0·8, relative to glutamic acid of 1·0) which could be explained either as analytical error due to difficulty of accurately assessing degradation during hydrolysis, or to the fact that these components possibly were present in the wall partially as derivatives that on hydrolysis did not liberate the free amino sugars (e.g. phosphorylated derivatives).

The above data strongly suggest that the mucopeptide component of bacterial walls is unlikely to vary substantially with changes in the growth condition. Our data, though preliminary, indicate that this conclusion is unwarranted.

A. COMPARATIVE EFFECTS OF LYSOZYME ON *Bacillus subtilis*

Bacillus subtilis W23 was grown in chemostat cultures with different essential nutrients limiting growth. Samples of culture were removed and the organisms immediately heated at 100° (for 20 min.) to inactivate the

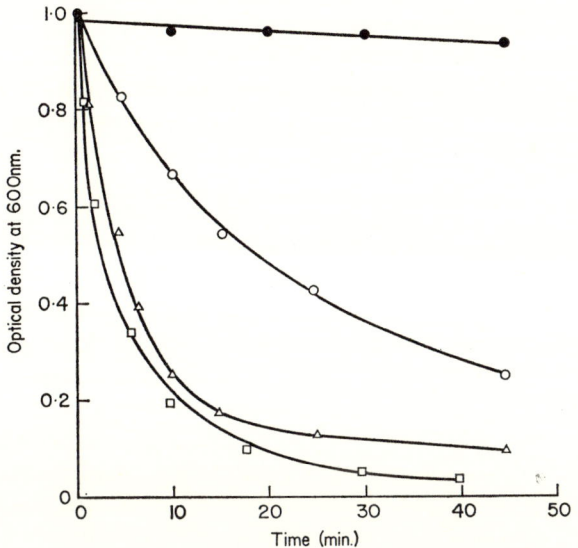

FIG. 8. Effect of lysozyme (20µg./ml.) on the optical density of heat-treated suspensions of *Bacillus subtilis* W23 organisms. The organisms were suspended in 0·05 M-sodium phosphate buffer (pH 6·8) to an optical density, at 600 nm., of about 1·0. Lysozyme was then added to the suspension and the change in optical density followed for up to 45 min. The organisms were from chemostat cultures (dilution rate of 0·3 hr.$^{-1}$; 35°; pH 7·0) that were either Mg^{2+}-limited (○), phosphate-limited (△) or ammonia-limited (□). The control suspension (●) contained Mg^{2+}-limited organisms but no lysozyme.

autolytic enzymes known to be present. After heating, the organisms were separated from the culture extracellular fluids and washed twice with 0·05 M-phosphate buffer (pH 7·0), using the centrifuge. The bacteria were then resuspended in the above buffer at a concentration sufficient to give an optical density of about 1·0 at 600 nm., in a cuvette of 1 cm. light path. Lysozyme (200 μg.) was added to 10 ml. of this suspension which was then incubated at 37° and the optical density measured at suitable time intervals. The results (Fig. 8) showed that Mg^{2+}-limited, PO_4^{3-}-limited and NH_3-limited organisms responded differently; Mg^{2+}-limited organisms lysed at a markedly slower rate than did organisms from either the PO_4^{3-}-limited or NH_3-limited cultures. The large difference in lysozyme sensitivity between the Mg^{2+}-limited organisms and the PO_4^{3-}-limited organisms possibly could be due to the different anionic polymers (teichoic acid as compared with teichuronic acid) being present in their walls. But this is unlikely to be the real cause of the differences since NH_3-limited *B. subtilis* organisms, which had teichoic acid in their walls, lysed even more rapidly than did the teichuronic acid-containing PO_4^{3-}-limited organisms. In fact no correlation could be found between the wall anionic polymer content and the rate of lysis, in the presence of lysozyme, suggesting that the difference in sensitivity lay in the mucopeptide content and composition. In this connexion it is worth noting that the walls of NH_3-limited bacteria were extremely prone to autolysis and therefore difficult to prepare.

B. Effect of Growth Condition

Although there are several other possible reasons for the differences in lysozyme sensitivity shown in Fig. 8, the most likely one is that the mucopeptide components of the different walls vary in some way. In order to investigate this further, walls were isolated, the anionic polymers removed with trichloroacetic acid, and the mucopeptide-containing residues hydrolysed with 4-N-HCl (18 hr. at 100° in an atmosphere of nitrogen gas). Qualitative chromatographic analyses of these hydrolysates showed each to be similar in containing substantial amounts of muramic acid, glucosamine, alanine, glutamic acid and diaminopimelic acid, together with traces of other amino acids (presumably derived from the small amounts of protein present in the walls of these bacilli; Table 3, p. 89). On the other hand, quantitative analyses, using a Technicon TMS amino-acid analyser, revealed marked differences in the muramic acid and glucosamine contents of the different mucopeptide preparations, although there was little variation in the molar proportions of alanine:glutamic acid:diaminopimelic acid (Table 8).

From the published data of Hughes *et al.* (1968) the expected molar

proportions of muramic acid:glucosamine:glutamic acid:alanine:diaminopimelic acid in the mucopeptide of *B. subtilis* W23 were 0·75: 0·75:1·0:1·6:1·0, respectively. Clearly (Table 8) the ratios found for the sugar components of the mucopeptides varied substantially from these values. Interestingly, the molar ratios of muramic acid and glucosamine to glutamic acid in the walls of NH_3-limited organisms (which were most

TABLE 8. Comparison of Mucopeptide Fractions from the Walls of *Bacillus subtilis* W23 Grown Under Different Conditions

Organisms were grown in a chemostat (pH 7·0; 35°) with different components of the medium limiting growth rate. Organisms were collected and walls isolated as described in Table 1. Purified wall fractions were extracted with 10%, w/v, trichloroacetic acid (37°, 18 hr.) to remove the teichoic acid and/or teichuronic acid and then washed and freeze-dried. Residues (2 mg.) were hydrolysed with 6N-HCl (1 ml.) in tubes sealed under nitrogen gas and heated at 100°. The release of amino compounds was followed by analysis of hydrolysates (using a Technicon TMS amino-acid analyser) after removal of the HCl by evaporation of the sample to dryness over NaOH and P_2O_5, twice. The optimum time for release of aminosugars was 6–8 hr., and for amino acids 15–22 hr. Consequently, routine analyses were carried out on samples that had been hydrolysed for 18 hr. and the results were corrected appropriately for partial destruction of amino sugars. The results below are expressed as molar ratios, assuming a glutamic acid value of 1·0.

Component	Growth-limiting substrate and dilution rate							
	Mg^{2+} 0·1 hr.$^{-1}$	Mg^{2+} 0·2 hr.$^{-1}$	Mg^{2+} 0·3 hr.$^{-1}$	K^+ 0·3 hr.$^{-1}$	Glucose 0·3 hr.$^{-1}$	NH_3 0·3 hr.$^{-1}$	SO_4^{3-} 0·3 hr.$^{-1}$	PO_4^{3-} 0·3 hr.$^{-1}$
Muramic acid	0·62	0·75	0·90	0·82	1·05	1·07	0·50	1·02
Glucosamine	0·79	0·92	1·04	0·67	0·70	1·03	0·70	0·70
Glutamic acid	1·00	1·00	1·00	1·00	1·00	1·00	1·00	1·00
Alanine	1·4	1·6	1·6	1·5	1·5	1·6	1·5	1·5
Diamino-pimelic acid	0·93	1·04	1·07	0·93	1·15	1·15	1·18	1·17

sensitive to lysozyme) were about unity, close to the theoretical value assuming none of the sugar constituents to be substituted. With Mg^{2+}-limited organisms, however, (and SO_4^{2-}-limited bacteria) the mucopeptides contained significantly more glucosamine than muramic acid, whereas the walls of K^+-limited, glucose-limited and PO_4^{3-}-limited *B. subtilis* W23 contained more muramic acid than glucosamine.

Comparison of the analytical data obtained from hydrolysates of *B. subtilis* W23 whole walls and mucopeptide fractions derived from them revealed no marked differences except in the alanine:glutamic acid

108 D. C. ELLWOOD AND D. W. TEMPEST

ratios (Table 9). The larger amount of alanine in the whole-wall preparations is consistent with the findings of Young (1965) with batch-grown
B. subtilis W23 and is not due to ester-bound alanine on the teichoic acid.
These Mg^{2+}-limited organisms were grown at pH 7·5 whereas the wall
teichoic acid only contains ester-bound alanine when grown in more
acidic environments (Table 9). But, in order to check that the
"additional" alanine in the unextracted walls was not ester-bound to
teichoic acid, we grew Mg^{2+}-limited cultures of *B. subtilis* W23 at two

TABLE 9. Influence of Culture pH Value on the Composition of the Mucopeptide
Fractions from Cell Walls of Mg^{2+}-Limited *Bacillus subtilis* W23
The organisms were grown at a dilution rate of 0·2 hr.$^{-1}$ in Mg^{2+}-limited
chemostat cultures with the temperature controlled at 35°. The steady-state
culture pH value was progressively decreased from 7·5 to 5·5 and samples removed
for analysis as described in Table 8. The table compares the molar ratios of muramic
acid, glucosamine, alanine and diaminopimelic acid to glutamic acid in unextracted
and trichloroacetic acid-extracted walls of the bacterium.

| | pH Value during growth | | | | | | |
| | Whole walls | | | Trichloroacetic acid-extracted walls | | | Walls of batch-grown organisms[a] |
Component	7·5	6·5	5·5	7·5	6·5	5·5	
Muramic acid	0·90	0·80	0·66	0·86	0·84	0·67	0·70
Glucosamine	0·87	0·83	0·77	0·85	0·77	0·77	1·10
Glutamic acid	1·00	1·00	1·00	1·00	1·00	1·00	1·00
Alanine	2·05	2·15	2·72[b]	1·61	1·53	1·53	2·26
Diaminopimelic acid	1·00	1·03	0·91	0·99	0·97	0·93	0·94

[a] Data of Young (1965).
[b] The teichoic acid present in the walls of these organisms contained ester-bound
alanine (see Table 7, p. 99).

lower pH values. A further marked increase in the wall alanine : glutamic
acid ratio was then observed (Table 9) which was consistent with the
formation of teichoic acid containing some ester-bound alanine (cf.
Table 7). One possible explanation for the increased alanine values of
whole walls is that the mucopeptide contains terminal D-alanyl-D-alanine
residues; that is, the mucopeptide is not fully cross-linked. In this
connexion, Hughes *et al.* (1968) found that the mucopeptide of *B.
subtilis* had only about 50% of its peptide chains cross-linked. The fact
that the extracted walls contained less alanine than the whole walls
suggests that this terminal alanine, if present, is acid-labile.

ENVIRONMENTAL EFFECTS ON BACTERIAL WALLS 109

TABLE 10. Effect of Incubation Temperature on the Composition of the Mucopeptide of Cell Walls of *Bacillus subtilis* W23

Organisms were grown in a Mg^{2+}-limited chemostat culture with the dilution rate fixed at 0·2 hr.$^{-1}$ and the pH value regulated at 7·0. The incubation temperature was varied progressively from 35° to 25° and samples removed and analysed as described in Table 8.

Component	Temperature of growth		
	25°	30°	35°
Muramic acid	1·18	0·80	0·84
Glucosamine	1·20	1·20	0·95
Glutamic acid	1·00	1·00	1·00
Alanine	2·20	2·20	1·99
Diaminopimelic acid	1·05	0·92	0·95

We repeated the above experiment but kept the pH value constant and varied the incubation temperature. Again (Table 10) it was found that the alanine : glutamic acid ratio was over 2·0 in the unextracted walls.

TABLE 11. Influence of Growth Condition on the Mucopeptide Composition of Cell Walls of *Micrococcus lysodeikticus* (NCTC 2665)

Organisms were grown in chemostat cultures with either the availability of Mg^{2+} or PO_4^{3-} limiting growth. The dilution rate was fixed at 0·2 hr.$^{-1}$ and the pH value and temperature regulated at 7·0 and 35°, respectively. The organism was also grown in batch culture in a tryptone-salts medium. Walls were prepared and analysed as described in Table 8.

Component	Condition of growth		
	Mg^{2+}-Limited[a]	PO_4^{3-}-Limited	Batch culture[b]
Muramic acid	1·03	0·79	0·66
Glucosamine	1·90	0·66	0·79
Glutamic acid	1·00	1·00	1·00
Alanine	4·24	2·09	1·92
Lysine	1·08	0·88	0·92
Glycine	2·72	0·81	1·11
Aspartic acid	0·32	0·08	—
Threonine	0·15	0·02	—
Serine	0·99	0·15	—

[a] The walls of Mg^{2+}-limited *M. lysodeikticus* contained a glycerol teichoic acid not previously found in this organism.

[b] Data derived from those of Anderson *et al.* (1969).

Surprisingly, incubation temperature also affected the muramic acid and glucosamine contents of walls, both of which increased substantially as the temperature was lowered. Thus, lowering the temperature of growth effected changes in mucopeptide composition similar to those occurring when the growth rate was increased; this is consistent with observations made on the cellular content of other components (Tempest and Hunter, 1965).

The walls of *Micrococcus lysodeikticus* (NCTC 2665) were similar in composition to those reported for batch-grown organisms (see Anderson *et al.*, 1969) when the cells were grown in a PO_4^{3-}-limited chemostat culture (Table 11). However, when grown in a Mg^{2+}-limited culture, the walls of this organism contained larger amounts of aspartic acid and considerably increased amounts of serine, glycine, alanine and glucosamine. The increased amounts of alanine and glucosamine could be explained by the fact that the walls also contained a glycerol teichoic acid that had glucosamine and ester-bound alanine associated with it. But the reason for the elevated levels of serine and glycine, and for the presence of aspartic acid, is unknown.

Clearly, many of the above data are of a preliminary nature and more definitive experiments remain to be done. Nevertheless the results so far obtained strongly suggest that, in Gram-positive bacteria at least, the mucopeptide component of the wall may change phenotypically. For the time being it would seem prudent, therefore, to treat sceptically data obtained from organisms grown in uncontrolled environments (that is, in batch-type cultures).

VI. Lipopolysaccharide

Though structurally very distinct, lipopolysaccharides (which are only found in Gram-negative bacteria) resemble teichoic acids (which are confined to Gram-positive species) in being anionic polymers that contain sugar phosphate residues (Fig. 9). But whether lipopolysaccharides and teichoic acids are functionally similar is for the moment a matter of speculation.

Among the many species of Gram-negative bacteria that have been examined, the lipopolysaccharide "core" structures are very similar; but the structure of the polysaccharide side-chains differs substantially between species and, indeed, between strains of a single species. In fact, mutants blocked in the synthesis of a particular nucleotide-sugar precursor produce a lipopolysaccharide that lacks the appropriate sugar component. And mutants can be obtained whose lipopolysaccharide contains the "core" structure only. Sugar-depleted lipopolysaccharides are presumably functionally competent since these mutants grow

normally and survive as well as the wild-type organisms under laboratory conditions.

The "core" structure contains two chemical components that do not occur elsewhere in the cell wall, that is KDO and heptose. Consequently it should be possible to determine the lipopolysaccharide content of bacterial walls simply by measuring the amount of either KDO or heptose in the wall preparation. When this experiment was carried out with isolated wall preparations from chemostat cultures of *Aerobacter*

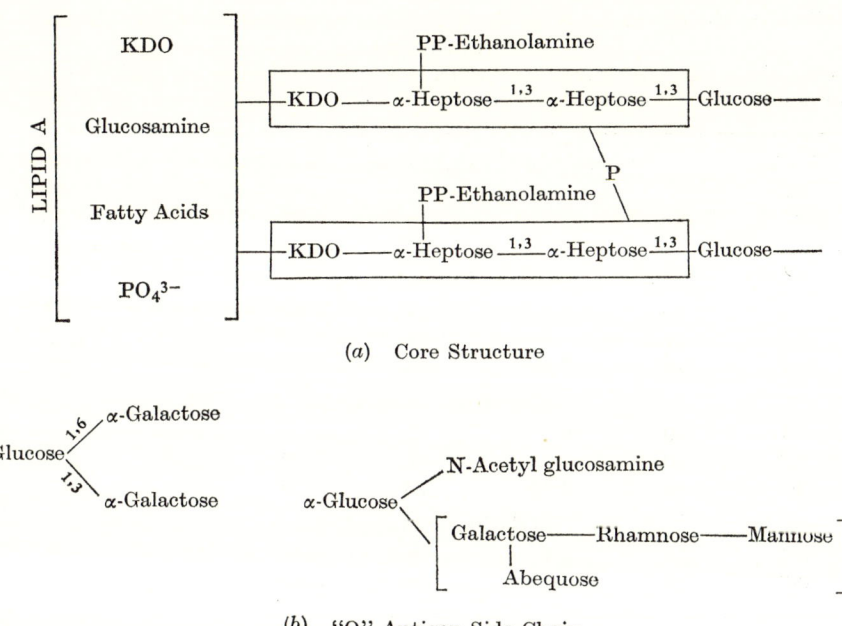

(*a*) Core Structure

(*b*) "O" Antigen Side Chain

FIG. 9. Structures of the "core" and "O-antigen" side chain of the lipopolysaccharide from *Salmonella typhimurium*. Details from Osborn (1969).

aerogenes, some unexpected results were obtained (Table 12). With glycerol-limited organisms, the wall KDO and heptose contents increased markedly with growth rate indicating a general increase in wall lipopolysaccharide content (and this conclusion was supported by the finding of a general increase in wall carbohydrate content). But Mg^{2+}-limited organisms behaved differently; the wall carbohydrate content increased only slightly and the KDO and heptose contents actually decreased with increasing growth rate.

Irrespective of changes in wall lipopolysaccharide content, the "core" structure should not vary, and hence the heptose:KDO ratio always should be constant. Clearly (Table 12) this assumption is not supported

112   D. C. ELLWOOD AND D. W. TEMPEST

by the facts. The heptose:KDO ratio varied from about 6 to 10, depending on the growth condition, being greatest in slowly-growing glycerol-limited and fast-growing Mg^{2+}-limited *A. aerogenes*. Again, the phosphorus content of the walls varied markedly and, although increases in wall heptose content were invariably accompanied by increase in wall phosphorus content, the ratios of heptose:phosphorus were far from uniform. Indeed, limiting growth by the availability of phosphate

TABLE 12. Influence of Growth Rate and Growth-Limiting Component of the Medium on the Composition of the Walls from *Aerobacter aerogenes* Grown in a Chemostat

Organisms were grown, and cell walls isolated, as described in Table 1 (p. 85). For details of the analytical procedures, see Tempest and Ellwood (1969).

Dilution rate (hr.$^{-1}$)	Content (g./100 g.) in dried cell walls of					Ratio of heptose: KDO
	Protein[a]	Phosphorus	Carbohydrate	KDO	Heptose[b]	
Glycerol-limited organisms						
0·1	90	0·5	13	0·43	3·9	9·1
0·3	71	—	19	0·54	5·7	10·5
0·7	69	0·6	27	1·07	6·7	6·3
Mg^{2+}-limited organisms						
0·1	80	0·7	18	0·92	6·0	6·5
0·3	60	—	21	0·52	4·4	8·5
0·7	58	0·45	22	0·50	4·6	9·1
PO_4^{3-}-limited organisms						
0·1	58	0·1	8	0·80	5·0	6·2
0·7	60	0·3	10	0·80	6·1	7·5

[a] Protein values were derived by assuming that the proteins have an overall composition similar to bovine serum albumen. This is clearly not so. Thus the figures may not be accurate but they allow a comparison to be made.

[b] The values for heptose are also included in the values for total carbohydrate.

effected a marked decrease in wall phosphorus content (c.f. Gram-positive organisms) and an enormous increase in the ratio of heptose: phosphorus (Table 12).

Once more it is necessary to point out that the data contained in Table 12 are of a preliminary nature; but plainly these data suggest that it is unwise to assume that the walls of Gram-negative bacteria do not vary phenotypically. Indeed, it has already been shown that the wall composition of a marine pseudomonad can vary enormously with the ionic strength of the growth medium (Brown, 1961). Our results fully support this conclusion, as well as that of Lacey (1961) who showed that growth

conditions greatly affected the antigenicity of Gram-negative bacteria (which is, presumably, a cell-wall phenomenon).

VII. Protein

Apart from work of Davies *et al.* (1968) on the adaptation of Group A haemolytic streptococci to produce M protein in a completely synthetic medium, we are not aware of any studies that have been made of phenotypic changes in bacterial wall protein content or composition.

The walls of Gram-negative bacteria are largely proteinaceous (Table 12) whereas proteins generally account for only a small proportion (about 10%, w/w) of the walls of Gram-positive bacteria (Table 3; Hughes, 1965). However, streptococci do have considerable amounts of protein in their walls, and the walls of *Staphylococcus aureus* can contain up to 30%, w/w protein.

By growing bacteria in a C-limited or N-limited environment, one would expect them to curtail severely the synthesis of non-essential proteins. But when cultures of *Bacillus subtilis* var. *niger* were thus cultured, the walls contained undiminished amounts of protein (Table 3, p. 89) suggesting that this protein was an indispensable component of the wall.

VIII. Lipids

Much work has been done on the effects of growth temperature (Gaughran, 1947; Marr and Ingraham, 1962; Shaw and Ingraham, 1965; Okuyama, 1969), and of other growth conditions (Marr and Ingraham, 1962; Knivett and Cullen, 1965, 1967; Hancock and Meadow, 1969) on the lipid content and composition of Gram-negative bacteria (principally *Escherichia coli*). In these investigations, no attempt was made to distinguish between wall and membrane-bound lipids, but the total extractable lipids were found to vary markedly. Temperature shifts affected the content and degree of saturation of the lipid fatty acid moieties; the lower the growth temperature the greater the proportion of unsaturated fatty acids in the bacterial lipids. Lipid content was also affected by the chemical environment; depletion of phosphate or utilizable nitrogen in a batch culture, or of utilizable nitrogen (ammonia) in a chemostat culture, caused the accumulation of much lipid in the cells.

Unlike Gram-negative bacteria, the walls of Gram-positive organisms generally contain little lipid (less than 5%, w/w). But *Bacillus polymyxa* has been shown to possess a lipoprotein coat which, when viewed by electron microscopy, is evident as an ordered surface structure (Goundry *et al.*, 1967).

We have grown a lipoprotein-containing strain of *B. subtilis* (designated *B. polymyxa*) in variously-limited chemostat cultures, isolated the walls and determined their compositions. Clearly (Table 13) these varied markedly; the walls of sulphate-limited and Mg^{2+}-limited organisms had a high content of lipid whereas glucose-limited organisms had an extremely high content of protein. The wall compositions of chemostat-grown organisms were markedly different from those of batch-grown *B. polymyxa* and, when examined in the electron microscope, failed to

TABLE 13. Effect of Different Growth Limitations on the Cell-Wall Composition of an Unidentified (Lipoprotein-Containing) Strain of *Bacillus subtilis* Grown in a Chemostat at a Dilution Rate of 0·3 hr.$^{-1}$, 35°, pH 7·0

Growth of organisms and isolation of wall preparations were as described in Table 1 (p. 85). The lipid content of freeze-dried wall samples was determined gravimetrically following extraction with $CHCl_3$/MeOH (2:1, by volume). Protein and total hexose were determined by the usual methods (see Ellwood, 1970) and lipoprotein by extraction with sodium dodecyl sulphate.

Growth-limiting nutrient	Content (g./100 g. dry weight) in purified cell walls of			
	Lipid	Protein	Total hexose	Lipoprotein
Magnesium	29	39	11	27
Glucose	17	62	20	63
Ammonia	11	20	30	15
Sulphate[a]	25	24	5	20
Phosphate	18	20	3	18
Batch-grown	5	12	23	18

[a] Dilution rate of 0·2 hr.$^{-1}$.

show the characteristic surface pattern. Similarly, chemostat-grown cultures of the organism used by Goundry *et al.* (1967) did not appear to have the ordered surface structure evident in batch-grown organisms.

IX. Concluding Remarks

In the Introduction, we emphasized the fact that, in order to study phenotypic variation in bacterial walls, it is essential to control rigidly the chemical environment in which the organisms are growing. Since, by their very nature, batch-type cultures can never fulfil this condition (see Tempest, 1970) one must inevitably resort to growing the organisms in an "open" (continuous-flow) culture system, and this we have done routinely.

From the preliminary data which have so far been obtained (and which are assembled in this article) it is clear that the walls of bacteria are

highly variable structures and capable of undergoing radical change in response to relatively small changes in the growth environment. But it is important to stress that, by choosing to use a chemostat to provide controlled growth conditions, we have necessarily had to culture the organisms at rates substantially below those expressed during the "logarithmic phase" of a batch culture; the rates at which the various bacteria were grown never exceeded one doubling of mass per hour and frequently were as little as 0·1 doubling per hour. Thus, caution must be exercised in comparing our data with those derived from organisms grown in batch cultures which were probably dividing, during the logarithmic growth phase, at a rate of 2–3 per hour. In this connexion, Donachie and Begg (1970) showed that organisms growing at a rate faster than one doubling per hour possessed two growth points whereas slower growing organisms had only one. Therefore, for this reason if for none other, one might expect substantial differences in cell wall structure between fast-growing ("batch") and slow-growing ("chemostat") bacteria.

Although it is clear that our work must not be compared directly with that carried out on batch-grown organisms, the data contained in this article show unequivocally that the walls of bacteria can *and do* vary phenotypically, and this fact can never be ignored when attempting to correlate the fine structural details of bacterial cell walls to the functioning of these organelles.

REFERENCES

Anderson, J. C., Archibald, A. R., Baddiley, J., Curtis, M. J. and Davey, N. B. (1969). *Biochem. J.* **113**, 183.

Armstrong, J. J., Baddiley, J. and Buchanan, J. G. (1961). *Biochem. J.* **80**, 254.

Baddiley, J. (1962). *J. Roy. Instit. Chem.* p. 366.

Baddiley, J. (1964). *Endeavour*, **23**, 33.

Baddiley, J. (1970). *Accounts Chem. Res.* **3**, 98.

Baddiley, J., Buchanan, J. G., RajBhandary, U. L. and Sanderson, A. R. (1962). *Biochem. J.* **82**, 438.

Brown, A. D. (1961). *Biochem. biophys. Acta* **49**, 585.

Burger, M. M. and Glaser, L. (1966). *J. biol. Chem.* **241**, 494.

Callow, D. S. and Pirt, S. J. (1956). *J. gen. Microbiol.* **14**, 661.

Chin, T., Burger, M. M. and Glaser, L. (1966). *Archs biochem. Biophys.* **116**, 358

Collins, F. M. (1964). *Aust. J. exp. Biol. med. Sci.* **42**, 255.

Davies, H. C., Karush, F. and Rudd, J. H. (1968). *J. Bact.* **95**, 162.

Donachie, W. D. and Begg, K. J. (1970). *Nature, Lond.* **227**, 1220.

Ellwood, D. C. (1970). *Biochem. J.* **118**, 367.

Ellwood, D. C. (1971). *Biochem. J.* **121**, 349.

Ellwood, D. C. and Tempest, D. W. (1968). *Biochem. J.* **108**, 40p.

Ellwood, D. C. and Tempest, D. W. (1969). *Biochem. J.* **111**, 1.

Evans, C. G. T., Herbert, D. and Tempest, D. W. (1970). "Methods in Microbiology," Vol. 2, p. 277. Academic Press, London.

Gaughran, E. R. L. (1947). *J. Bact.* **53**, 506.

Ghuysen, J. M. (1964). *Biochim. biophys. Acta* **83**, 132.

Ghuysen, J. M. (1968). *Bact. Rev.* **32**, 425.

Goundry, J., Davison, A. L., Archibald, A. R. and Baddiley, J. (1967). *Biochem. J.* **104**, 1c.

Gould, G. W. (1958). M.Sc. Thesis: University of Bristol.

Hancock, I. and Meadow, P. M. (1969). *Biochim. biophys. Acta* **187**, 366.

Heptinstall, S., Archibald, A. R. and Baddiley, J. (1970). *Nature, Lond.* **225**, 519.

Herbert, D. (1958). "Recent Progress in Microbiology", *7th Intern. Congr. Microbiol.* p. 381.

Herbert, D. (1961a). "Microbial Reaction to Environment", *Symp. Soc. gen. Microbiol.* **11**, 391.

Herbert, D. (1961b). "Continuous Cultivation of Microorganisms", *Symp. Soc. Chem. Ind. Monograph* No. 12, p. 21.

Herbert, D., Phipps, P. J. and Tempest, D. W. (1965). *Lab. Pract,* **14**, 1150.

Hughes, R. C. (1965). *Biochem. J.* **96**, 700.

Hughes, R. C. (1966). *Biochem. J.* **101**, 693.

Hughes, R. C. and Thurman, P. F. (1970). *Biochem. J.* **117**, 441.

Hughes, R. C., Parlik, J. G., Rogers, H. J. and Tanner, P. J. (1968). *Nature, Lond.* **219**, 642.

Houtsmuller, U. M. T. and van Deenen, L. L. M. (1964). *Biochim. biophys. Acta* **84**, 96.

Houtsmuller, U. M. T. and van Deenen, L. L. M. (1965). *Biochim. biophys. Acta* **106**, 564.

Janczura, E., Perkins, H. R. and Rogers, H. J. (1961). *Biochem. J.* **80**, 82.

Knivett, V. A. and Cullen, J. (1965). *Biochem. J.* **96**, 771.

Knivett, V. A. and Cullen, J. (1967). *Biochem. J.* **103**, 299.

Lacey, B. W. (1961). "Microbial Reaction to Environment", *Symp. Soc. gen. Microbiol.* **11**, 343.

Marr, A. G. and Ingraham, J. L. (1962). *J. Bact.* **84**, 1260.

Mauck, J. and Glaser, L. (1970). *Biochem. biophys. Res. Commun.* **39**, 699.

Meers, J. L. and Tempest, D. W. (1970). *J. gen. Microbiol.* **63**, 325.

Neidhardt, F. C. (1963). *A. Rev. Microbiol.* **17**, 61.

Novick, A. and Szilard, L. (1950). *Science, N.Y.* **112**, 715.

Okyama, H. (1969). *Biochim. biophys. Acta* **176**, 125.

Op den Kamp, J. A. F., Houtsmuller, U. M. T. and van Deenen, L. L. M. (1965). *Biochim. biophys. Acta* **106**, 438.

Osborn, M. J. (1969). *A. Rev. Biochem.* **38**, 501.

Perkins, H. R. (1963). *Biochem. J.* **86**, 475.

Petit, J. F., Adam, A., Wutzerbin-Falszpan, J., Lederer, E. and Ghuysen, J. M. (1969). *Biochem. biophys. Res. Commun.* **35**, 478.

Prasad, A. and Litwack, G. (1965). *Biochemistry, N.Y.* **4**, 496.

Salton, M. R. J. (1953). *Biochim. biophys. Acta* **10**, 512.

Schleifer, K. H., Huss, L. and Kandler, O. (1969). *Arch. Mikrobiol.* **68**, 387.

Shaw, M. K. and Ingraham, J. L. (1965). *J. Bact.* **90**, 141.

Strange, R. E. and Shon, M. (1964). *J. gen. Microbiol.* **34**, 99.

Sud, I. J. and Schaechter, M. (1964). *J. Bact.* **88**, 1612.

Tempest, D. W. (1970). *Adv. microbial Physiol.* **4**, 223.

Tempest, D. W. and Ellwood, D. C. (1969). *Biotechnol. bioengng.* **11**, 775.

Tempest, D. W. and Hunter, J. R. (1965). *J. gen. Microbiol.* **41**, 267.

Tempest, D. W. and Strange, R. E. (1966). *J. gen. Microbiol.* **44**, 273.

Tempest, D. W., Hunter, J. R. and Sykes, J. (1965). *J. gen. Microbiol.* **39**, 355.

Tempest, D. W., Meers, J. L. and Brown, C. M. (1970). *J. gen. Microbiol.* **64**, 171.

Tipper, D. J. and Strominger, J. L. (1966). *Biochem. biophys. Res. Commun.* **22**, 48.

Young, F. E. (1965). *Nature, Lond.* **207**, 104.

Young, F. E., Tipper, D. J. and Strominger, J. L. (1964). *J. biol. Chem.* **239**, PC 3600.

The Metabolism of One-Carbon Compounds by Micro-Organisms

J. R. QUAYLE

Department of Microbiology, University of Sheffield, Sheffield, England

I. Introduction 119
II. Description and Physiology of One-Carbon-Utilizing Micro-Organisms 120
    A. Aerobic Organisms 120
    B. Photosynthetic Bacteria 138
    C. Non-Photosynthetic Anaerobic Bacteria 140
III. Energy Metabolism 143
    A. Aerobic Oxidation 143
    B. Anaerobic Dismutation 161
    C. Photometabolism 168
IV. Carbon Assimilation 169
    A. Assimilation of Carbon Dioxide 169
    B. Assimilation of Reduced One-Carbon Compounds . . 177
V. Acknowledgements 197
    References 197

I. Introduction

One-carbon compounds at all oxidation levels between methane and carbon dioxide occur abundantly throughout Nature. Methane occurs in coal and oil deposits and is also evolved on a large scale as an end-product of many fermentations, e.g. during sewage production or ruminant digestion. Indeed, the cattle population (1967) of the United Kingdom eructating in concert, could have filled the airship "Hindenburg" with methane in 114 minutes. Methanol arises from breakdown of such natural products as methyl esters or ethers, e.g. pectin. Methylated amines or amine oxides occur widely in tissues of plants, animals and fishes. Formic acid occurs in plant and animal tissue and is frequently encountered as an end product of carbohydrate fermentation. Carbon dioxide is present in the atmosphere, in natural waters, and as carbonates in the earth. It is not surprising, therefore, that a considerable number of micro-organisms have developed the ability to utilize such compounds as carbon and/or energy sources. It is the purpose of this article to survey

120   J. R. QUAYLE

these organisms and to examine the biochemical problems which are
posed by energy transduction and biosynthesis of cell constituents from
the one-carbon substrate. More attention will be paid to organic one-
carbon compounds than to carbon dioxide, because the physiology and
biochemistry of autotrophic organisms is often conveniently reviewed
as a single topic (Peck, 1968; Pfennig, 1967; Rittenberg, 1969; Walker
and Crofts, 1970; Gibbs, 1967; Kiesow, 1967; Kelly, 1967). General
aspects of one-carbon metabolism common to many kinds of living
tissue, e.g. one-carbon transfer reactions in biosynthesis of purines,
pyrimidines, methionine, and choline, which have been presented elsewhere (Blakley, 1969; Stokstad and Koch, 1967; Barker, 1967; Weissbach and Taylor, 1968) will not be dealt with except in so far as they bear
directly on the problems unique to microbial growth on one-carbon
compounds. A somewhat similar review has been published recently by
Ribbons et al. (1970).

II. Description and Physiology of One-Carbon-Utilizing
Micro-organisms

A. AEROBIC ORGANISMS

Much progress has been made during the last ten years on the isolation
of aerobic one-carbon-utilizing micro-organisms; in particular, the
microbiology of methane utilizers has been opened up and now promises
to be an exciting field for both microbial physiologists and biochemists.

Utilization of one-carbon compounds appears to be largely confined
to prokaryotic organisms. This reviewer is not aware of any case where a
eukaryotic organism in pure culture has been shown unequivocally to
be growing on a one-carbon compound as a major carbon and energy
source.

1. Methane and Methanol Utilizers

The first well-described methane-utilizing micro-organism was
isolated in 1906 by Söhngen (Söhngen, 1906) and named by him, Bacillus
methanicus. Considering the widespread natural occurrence of methane
it is surprising that so few other methane-utilizing micro-organisms were
isolated in the succeeding sixty years. By 1966, Bacillus methanicus had
been re-isolated by Dworkin and Foster (1956) who renamed it Pseudomonas methanica (Söhngen), and only three other new species had been
isolated and described reasonably fully, namely Pseudomonas methanitrificans (Davis et al., 1964), Methanomonas methanooxidans (Stocks and
McCleskey, 1964b), and Methylococcus capsulatus (Foster and Davis,
1966). Since 1966, the field has been transformed by the isolation of over
100 strains of methane-utilizers by Whittenbury, Wilkinson and their

METABOLISM OF ONE-CARBON COMPOUNDS BY MICRO-ORGANISMS 121

colleagues at Edinburgh (Whittenbury, 1969; Whittenbury *et al.*, 1970b; Wilkinson, 1971). One of the most remarkable findings of the above studies is that, as far as has been tested, all the methane-utilizers grow on methane, methanol and dimethyl ether, but on no other substrate. They appear to be obligate methylotrophs. For this reason they have been brought together in this section of the present review, although, as the following detailed examination will show, they actually comprise many apparently quite different species.

a. *Organisms Isolated Prior to* 1966. In order to view the classic methane utilizers in the context of the newer findings by the Edinburgh group, the four organisms *Ps. methanica, Ps. methanitrificans, Mtn. methanooxidans* and *Mtlc. capsulatus* will first be described. Particular attention will be paid to the method of isolation that was used in each case.

i. *Pseudomonas methanica* (Söhngen). Following its initial isolation by Söhngen (1906), this organism was re-isolated and extensively investigated by Dworkin and Foster (1956) and Leadbetter and Foster (1958). It is a pink-coloured, Gram-negative rod with a single polar flagellum, unable to grow on any other substrate except methane or methanol. It forms colonies of 2–4 mm. diam. when growing on agar plates. Harrington and Kallio (1960) isolated a pink, methanol-utilizing bacterium which they identified as a strain of *Ps. methanica*. In view of its inability to grow on methane its identification should be viewed with caution.

Foster and his colleagues investigated a variety of inocula and experimental conditions for the elective culture of aerobic methane-utilizers. They concluded that the conditions necessary for isolation of *Ps. methanica* were rather narrowly selective. Aquatic environments, such as ponds, ditches or weeds, provided the best sources of inocula. Such inocula were transferred to liquid salts media and incubated under methane-air at 30° without shaking. *Pseudomonas methanica* formed a pink pellicle after about six days growth and loopfuls of the pellicle were then streaked out on to agar plates containing mineral salts and incubated under methane-air at 30°. After 5–7 days, pink colonies of *Ps. methanica* appeared. Using these conditions, Foster was able to isolate routinely *Ps. methanica* from many sources. If the conditions were varied, e.g. cultures shaken instead of held stationary, incubated at 37° instead of 30° or Petri plates inoculated directly with soil particles, the emergence of *Ps. methanica* was diminished or abolished. Instead, a variety of yellow, brown or colourless bacteria were obtained. These were apparently indistinguishable from *Ps. methanica* except for their pigmentation; they were all obligate methylotrophs. Their close relationship to each other was further indicated by the finding that members of each of the differently pigmented groups mutated spontaneously to each

122 J. R. QUAYLE

other. Leadbetter and Foster (1958) concluded that non-pink varieties were predominant in nature. They raised two rather profound questions:

    (a) why had the many different sources of inocula and elective conditions of culture only resulted in the isolation of one type of pseudomonad?

    (b) why had their wide-ranging study failed to isolate methane-utilizing organisms that also were able to grow on a variety of other substrates?

These questions have loomed large in this field of work ever since.

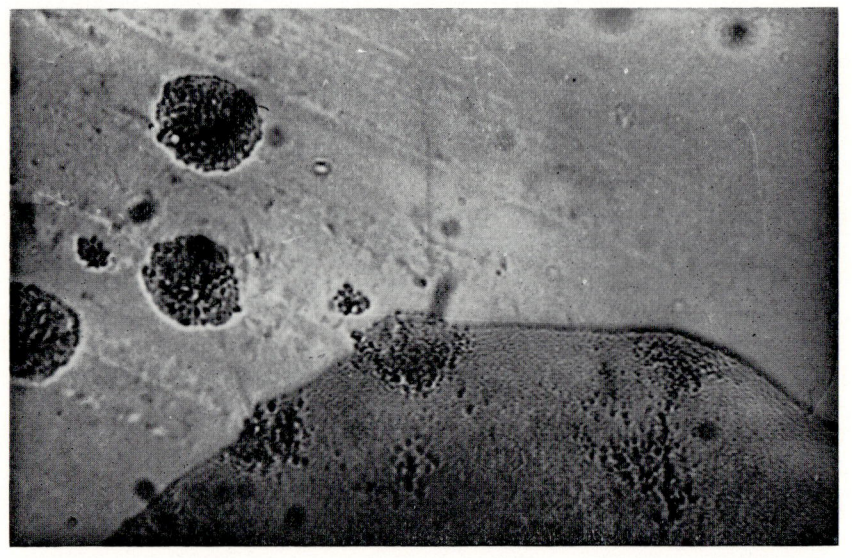

Fig. 1. Photomicrograph of the surface of an agar plate showing engulfment of microcolonies of methane-utilizing bacteria by a large contaminating colony. From Ribbons (1968).

ii. *Pseudomonas methanitrificans*. This organism was isolated by Davis *et al.* (1964) from a variety of natural sources, including soils exposed to methane or natural gas. Enrichment was performed by successive transfer in liquid media lacking a nitrogen source, the cultures being incubated at 30° with shaking under methane-air. Various isolates of methane-oxidizing nitrogen-fixing bacteria were obtained which were white or yellow in colour. The most common isolate was a Gram-negative motile rod forming pale-yellow colonies on agar. The organism did not grow on any other hydrocarbon tested although mention was made (without experimental detail) of its growth on conventional media.

iii. *Methanomonas methano-oxidans*. The first full reports of the isolation and description of this organism were given by Brown *et al.* (1964) and

METABOLISM OF ONE-CARBON COMPOUNDS BY MICRO-ORGANISMS   123

Stocks and McCleskey (1964b). Inocula were obtained from a variety of natural sources including soils, muds, coal-mine water and rumen contents. Elective growth was carried out in liquid media at 28–30°, shaken under methane-air. Serial transfers were made through five or six successive growths in liquid media; immediately after the last transfer showed detectable turbidity, a sample was withdrawn and serially diluted by 10^{-10}. Agar plates were inoculated from this final dilution and incubated under methane-air at 28°. After 10 days incubation, colourless micro-colonies had formed and these contained a Gram-negative, motile, pleomorphic bacterium, often associated in character-istic rosettes. The colonies never grew bigger than 0·1 mm. in diameter even after prolonged incubation, and microscopic examination of plates was necessary for their detection. The organism, named *Methanomonas methano-oxidans*, was found to grow only on methane or methanol. The slow growth of the organism, the invisibility of its colonies to the naked eye, and its tendency to grow in rosette-like clusters leads to difficulty in isolating it in pure culture. Other methanol-utilizing organisms are entangled in these clusters and indeed there may be some symbiotic growth on the methanol which is provided by oxidation of the methane. Close microscopic examination of colonies of methanol-utilizing bacteria often reveals the presence inside of colonies of *Mtn. methano-oxidans* (Fig. 1). Hence the initial successes in isolation of pure cultures of *Mtn. methano-oxidans* (Brown *et al.*, 1964; Stocks and McCleskey, 1964b; Chapman and Ribbons, 1968) were due to dilution to single-cell level before plating out, and use of the microscope to examine the surface of agar plates for micro-colonies.

iv. *Methylococcus capsulatus*. This organism was isolated from sewage by Foster and Davis (1966) by shake culture in liquid medium under methane-air at 50–55°. Streaks were made from the secondary enrich-ment culture onto mineral salts agar. Colourless colonies of 1 mm. diameter formed in 10–14 days. *Methylococcus capsulatus* is a non-motile, encapsulated, Gram-negative coccus mostly adopting a diplococcoid arrangement. Its optimum growth temperature is 37° and hence the organism is thermotolerant rather than thermophilic. Only methane or methanol supports significant growth.

v. Organism JOB5. This organism was isolated by Ooyama and Foster (1965) by elective culture on 2-methylbutane. It is a small, Gram-negative, non-motile, rod-shaped bacterium, pale-yellow in colour and capable of growth on methane. It has not been described as fully as the above four organisms, but it is remarkable in that it can apparently grow on *n*-alkanes of chain length C_1–C_{22}. The figures given by Ooyama and Foster (1965) suggest that its growth on methane was only relatively slight; however this property was again reported by Perry (1968). This

124   J. R. QUAYLE

organism may therefore be the only well authenticated case, so far, of a
facultative methylotroph and thus obviously warrants a careful biochemical investigation.

b. *Organisms Isolated Subsequent to* 1966. By 1966 only the above four
or five well defined species of methane-utilizer had been isolated in pure
culture. At this time they were generally considered difficult organisms
to isolate and maintain, particularly *Ps. methanica*. Some workers had
tried to re-isolate this organism without success, and there were many
verbal reports of failure to maintain cultures of the original isolate of
Dworkin and Foster (1956). This situation has now been transformed by
the achievements of Whittenbury, Wilkinson and their colleagues at
Edinburgh as a result of their comprehensive study of methods of
enrichment, isolation and culture of methane utilizers (Whittenbury,
1969; Whittenbury *et al.*, 1970b; Wilkinson, 1971). These results will
now be examined in some detail.

i. *Enrichment and isolation methods.* A wide variety of sources of inocula
was used encompassing mud, water and soil from different parts of the
world. Inocula were placed in 25 ml. of liquid medium and incubated at
30°, 45° and 55° under methane-air. After 3 days of incubation, turbidity,
often accompanied by a pellicle, indicated the presence of methane
utilizers. At this stage, such cultures were serially diluted in sterile tap
water and spread on to salts-agar plates which were incubated under
methane-air for 3–4 weeks. The plates were examined under a plate
microscope at 3-day intervals. Colonies of non-methane-utilizers reached
their maximum size in about 3 days and remained relatively transparent.
Methane-utilizing colonies began to appear after 5–7 days and were often
50–100-fold fewer in number than the non-methane-utilizing colonies.
They continued to increase their size over 2–3 weeks and were opaque.
Isolation was most successful while the colonies were still small (about
0·2 mm. diam.); they were picked off under the microscope and transferred to agar slopes for further subculture and single-colony isolation.
Use of these methods resulted in isolation of over 100 strains of methane-
utilizing bacteria many of which were totally new; also, organisms
identical or closely similar to those isolated prior to 1966 were re-isolated
(Fig. 2). It is thus clear that the methods used by the Edinburgh group

---

FIG. 2. Phase-contrast micrographs of methane-utilizing bacteria, magnification
×1900. From Whittenbury *et al.* (1970a, b). (A) "*Methylosinus sporium*", showing
occasional vegetative cells and many sporulating organisms and spores; (B) Indian
ink preparation of "*Methylosinus trichosporium*", showing a vegetative organism
(lower right) and sporulating organisms budding-off spores; (C) "*Methylocystis
parvus*"; (D) "*Methylomonas methanica*"; (E) "*Methylobacter chroococcum*";
(F) "*Methylobacter bovis*", showing germination of cysts; (G) "*Methylococcus
minimus*"; (H) "*Methylococcus capsulatus*".

METABOLISM OF ONE-CARBON COMPOUNDS BY MICRO-ORGANISMS 125

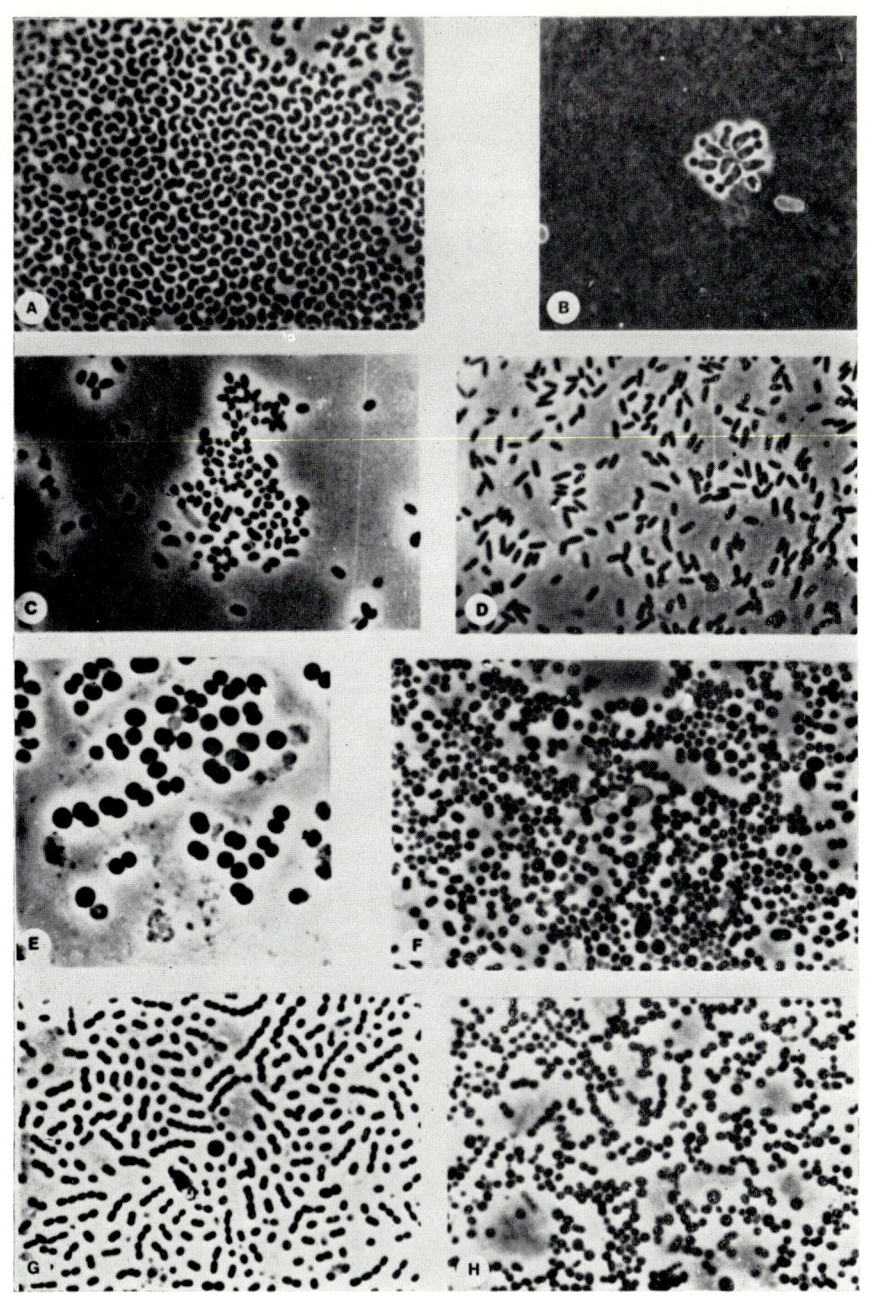

126   J. R. QUAYLE

constitute a major advance in a hitherto rather intractable field. Comparison of their methods with those used previously indicates the crucial improvements in technique:

(a) methane utilizers are few in number and slow growing relative to non-methane-utilizers in the initial enrichment cultures. It is therefore necessary to streak out on to plates *without further serial culture*. This also decreases losses due to the activities of predacious organisms;

(b) plates must be examined under a microscope at frequent intervals. Experience of previous investigators in respect of *Mtn. methanooxidans* had demonstrated the likelihood of micro-colony formation and the necessity for plating out such organisms from highly dilute suspensions. The Edinburgh group were able to make use of this, and they found that isolation of many organisms, not necessarily micro-colony formers, was best done using colonies still visible only under the microscope.

It seems that the answer to the first of the two questions posed by Leadbetter and Foster (1958; see p. 122) has now been answered. Their second question, however, assumes even greater importance since all of the new methane-utilizing organisms isolated by the Edinburgh group are obligate methylotrophs, growing only on methane, methanol and dimethyl ether.

ii. *Description and classification of organisms.* Many of the newly isolated organisms are capable of existing in either heat-resistant or desiccation-resistant forms. Exospores and different types of cysts are associated with such forms (Plate 2) (Whittenbury *et al.*, 1970a). Also, all the methane-utilizing organisms so far examined under the electron microscope possess elaborate internal membrane structures, resembling those found in nitrifying and photosynthetic bacteria (Ribbons, 1968; Proctor *et al.*, 1969; Whittenbury, 1969; Davies and Whittenbury, 1970; Smith and Ribbons, 1970; Smith *et al.*, 1971). The types of membrane structures encountered have been subdivided into two basic patterns: a system of paired membranes running throughout the cell or concentrated round its periphery (Type I; Figs. 3 and 4); bundles of disc-shaped membrane vesicles distributed throughout the cell (Type I; Figs. 5 and 6). Tables 1 and 2 show a tentative classification into groups and sub-groups which has been drawn up by Whittenbury *et al.* (1970b). It should be noted that the most frequently encountered methane-utilizing organism was found to be "*Methylomonas methanica*", considered to be identical with *Pseudomonas methanica* (Dworkin and Foster); out of the 99 isolates which were subdivided into 14 sub-groups, 30 isolates belonged to the sub-group "*Methylomonas methanica*".

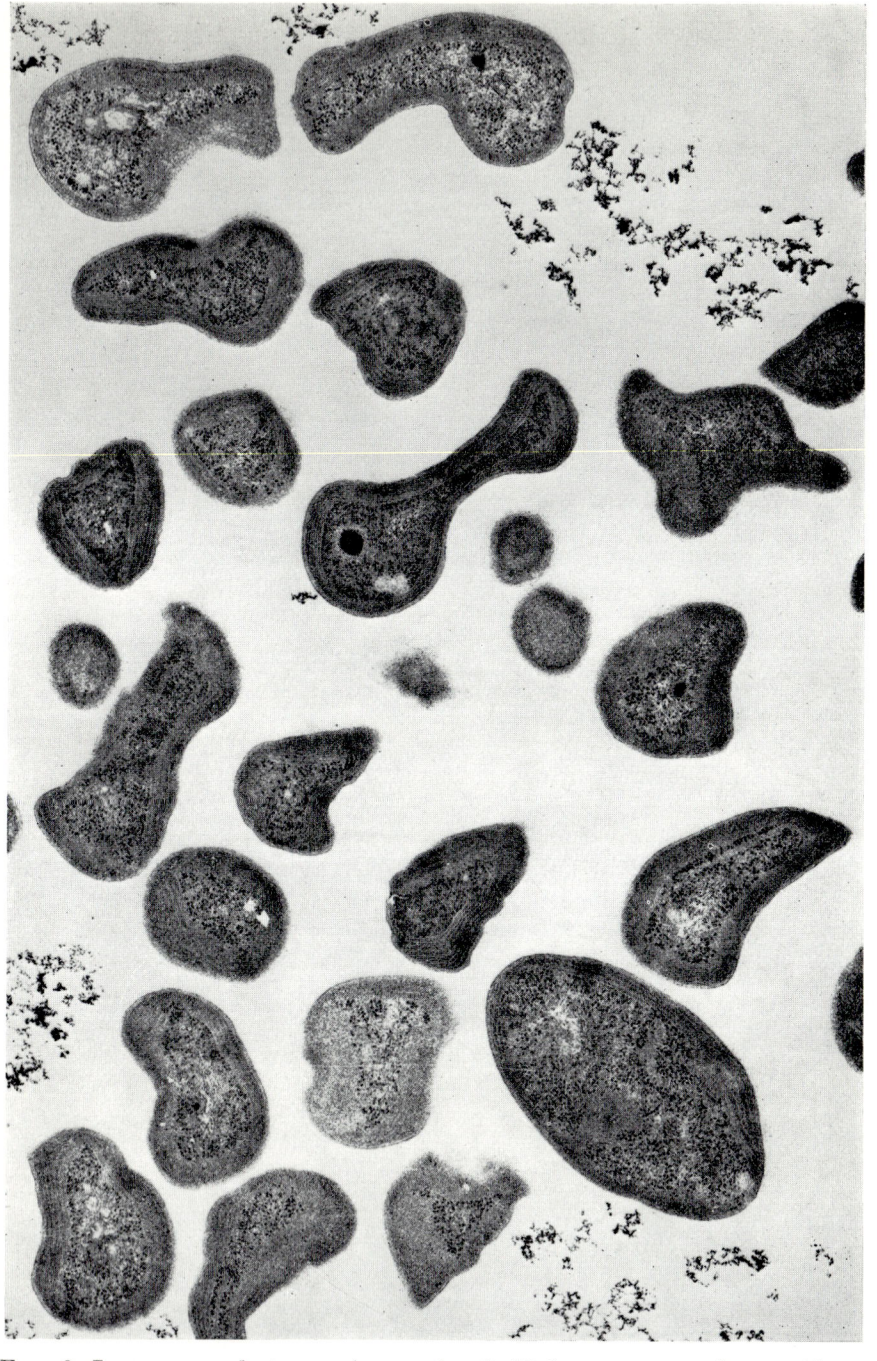

FIG. 3. Low-power electron micrograph of *Methanomonas methano-oxidans*; magnification ×24,000. From Smith and Ribbons (1970).

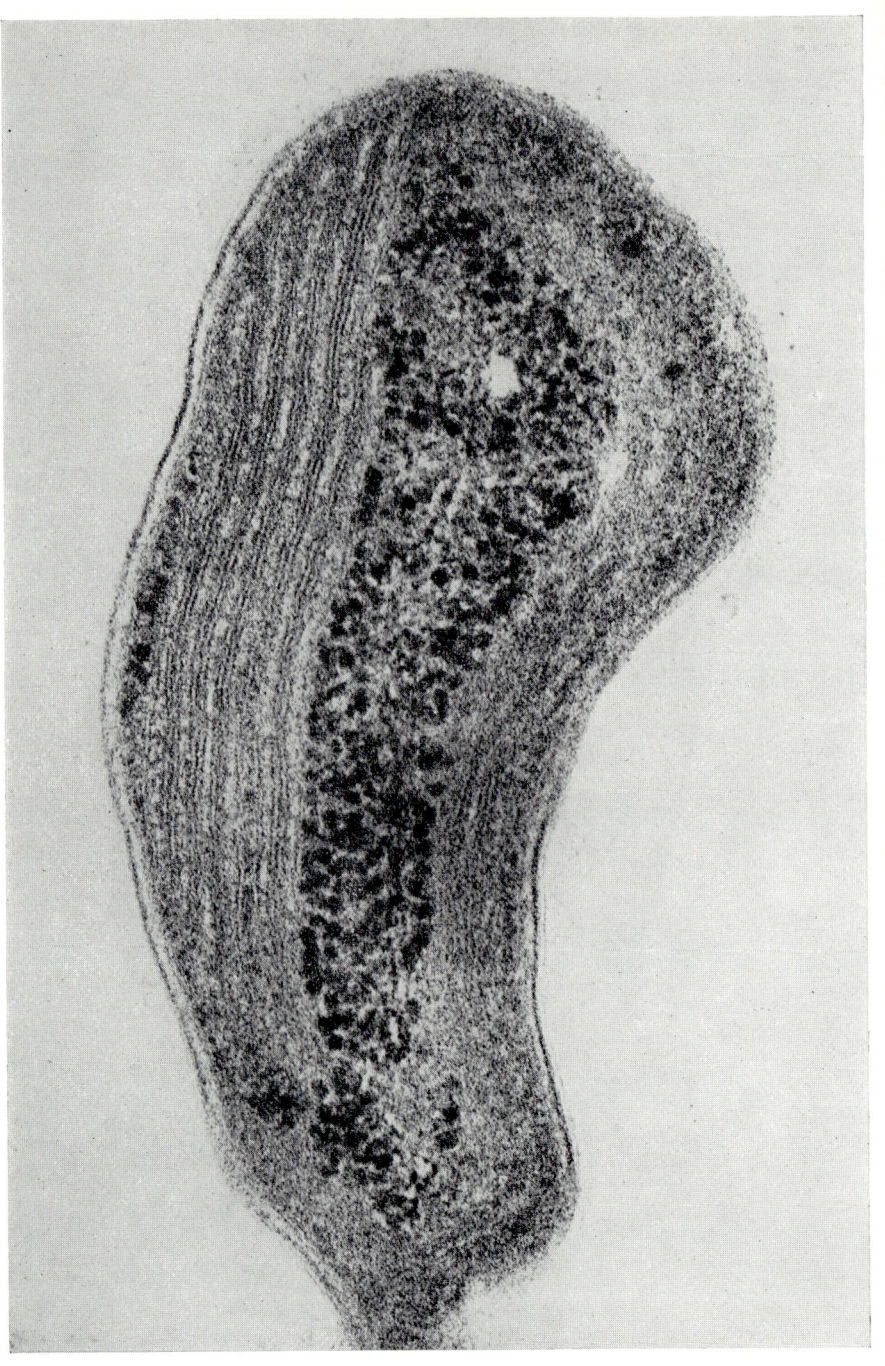

FIG. 4. Electron micrograph of single cell of *Methanomonas methano-oxidans* showing multilayered peripheral array of internal membranes; magnification ×132,000. From Smith and Ribbons (1970).

METABOLISM OF ONE-CARBON COMPOUNDS BY MICRO-ORGANISMS   129

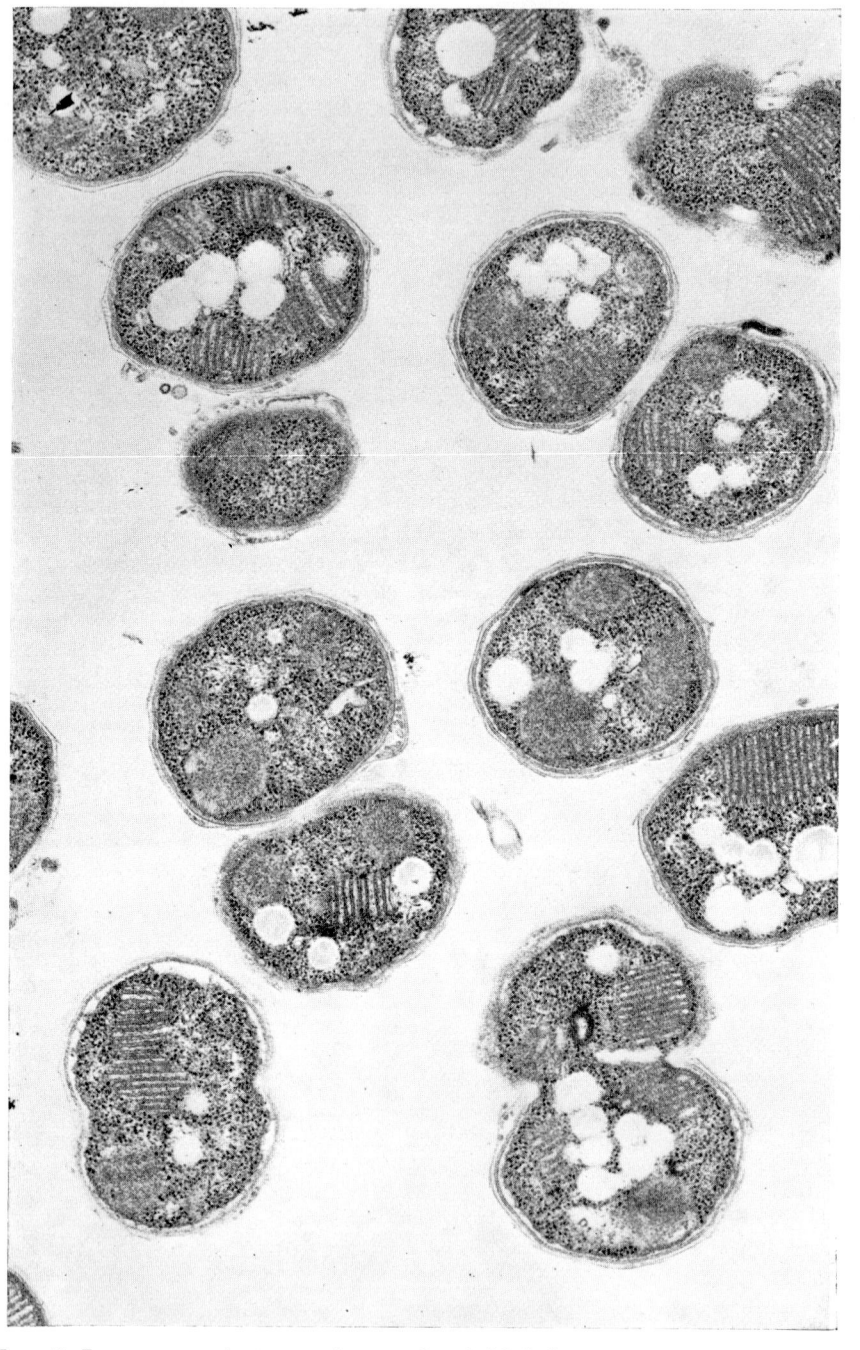

FIG. 5. Low-power electron micrograph of *Methylococcus capsulatus* showing bundles of disc-shaped membrane vesicles distributed throughout the cell; magnification ×25,000. From Smith *et al.* (1971).

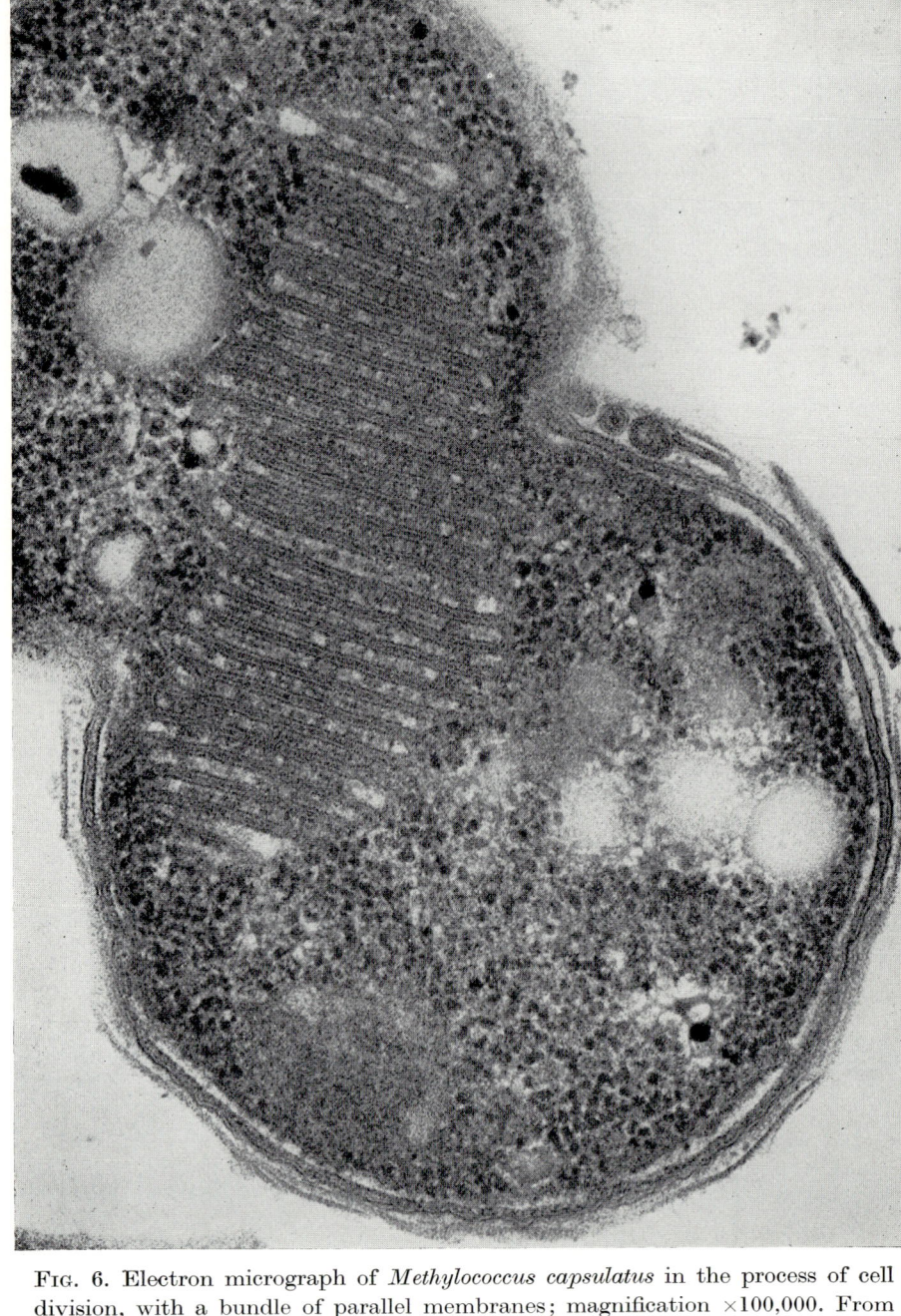

FIG. 6. Electron micrograph of *Methylococcus capsulatus* in the process of cell division, with a bundle of parallel membranes; magnification ×100,000. From Smith *et al.* (1971).

iii. *Miscellaneous organisms*. Various reports have been made of the isolation of methane-utilizing bacteria which are not obligate methylotrophs (Elizarova, 1963; Kersten, 1964; Kozlova *et al.*, 1969). These reports need careful checking in order to ascertain that such critical factors as the purity of methane used and the absence of contaminating symbiotic methanol-utilizers have been fully taken care of. There would clearly be considerable biochemical interest in a comparison of obligate and facultative methylotrophs.

TABLE 1. Division of Methane-Utilizing Bacteria into Groups.
Based on Whittenbury *et al.* (1970b)

Group	Resting stage	Membrane type	Morphology	Rosette formation
"Methylosinus"	Exospore	II	Rod or pear-shaped	+
"Methylocystis"	"Lipid" cyst	II	Rod or vibroid	+
"Methylomonas"	"Immature Azoto-bacter-type" cyst[a]	I	Rod	−
"Methylobacter"	Azotobacter-type cyst	I	Rod	−
"Methylococcus"	"Immature Azoto-bacter-type" cyst[a]	I	Coccus	−

[a] Not all organisms form an identifiable resting stage.

Reports of growth on methane by eukaryotic organisms are scanty. Erebo (1967) described a series of experiments with a species of *Chlorella* in which addition of methane to the gas phase caused an increase in growth of 35–40%. Although the algal culture was contaminated with bacteria, the extent of the bacterial contamination is claimed to be insufficient to account for the methane-dependent increase in cell yield. Zajic *et al.* (1969) studied growth of a species of *Graphium* on a natural gas mixture. This gas mixture contained methane (90·5%), ethane (6·0%) and propane (0·21%) together with traces of higher hydrocarbons; the relative contributions of these hydrocarbons towards the growth of the fungus remains to be ascertained.

2. *Methylamine, Methanol and Formate Utilizers*

Elective culture on one-carbon compounds other than methane yields a wide variety of bacteria which are able to grow on organic compounds containing one or more carbon atoms. Many of these organisms are also isolated along with methane utilizers during elective culture on methane; no doubt this is due to excretion of catabolites of methane such as

TABLE 2. Division of Methane-Utilizing Bacteria into Sub-Groups. Based on Whittenbury et al. (1970b)

Group and sub-group	Colony colour	Capsule formation	Flagella	Shortest generation time (hr.)	Identity with organisms isolated prior to 1966
"Methylosinus"					
"trichosporium"	White to yellow	+	Polar tufts	5	*Pseudomonas methanitrificans* may be a strain of "*Methylosinus*".
"sporium"	White to buff	+	Polar tufts	5	*Methanomonas methano-oxidans* (Stocks and McCleskey) is generally similar to members of these groups.[a]
"Methylocystis"					
"parvus"	White	+	–(Non-motile)	5	
"Methylomonas"					
"methanica"[b]	Yellow ochre to pink	+	Polar	3·5	*Pseudomonas methanica* (Dworkin and Foster)
"albus"	White	–	Polar	3	
"streptobacterium"	White	+	–(Non-motile)	4	
"agile"	White	–	Polar	3·5	
"rubrum"	Red	+	Polar	4	
"rosaceus"	Pale pink	+	Polar	4	
"Methylobacter"					
"chroococcum"	Pale pink	+	–(Non-motile)	5	Some strains of "*Methylobacter*" resemble the organisms isolated at 37° by Leadbetter and Foster (1958).
"bovis"	White (brown if cysts present)	–	–(Non-motile)	4	
"capsulatus"	White (brown if cysts present)	+	Polar	4	
"Methylococcus"					
"capsulatus"	White	+	–(Non-motile)	3·5	*Methylococcus capsulatus* (Foster and Davis)
"minimus"	White	+	–(Non-motile)	3·5	

[a] R. J. Whittenbury (personal communication).
[b] Most commonly encountered species.

METABOLISM OF ONE-CARBON COMPOUNDS BY MICRO-ORGANISMS   133

methanol, formaldehyde or formate by the methane utilizers. The non-methane-utilizing bacteria have variously been assigned to such genera as *Pseudomonas*, *Vibrio*, *Bacillus*, *Protaminobacter*, *Achromobacter* and *Hyphomicrobium*. A selection of well authenticated such organisms is given in Table 3. This list is not an exhaustive one but it hopefully includes all those organisms which have been used in significant biochemical studies. Included in this list is the organism identified by Harrington and Kallio (1960) as a strain of *Ps. methanica*. Since the organism could not oxidize methane, the possibility of its assignment to the organisms of Table 3 should be kept open.

There is a need for a comprehensive and detailed taxonomic survey of these organisms. Many of them appear to be species of *Pseudomonas* but their ability to grow on one-carbon compounds sets them apart from a large number of aerobic pseudomonads. Stanier *et al.* (1966) have made a detailed taxonomic study of 267 individual strains of *Pseudomonas* and tested their ability to grow on 146 different organic compounds; included in the substrates tested were oxalate and two one-carbon compounds, methanol and methylamine. None of the 267 strains examined could grow on oxalate, methanol or methylamine. Thus the ability to utilize these compounds, although possessed by some species of *Pseudomonas*, must be confined to a restricted segment of the genus.

Stocks and McCleskey (1964a) compared 15 isolates of methanol-utilizing bacteria obtained either from their own elective cultures or as authenticated methanol utilizers from other laboratories. The fifteen strains were morphologically and nutritionally very similar; all were pink-pigmented, Gram-negative, motile, non-sporeforming rods with a single polar flagellum; capsules were not observed. All strains grew well on methanol, glycerol or oxalate; all except *Protaminobacter ruber* grew almost as well on formate. None of the other carbon sources tested was satisfactory for all the strains, and no strain grew on methane. Serological studies were made using antisera prepared from rabbits. These showed that the strains were not antigenically homogeneous; no single antiserum agglutinated more than three heterologous strains. Stocks and McCleskey concluded that all of the organisms they had studied were sufficiently alike to be considered as strains of one species and they suggested that, pending further investigations, *Vibrio extorquens* be considered the type species. Their choice of type species was based on the fact that the first published description of an organism of this group was that of *Bacillus extorquens* by Bassalik (1913). This species was later placed in the genus *Vibrio* by Bhat and Barker (1948) and in the genus *Pseudomonas* by Janota (1950). It has been recorded as *Vibrio extorquens* by Breed *et al.* (1957). The assignment to the genus *Vibrio* rather than *Pseudomonas* must however be regarded as only tentative.

134   J. R. QUAYLE

TABLE 3. Aerobic Organisms Which Utilize Methylamines, Methanol or Formate as Sole Carbon Source

Name	Remarks	References to isolation/ description of organism
Bacillus PM6	Aerobic, spore-forming, Gram-positive rod. Grows on trimethylamine N-oxide, tri-methylamine, dimethylamine or methyl-amine.	Myers and Zatman (1971)
Bacillus sphaericus	Gram-positive, spore-forming rod. Grows on N-monomethylurea.	Iyer and Kallio (1958)
Bacterium 4B6	Gram-negative, non-motile rod. Grows on trimethylamine, dimethylamine and methylamine.	Colby and Zatman (1971)
Bacterium formicum	Grows on formate aerobically or a mixture of formate plus C_4 dicarboxylic acid anaerobically.	Shaposhnikov and Loriya (1964, 1967)
Bacterium formoxidans	Gram-negative, non-motile, non-flagellate rod/coccoid. Grows on formate but not on methanol or oxalate.	Sorokin (1961)
Hyphomicrobium spp.	Colourless, stalked bacteria growing by budding from ends of hyphae. Grow on methylamine, methanol or formate but not on oxalate.	Hirsch and Conti (1964a, b; 1968)
Nitrobacter winogradskyi	Grows very slowly on formate.	van Gool and Laudelot (1966)
PAR	Gram-negative, non-motile diplococcus. Grows on methylamine.	Leadbetter and Gottlieb (1967)
Protaminobacter alboflavus	Colourless or yellow bacterium. Grows on formate or formamide but not on methanol or methylamine.	den Dooren de Jong (1926, 1927)
Protaminobacter ruber	Red bacterium. Grows on methanol, methylamine and formate.	den Dooren de Jong (1926, 1927); Stocks and McCleskey (1964a)
Pseudomonas AM1	Pink, Gram-negative, motile rod with single polar flagellum. Grows on methylamine, methanol, formate and oxalate.	Peel and Quayle (1961); Stocks and McCleskey (1964a)
Pseudomonas aminovorans	Colourless or yellow, motile bacterium with single polar flagellum. Different strains have different substrate specificities.	den Dooren de Jong (1927); Eady and Large (1969)

METABOLISM OF ONE-CARBON COMPOUNDS BY MICRO-ORGANISMS 135

TABLE 3—*continued*

Name	Remarks	References to isolation/ description of organism
Pseudomonas M27	Pink, Gram-negative, motile rod with single polar flagellum. Grows on methyl-amine, methanol, formate or oxalate.	Anthony and Zatman (1964a)
"*Pseudomonas methanica*"	Pink, Gram-negative, non-motile rod. Grows on methanol but not on any other substrate tested, including methane. Claimed to be a strain of *Pseudomonas methanica*.	Harrington and Kallio (1960)
Pseudomonas MS	Brown, Gram-negative, motile bacterium. Grows on trimethylsulphonium chloride and trimethyl-, dimethyl- and mono-methylamine but not on methanol or formate.	Wagner (1964); Kung and Wagner (1970b)
Pseudomonas oxalaticus	Colourless, Gram-negative, motile rod. Grows on formate but not on methyl-amine or methanol.	Khambata and Bhat (1953)
Pseudomonas PP	Pink organism resembling *Pseudomonas* AM1 and *Pseudomonas* M27.	Ladner and Zatman (1969)
Pseudomonas PRL-W4	Pink, Gram-negative, motile rod with single polar flagellum. Grows on methanol but not on formate or oxalate. Needs biotin as growth factor.	Kaneda and Roxburgh (1959a)
Pseudomonas sp.	Colourless, Gram-negative, motile rod with single polar flagellum. Grows on methyl-amine but not on methanol, formate or oxalate.	Shaw *et al.* (1966)
Vibrio(Bacillus) extorquens	Red, motile, curved rod with single polar flagellum. Grows on methanol, formate and oxalate.	Bassalik (1913)
Vibrio oxaliticus	Colourless, Gram-negative, curved rod with single polar flagellum. Grows on formate (in presence of yeast extract) or oxalate but not on methanol.	Bhat and Barker (1948)

While many of the methanol-utilizing organisms appear similar to *V. extorquens* there are others which clearly are quite different, for example, the stalked bacteria belonging to the genus *Hyphomicrobium*. These have long been difficult organisms to grow but the studies of Hirsch and Conti (1964a, b), which are outlined below, have resulted in a much clearer picture of their remarkable growth physiology and

136   J. R. QUAYLE

should facilitate more meaningful biochemical work than has hitherto
been possible.

Hyphomicrobium species are found to be ubiquitous but are easily out-
grown by other micro-organisms in the primary enrichments. However,
they have the property of growing "oligocarbophically", i.e. growing,
albeit slowly, in a mineral medium without added carbon source, the
carbon being obtained in trace quantities from the air. This forms the
basis of the best primary enrichment procedure. Since the organisms
often grow attached to solid surfaces it can be advantageous to introduce
glass slides into the medium (Hirsch and Rheinheimer, 1968). This is
then followed by isolation from the primary enrichments using methylamine as carbon source. It is interesting to note the parallel between the
problems of enrichment of these organisms and methane-utilizing
bacteria. In both cases the main problem is to avoid overgrowth of the
one-carbon-utilizer in the *initial* enrichment. Whittenbury *et al.* (1970b)
accomplished this by plating out immediately after the first detectable
growth, whereas Hirsch and Conti were able to use starvation media to
suppress growth of other organisms. Using these methods, Hirsch and
Conti (1964a, b) have isolated 11 new strains of *Hyphomicrobium* and
compared them with cultures of the strain isolated in Germany by Mevius
(1953); the organisms have been divided into four groups based on
morphology and enrichment procedure. Hirsch and his colleagues have
now extended these studies to 89 isolates with respect to 250 characteristics including 100 related to growth on different carbon sources. It is
expected that this work will be published shortly (P. Hirsch, personal
communication).

Hyphomicrobia develop hyphae of a fairly constant diameter (0·2–0·3
μm.) but of varying lengths. Daughter cells ("swarmers") develop from
the end of the hyphae and sometimes break away as motile organisms
each with a single flagellum (Fig. 7). In other organisms the daughter
cell remains attached and chain or cross formation can occur. The cells
tend to attach to surfaces and, for reasons unknown, this attachment
process is inhibited by light. The shape and size of the rod is relatively
constant but the length and degree of branching of the hyphae are dependent on external factors, particularly the nature of the growth substrate
and its concentration. Slow growth results in longer hyphae but fewer
swarmers, and *vice versa*. This might be related to its ability to survive in
media containing extremely low concentrations of substrate if the
elongated hyphae can serve as "root surfaces" to facilitate absorption
of nutrients (Hirsch and Rheinheimer, 1968).

Hyphomicrobia grow little, or not at all, in rich organic media such
as nutrient broth; their range of single growth substrates is also limited.
Utilization of one-carbon compounds is in the order methanol > urea >

METABOLISM OF ONE-CARBON COMPOUNDS BY MICRO-ORGANISMS 137

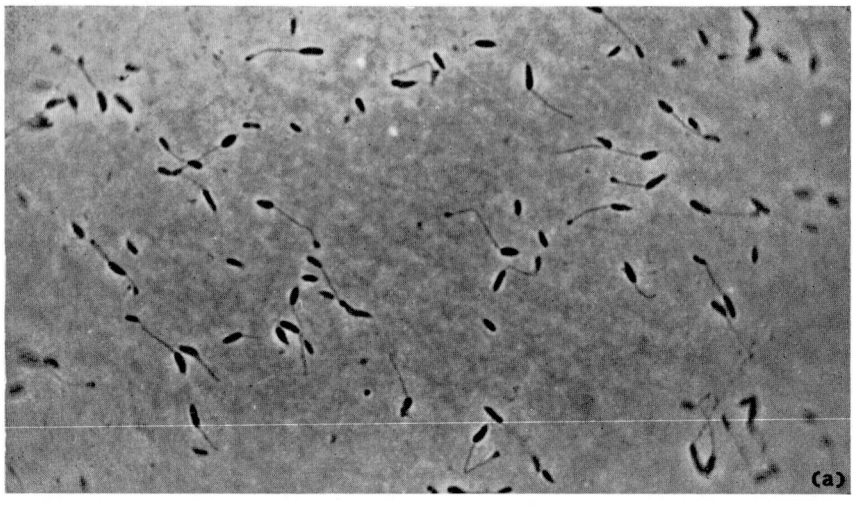

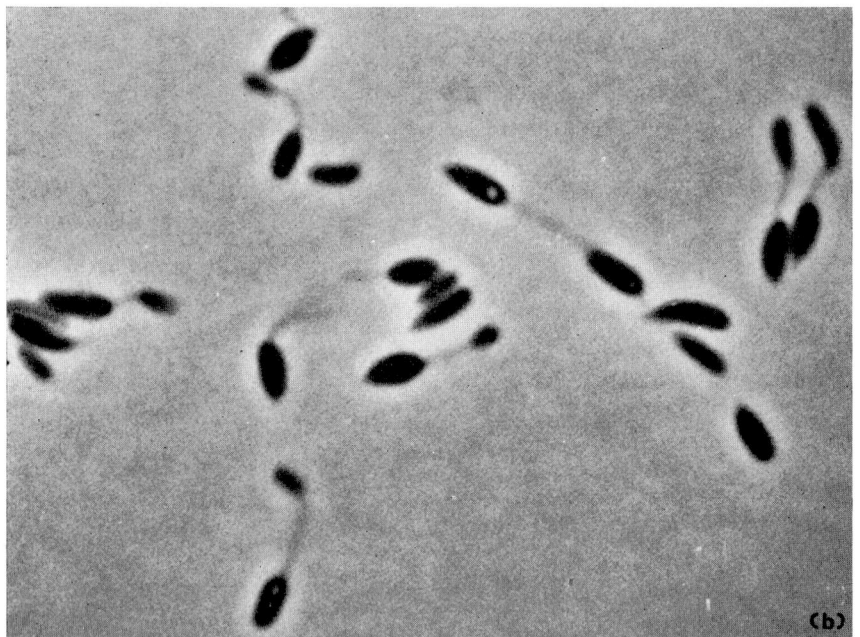

FIG. 7. Phase-contrast micrographs of growing cells of *Hyphomicrobium*, strain B-522. (Reproduced by courtesy of Prof. P. Hirsch, University of Kiel, West Germany.) (a) Magnification ×1200; (b) Magnification ×4500.

methylamine > formamide > formate. Growth was not observed with oxalate or methane. The best conditions for growth of *H. vulgare* were found to be vigorous aeration in the dark at 30° in a neutral, well-buffered medium containing Ca^{2+}, trace elements, either nitrate or ammonia as nitrogen source, and methanol, urea or methylamine as carbon source. Using methylamine as carbon source under these conditions resulted in a yield of 100 mg. of bacterial protein/100 ml. of medium in three days, representing a doubling time of approximately 12 hours. The majority of strains of *Hyphomicrobium* possess well developed internal membrane systems which appear to be derived from invagination of the cytoplasmic membrane (Conti and Hirsch, 1965).

3. *Carbon Monoxide Utilizers*

To search the literature for an authenticated, aerobic, carbon monoxide utilizer is to pursue a microbial Jack O'Lantern. Early work suggested that an organism, capable of growth in mineral medium without a carbon source other than that present in the air, could also utilize carbon monoxide; the organism was named *Bacillus oligocarbophilus* or *Carboxydomonas oligocarbophila* (Beijerinck and van Delden, 1903; Kaserer, 1906; Orla-Jensen, 1909). Later work by Kistner (1953, 1954) suggested that the organism could oxidize hydrogen and that the ability to utilize carbon monoxide was inducible; he renamed it *Hydrogenomonas carboxydovorans*. Collins (1969) has also found, using the medium of Kistner (1953), several isolates which oxidized both carbon monoxide and hydrogen. The field is thus rather confused and calls for a careful re-examination of the isolation and identity of carbon monoxide utilizers. Such a study might be of more than academic interest in view of the increasing concern over atmospheric pollution.

4. *Cyanide Utilizers*

A considerable amount of work has been done on metabolism of cyanide by plant, animal and microbial tissue but, as far as this reviewer is aware, there is only one report of a micro-organism growing on cyanide as sole source of carbon (Ware and Painter, 1955). The organism was isolated on silica gel plates containing mineral salts and potassium cyanide as sole source of carbon and nitrogen. It grew in the form of branching filaments and was Gram-positive. It was provisionally classified as an actinomycete.

B. PHOTOSYNTHETIC BACTERIA

There are several instances of photosynthetic bacteria which can utilize one-carbon compounds other than carbon dioxide; however, the

METABOLISM OF ONE-CARBON COMPOUNDS BY MICRO-ORGANISMS 139

number which have been shown unequivocally to use the one-carbon compound as sole carbon source is very small indeed. The range of one-carbon compounds reported to be utilized appears to be limited to methane, formate and carbon monoxide and hence, compared to aerobic organisms, the field is very restricted.

1. *Methane Utilizers*

From an ecological point of view one might expect methane to be a popular source of carbon for photosynthetic bacteria. Large quantities of methane are generated in aqueous environments through the action of methane fermenters and these aqueous environments often support large populations of photosynthetic bacteria. There is however only one report of methane utilization by such an organism (Wertlieb and Vishniac, 1967). These authors set up enrichment cultures for organisms capable of growing anaerobically in the light on methane plus carbon dioxide. They isolated a strain of *Rhodopseudomonas gelatinosa* which was able to oxidize methane to carbon dioxide and to incorporate small amounts of carbon from methane into cellular components. They were not able to demonstrate that methane was the sole electron donor for the organism.

There may, of course, be a biochemical reason for methane being a refractory substrate for an anaerobic organism. Oxidation of methane to most intermediary metabolites must involve introduction of oxygen into the molecule and since there is evidence (see p. 144) that in the initial oxidation of methane to methanol by aerobic bacteria the source of the oxygen atom is molecular oxygen, this mechanism would not be possible anaerobically. Neither would it be possible to introduce a double bond into this particular molecule in order to facilitate a hydroxylation with water. Clearly an anaerobic hydroxylation of methane would demand, at the least, a rather special kind of biochemical transformation.

2. *Formate Utilizers*

A species of green sulphur bacterium, *Chloropseudomonas* spp., has been reported to oxidize formate to carbon dioxide (Kondrat'eva and Trotsenko, 1969); the authors did not exclude the possibility that some of the formate itself was incorporated into cell constituents.

The Athiorhodaceae are able to grow photosynthetically on a wide variety of simple organic compounds including formate (van Niel, 1944). There have been several later reports of photometabolism of formate by members of this group, usually however with organisms grown in media in which formate was not the carbon source. In general, their ability to *grow* photosynthetically on one-carbon compounds has not been well established.

140   J. R. QUAYLE

Recent work by Rolls and Lindstrom (1967) and Qadri and Hoare (1968) has defined the capabilities of strains of *Rhodopseudomonas palustris* for photosynthetic growth on formate. Rolls and Lindstrom (1967) showed that the extent of growth of *Rh. palustris* (strain ATCC 11168) in a mineral medium containing formate as carbon source was doubled when thiosulphate was added as an extra source of electrons. Qadri and Hoare (1968) isolated a new strain of *Rh. palustris* by using an elective medium containing mineral salts, sodium sulphide, sodium formate and yeast extract; the enrichment was carried out anaerobically in the light. Although the organism was isolated by enrichment in a formate-containing medium, it grew only slowly in this medium (mean generation time of 18 hours) whereas many other organic compounds supported more rapid growth. It was unable to grow photosynthetically on formate in the absence of yeast extract. Neither methanol nor oxalate supported growth, even in the presence of yeast extract.

3. *Carbon Monoxide Utilizers*

Hirsch (1968) reported the isolation of a photosynthetic bacterium, tentatively identified as a species of *Rhodopseudomonas*, which grew anaerobically in the light in a mineral medium maintained under carbon monoxide. Under these conditions, light, carbon monoxide and ammonium ions apparently were the only possible sources of energy. Since oxidation of ammonia was not detected, Hirsch assumed that water functioned as an electron source for the organism; if so, this would raise several interesting questions in comparative biochemistry.

C. NON-PHOTOSYNTHETIC ANAEROBIC BACTERIA

Non-photosynthetic anaerobic bacteria which are able to grow on one-carbon compounds, in otherwise mineral media, belong to the group of methane-producing bacteria. The possibility should not be excluded however that non-methanogenic anaerobes may exist which are capable of growth on one-carbon compounds. For example, Wieringa (1940) isolated *Clostridium aceticum* which carried out a fermentation of hydrogen and carbon dioxide to give acetate. The organism needed a supplement of mud extract in order to grow. The culture has since been lost. Its exact nutritional requirements were not determined (Karlsson *et al.*, 1948).

The methanogenic bacteria are strictly anaerobic bacteria found in the rumen, in aquatic mud and in sewage. For many years these bacteria were considered to grow mainly on short-chain alcohols and fatty acids, in some cases on methanol, formate or possibly carbon monoxide and hydrogen plus carbon dioxide (Barker, 1956). The organic substrates

METABOLISM OF ONE-CARBON COMPOUNDS BY MICRO-ORGANISMS   141

were fermented to a mixture of carbon dioxide and methane. One of the most studied organisms was *"Methanobacterium omelianskii"* isolated by Barker (1940); however, this isolate has now been shown (Bryant *et al.*, 1967) to be a symbiotic mixture of two organisms, namely a non-methanogenic organism, "S", which ferments ethanol to a mixture of hydrogen and acetate and a methanogenic organism (*Methanobacterium* strain M.o.H.) which ferments hydrogen plus carbon dioxide to methane. This finding has cast doubt as to whether alcohols other than methanol and fatty acids other than acetate or formate are in fact attacked by methanogenic bacteria. Fermentation of one-carbon substrates does not seem in question and the well authenticated species capable of utilizing these compounds will be briefly described.

1. *Methanol Utilization*

Enrichment culture in mineral medium using methanol plus sodium carbonate as carbon sources led to the isolation of *Methanosarcina barkerii* (Schnellen, 1947; Kluyver and Schnellen, 1947). Of the many other organic compounds tested, only acetate was fermented, and that slowly.

2. *Formate Utilization*

Two species of bacteria have been isolated which are able to grow in mineral medium containing formate plus bicarbonate as sole carbon sources, namely *Methanobacterium formicicum* (Schnellen, 1947; Kluyver and Schnellen, 1947) and *Methanococcus vannielii* (Stadtman and Barker, 1951).

Methanobacterium formicicum is a rod-shaped organism apparently restricted to formate as the only fermentable organic substrate.

Methanococcus vannielii is a motile coccus that was isolated once from mud in San Francisco Bay but not elsewhere. It grew vigorously in a mineral medium containing formate and bicarbonate, and growth was stimulated by yeast extract. No other organic compound besides formate was found to support growth.

In addition to the above two bacteria which are able to grow on formate, in a mineral medium, the following two species of bacteria, isolated from rumen contents, are able to ferment formate. In order to grow appreciably, they need growth factors supplied from complex media such as rumen fluid or yeast extract.

Methanobacterium ruminantium, isolated by Smith and Hungate (1958), in a short, Gram-positive rod, probably non-motile, and is able to utilize formate but none of the other organic substrates tested.

Methanobacterium mobilis, isolated by Paynter and Hungate (1968), is a Gram-negative, motile rod with single polar flagellum. The only

142   J. R. QUAYLE

organic compound tested which produced methane in appreciable
amounts was formate. The organism required a supplement of rumen
fluid in order to grow well; yeast extract in place of rumen fluid also
allowed limited growth. *Methanobacterium mobilis* closely resembles
Methanobacterium ruminantium but differs in morphology, Gram stain
reaction and motility.

3. *Carbon Monoxide Utilization*

It has been shown by Kluyver and Schnellen (1947) that *Methanosarcina barkerii* and *Methanobacterium formicicum* are able to ferment
carbon monoxide to a mixture of carbon dioxide and hydrogen, the
necessary hydrogen and extra oxygen atoms presumably coming from
water. Their experiments were performed on suspensions of the organisms
and it is not stated whether the organisms would grow anaerobically in a
mineral medium, on carbon monoxide; it seems very likely that the
organisms should be able to do so. It is not clear to what extent fermentation of carbon monoxide is a general property of methanogenic bacteria.

4. *Utilization of Hydrogen and Carbon Dioxide*

Fermentation of hydrogen plus carbon dioxide to methane plus water
is a general property of methanogenic bacteria; all the species mentioned
in this section are capable of carrying out this dismutation. Indeed,
Smith and Hungate (1958) suggest that hydrogen and carbon dioxide
may be the chief substrates giving rise to methane in the rumen.

Methanosarcina barkerii and *Methanobacterium formicicum* can grow
in simple mineral medium on hydrogen plus carbon dioxide (Kluyver
and Schnellen, 1947), *Methanobacterium ruminantium* (Smith and
Hungate, 1958) and *Methanobacterium mobilis* (Paynter and Hungate,
1968) grow on this gas mixture but require added growth factors.

A detailed study has been made of techniques for growth on hydrogen
plus carbon dioxide of *Methanobacterium* strain M.o.H. and another
isolate called *Methanobacterium* strain B.B. (Bryant *et al.*, 1968).
Growth was accomplished in a modified 12-litre "Microferm" fermenter
(New Brunswick Scientific Co. Inc., U.S.A.) containing a complex
medium gassed with a hydrogen-carbon dioxide mixture. Cell yields of
60 g. wet weight of *Methanobacterium* M.o.H. per 12 litres were reported
after 60–70 hours growth from a 200 ml. initial inoculum. Although a
complex medium which included rumen fluid was used for large-scale
growth, the Illinois group have reported (Bryant *et al.*, 1967) that
Methanobacterium M.o.H. can be grown on hydrogen-carbon dioxide in a
defined medium containing mineral salts, B-vitamins, volatile fatty
acids, cysteine and sulphide.

METABOLISM OF ONE-CARBON COMPOUNDS BY MICRO-ORGANISMS   143

An examination under the electron microscope of *Methanobacterium* strain M.o.H., *Methanobacterium formicicum*, and *Methanobacterium ruminantium* (Langenberg *et al.*, 1968) revealed a complex pattern of intracytoplasmic membraneous elements in the first two organisms. No such elements were observed in *Methanobacterium ruminantium*.

III. Energy Metabolism

A. Aerobic Oxidation

Many enzymes which catalyse the oxidation of one-carbon compounds at a variety of oxidation levels have now been purified and their properties studied. A notable gap exists in respect of methane, for which, until very recently, no active enzymic system had been reported. The mechanism of oxidation of methane also raises some interesting problems of energetics and these will be referred to in more detail under the appropriate section below.

1. *Methane*

The route of microbial methane oxidation is generally thought to be:

$$CH_4 \rightarrow CH_3OH \rightarrow HCHO \rightarrow HCO_2H \rightarrow CO_2 \qquad (1)$$

Evidence in support of this is as follows:

(a) Methane-utilizing bacteria are capable of oxidizing methane and methanol, and, where tested, formaldehyde and formate (Dworkin and Foster, 1956; Leadbetter and Foster, 1958; Brown *et al.*, 1964; Stocks and McCleskey, 1964b; Foster and Davis, 1966; Whittenbury *et al.*, 1970a).

(b) Some of the postulated intermediary metabolites have been detected during the oxidation of methane or methanol, e.g. formate during the oxidation of methanol by *Ps. methanica* (Leadbetter and Foster, 1958) or during the oxidation of methane, methanol or formaldehyde by *Mtn. methano-oxidans* (Brown *et al.*, 1964); methanol during the oxidation of methane by *Ps. methanica* or *Mtn. methano-oxidans* (Higgins and Quayle, 1970) or *Mtlc. capsulatus* (Harwood, 1970).

(c) Some of the postulated intermediates have been accumulated in the presence of inhibitors or trapping agents. For example, Brown *et al.* (1964) reported that methanol and formaldehyde accumulated during methane oxidation by suspensions of *Mtn. methano-oxidans* in the presence of iodoacetate or sodium sulphite, respectively, as inhibitors. It should be noted, however, that

144   J. R. QUAYLE

Higgins and Quayle (1970) were not able to duplicate these findings in respect of methanol accumulation in the presence of iodoacetate (see p. 145).

(d) Enzymes catalysing the oxidation of methanol, formaldehyde and formate have been demonstrated in extracts of methanol-grown *Ps. methanica* (Johnson and Quayle, 1964) and methane-grown "*Methylosinus sporium*" (strain 5) (J. R. Quayle, unpublished results).

It can be seen that the evidence in favour of the stepwise breakdown of methane to carbon dioxide is not complete for any one organism, but the scheme is the simplest one to account for the known facts. Bryan-Jones and Wilkinson (personal communication) have made the interesting observation that dimethyl ether is oxidized, and used as a growth substrate, by methane-utilizing bacteria and it has been detected in the medium when the organisms have been grown on methane under conditions of oxygen limitation. This raises the possibility of dimethyl ether also being an intermediate in methane oxidation.

If methane is converted directly to methanol, two possibilities for such a reaction suggest themselves. By analogy with oxidation of higher alkanes (Peterson *et al.*, 1967) a possible reaction might be one which is mediated by a mixed function oxidase (mono-oxygenase):

$$CH_4 + O_2 + XH_2 \rightarrow CH_3OH + H_2O + X \qquad (2)$$

where XH_2 is a reducing agent. Another possibility might be an oxygenase system in which two molecules of methane are oxygenated in the overall reaction:

$$2CH_4 + O_2 \rightarrow 2CH_3OH \qquad (3)$$

Indirect evidence for involvement of oxygen was first obtained by Leadbetter and Foster (1959) who grew *Ps. methanica* in the presence of $^{18}O_2$ and found that if the carbon source was methane, the incorporation of ^{18}O into cell material was 16 times that found when the micro-organism was grown on methanol. There is now direct evidence which establishes that the oxygen atom in methanol is derived exclusively from molecular oxygen during oxidation of methane by *Ps. methanica*. Higgins and Quayle (1970) have shown that, during the oxidation of methane by *Ps. methanica* or *Mtn. methano-oxidans* in the presence of $^{18}O_2$, the $^{18}O/^{16}O$ abundance ratio in the methanol was, within the limits of experimental error, the same as that in the molecular oxygen in the gas phase. There was negligible incorporation into methanol of oxygen derived from water when the experiments were repeated with $^{16}O_2$ in the gas phase but $[^{18}O]$ water in the incubation medium.

METABOLISM OF ONE-CARBON COMPOUNDS BY MICRO-ORGANISMS 145

The success of this isotope experiment depended on obtaining sufficient methanol from the bacterial oxidation of methane to permit the necessary isotopic analyses. Methanol could not be detected when washed suspensions of the organisms were incubated in Warburg cups at cell densities of 1 mg. dry weight/ml. in 20–40 mM-phosphate buffer at pH 7·0. Following the work of Brown $et\ al.$ (1964), preliminary attempts were made to cause methanol accumulation from methane, by suspensions of $Mtn.\ methano-oxidans$, by using iodoacetate as an inhibitor. In contrast to the results reported by Brown $et\ al.$ (1964), methanol could not be detected under a wide range of conditions using this inhibitor. In all experiments, uptake of a methane-oxygen mixture by the bacteria was actually $more$ sensitive to iodoacetate than was methanol-dependent oxygen uptake. However, during the course of this work, it was found that, in phosphate buffer under strictly controlled conditions of phosphate concentration, pH value and cell density, methanol accumulated to concentrations of up to 2·5 mm. This was sufficient to enable the methanol to be separated from water by fractional distillation and gas-liquid chromatography, followed by mass spectrometric analysis. It is interesting to note that Anthony (1970) observed that phosphate inhibited the oxidation of methanol and formaldehyde by $Pseudomonas$ AM1. Accumulation of methanol during growth of $Mtlc.\ capsulatus$ on methane under oxygen limitation has been observed by Harwood (1970).

As referred to previously, Brown $et\ al.$ (1964) reported that the presence of 3 mm-iodoacetate during oxidation of methane by $Mtn.$ $methano-oxidans$ caused at least 75% of the methane consumed to accumulate as methanol, the iodoacetate inhibiting methanol oxidation without inhibiting methane oxidation. Whittenbury (1969) has pointed out that, since the further oxidation of methanol would be the only source of reducing power for a mono-oxygenase reaction, the results of Brown $et\ al.$ (1964) could be inconsistent with the involvement of such an enzyme during methane oxidation. Furthermore, the consumption of reducing power by a mono-oxygenase might result in the cell yield of a methane-grown organism being less than that of the same organism grown on methanol. Measurements by Whittenbury $et\ al.$ (1970b) showed that the growth yields of two methane-utilizing bacteria were 25% higher on equimolar amounts of methane as compared to methanol. These two sets of data led to the suggestion by Whittenbury (1969) that the transformation of methane into methanol might be energy-yielding, e.g. a hydroxylation reaction in which methanol was formed together with two electrons for energy transduction:

$$CH_4 + H_2O \rightarrow CH_3OH + 2[H] \qquad (4)$$

146   J. R. QUAYLE

Since the later isotopic experiments of Higgins and Quayle (1970) rule out such a reaction, it is necessary to examine further the considerations which led to its proposal.

The selective inhibition by iodoacetate of methanol oxidation as compared to methane oxidation, reported by Brown *et al.* (1964) for *Mtn. methano-oxidans*, could not be repeated by Higgins and Quayle (1970). Unless the results are reproducible the deduction from them is clearly not valid.

The question of relative growth yields on methane and methanol is a rather complex one at the present time. There is considerable variation in the reported growth yields on methane alone. Table 4 lists such values, together with those for the relative growth yields on methane and methanol, which have been obtained with pure cultures. Similar measurements have also been made with mixed cultures (Vary and Johnson, 1967; Haggstrom, 1969) but these have not been included in Table 4 lest they confuse rather than clarify.

The results of Brown *et al.* (1964) and Whittenbury *et al.* (1970b) for three organisms are closely similar, suggesting the overall stoicheiometry for a growing culture to be:

$$CH_4 + 1 \cdot 0 - 1 \cdot 1 \ O_2 \ \rightarrow \ 0 \cdot 19 - 0 \cdot 3 \ CO_2 + cells \qquad (5)$$

These results are different from those of Dworkin and Foster (1956), Leadbetter and Foster (1958) and Foster and Davis (1966) for two organisms. These results indicate:

$$CH_4 + 0 \cdot 4 - 0 \cdot 58 \ O_2 \ \rightarrow \ 0 \cdot 18 - 0 \cdot 23 \ CO_2 + cells + \text{``soluble slime''} \qquad (6)$$

Both equations indicate a very high efficiency of conversion of methane-carbon to cellular-carbon, viz. 70–82%. However, equation (6), while it might be compatible with hydroxylation of methane with water (reaction 4), is scarcely compatible with oxygenation of methane by either reactions (2) or (3). In the case of (2) or (3) operating, the entire oxygen uptake (or more!) of the organism would be accounted for merely by the conversion of methane to methanol.

Harwood (1970) has studied the growth of *Mtlc. capsulatus* on methane under conditions of methane or oxygen limitation. Under oxygen limitation, the entire oxygen uptake would be accounted for in the conversion of 0·25 mole of methane to 0·2 mole of methanol (in the products) and 0·05 mole of carbon dioxide, if reaction (2) operated. This would leave no oxygen available for conversion of the remaining 0·75 mole of methane to cell carbon and products other than methanol. If reaction (3) operated, 0·05 mole of oxygen would be available for the latter conversion.

The incomplete results quoted by Johnson (1967) are based on indirect measurements of oxygen uptake which are open to criticism (Stouthamer,

Stoicheiometry of $CH_4/CH_3OH + O_2 \rightarrow CO_2 +$ Cells[a]

Organism	Reference	CH_4 (moles)	CH_3OH (moles)	O_2 (moles)	CO_2 (moles)	Cell carbon (moles)	Further data
Pseudomonas methanica	Dworkin and Foster (1956)	1		0·5	0·23	0·71	In a separate experiment the cell carbon was split into washed cells (47%, w/w) and soluble slime (53%, w/w).
Pseudomonas methanica	Leadbetter and Foster (1958)	1		0·40	0·21		
Methanomonas methano-oxidans	Brown et al. (1964)	1		1·1	0·19		
Methylococcus capsulatus	Foster and Davis (1966)	1		0·58	0·18	0·95	In the same experiment the cell carbon was split into 0·69 moles present as cells and 0·26 moles as "soluble C, mainly polysaccharide".
Methylococcus capsulatus	Harwood (1970)	1		1·8	0·34	0·66	Culture growing in chemostat under CH_4 limitation. No other products detected.
		1		0·3	0·05	0·20	Culture growing in chemostat under O_2 limitation. Other products contained 0·75 mole of carbon, of which 25–30% was methanol.
Pseudomonas sp.	Unpublished experiments; results quoted by Johnson (1967)	1		0·16[b]		1·07[c]	
Two strains of non-capsulate, non-slime-forming bacteria	Whittenbury et al. (1970b)	1 1	1	1·0–1·1 1·0–1·1	0·2–0·3 0·5–0·6		1 g. CH_4 yielded 1·0–1·1 g. cells. 1 g. CH_3OH yielded 0·4 g. cells. Carbon balances estimated to be approx. 100%.

[a] Where necessary, the results from different authors have been recalculated to bring them all to molar units.
[b] Calculated from the difference between the amount of oxygen required to burn the utilized substrate and the amount required to burn the product.
[c] Calculated from the dry weight of organism, assuming it contains 47% (w/w) of carbon.

148   J. R. QUAYLE

1969). The results indicate more carbon being fixed into cellular material than can be supplied from methane.

Whittenbury *et al.* (1970b) stress that their results on relative efficiencies of methane and methanol utilization apply to two strains which were non-capsulate, non slime-forming, did not form detectable lipid inclusions, and were not inhibited by methanol at the concentrations used. Other strains were less efficient. Variations in the stoicheiometry were observed under conditions of nitrogen limitation and with non-growing suspensions. In the former case, oxygen consumption and carbon dioxide production were lower per mole of methane utilized and poly-β-hydroxybutyrate accumulated. In the latter case, oxygen consumption and carbon dioxide evolution increased per mole of methane utilized, no doubt due to uncoupling of energy production. The results of Harwood (1970) also clearly show the very wide variation in stoicheiometry which is encountered under different growth conditions. More investigation is needed, particularly with known organisms under very carefully controlled conditions of growth, in order to establish the stoicheiometries and relative efficiencies of growth on methane and methanol. This is an area of prime importance with respect to the feasibility or otherwise of producing biomass from methane or methanol on a commercial basis.

Formation of more cellular material from growth on methane than from growth on equimolar amounts of methanol could be interpreted in more than one way without invoking reaction (4) in place of either reaction (2) or (3).

Two such interpretations are:

(a) Oxygenation of methane to methanol might be energy-yielding, i.e. coupled to ATP synthesis. There is ample free energy release from this to be thermodynamically possible for the standard free energy changes at pH 7 for reactions (2) and (3) are $-82 \cdot 1$ and $-59 \cdot 5$ kcal/mol., respectively, where X = a reduced nicotinamide nucleotide (Higgins and Quayle, 1970). It should nevertheless be pointed out that this would be a novel reaction as no monooxygenase system has yet been found which is coupled to ATP synthesis.

(b) Utilization of methanol might proceed by an energy-dependent reaction, e.g. uptake of substrate, which is not involved when methane is the initial substrate (Ribbons *et al.*, 1970).

The possible involvement of dimethyl ether as an intermediate between methane and methanol is difficult to rationalize in terms of either reaction (2) or (3). Rather would it indicate a novel reaction of the type:

$$2CH_4 + O_2 \rightarrow CH_3.O.CH_3 + H_2O \tag{7}$$

METABOLISM OF ONE-CARBON COMPOUNDS BY MICRO-ORGANISMS   149

The further oxidation of dimethyl ether might take place by a reaction catalysed by a mono-oxygenase:

$$CH_3.O.CH_3 + XH_2 + O_2 \rightarrow CH_3OH + HCHO + X + H_2O \qquad (8)$$

analogous to the O-demethylase discovered by Ribbons (1970):

$$\text{(structure) } O.CH_3 + O_2 + NADH_2 \longrightarrow \text{(structure) } OH + HCHO + NAD + H_2O \qquad (9)$$

or, as suggested by Bryan-Jones and Wilkinson (personal communication), a cleavage reaction:

$$CH_3.O.CH_3 + XH \rightarrow CH_3OH + CH_3.X \qquad (10)$$

Both reactions (8) and (10) would result in the oxygen atom of methanol arising solely from molecular oxygen and not from water, and would thus be consistent with the whole-cell isotope data of Higgins and Quayle (1970). However, it may be predicted that during growth of an organism on dimethyl ether in the presence of $^{18}O_2$, operation of reaction (8) would lead to half as much ^{18}O being incorporated into cell material as compared with growth on methane, whereas operation of reaction (10) would lead to very little incorporation of ^{18}O by the cells. Bryan-Jones and Wilkinson (personal communication) have shown that the latter situation obtains in the case of "*Methylomonas albus*" and "*Methylomonas agile*". Thus, the evidence, so far, favours operation of a reaction of type (10) rather than type (8) for the conversion of dimethyl ether to methanol. Further work is needed to establish whether dimethyl ether is indeed an obligatory intermediate between methane and methanol.

Study of the mechanism of methane oxidation has long been held up by inability to prepare a cell-free extract capable of oxidizing methane. It is thus of great interest that Ribbons and his colleagues have obtained an extract of *Mtlc. capsulatus* which exhibits a methane-stimulated $NADH_2$ oxidase activity (Smith *et al.*, 1971; Ribbons and Michalover, 1970). The extract was prepared by disruption of the organism in a French pressure cell and the active fraction was contained in particles which sedimented at 3,000–40,000 g in 15 minutes; this fraction contained fragmented intracytoplasmic membranes (see Fig. 4) and some cell-wall material. The active fraction showed $NADH_2$ (but not $NADPH_2$) oxidase activity which was stimulated by the presence of methane or ethane, but not propane. Ribbons and Michalover (1970) were not able to establish a complete stoicheiometry for the reaction owing to the difficulty of assessing the contribution made by methane-independent $NADH_2$ oxidase activity to the overall reaction in the

150   J. R. QUAYLE

presence of methane. Nevertheless, oxygen and $NADH_2$ were consumed in equimolar proportions as would be demanded by a mono-oxygenase reaction (2). The results of further work with this system are eagerly awaited as these may furnish a direct answer to many of the perplexing questions raised above.

Finally, it is interesting to note that Urushibara and Forrest (1970) have isolated four pteridine derivatives from each of the two methane-grown bacteria *"Methylosinus sporium"* (strain 5) and *"Methylosinus trichosporium"* (strain PG). These derivatives were characterized as 2 - amino - 4 - hydroxy - 6 - carboxypteridine, 2 - amino - 4 - hydroxy - 6 - (*D* - erythro*-1,2,3-trihydroxypropyl)pteridine (with the trivial name, neopterin), 2-amino-4-hydroxypteridine and 2-amino-4-hydroxy-6-methyl-pteridine. The last pteridine derivative has been used in its synthetic, reduced form as an artificial cofactor for phenylalanine hydroxylase (Kaufman, 1969; Kaufman and Levenberg, 1969).

2. *Methylated Amines and Methylated Thiols*

Table 3 (pp. 134–135) contains a list of well authenticated organisms which are capable of growing on trimethylamine, dimethylamine, mono-methylamine, trimethylamine-N-oxide, trimethylsulphonium chloride and N-monomethylurea. Since the carbon atoms in these substrates are not joined together but are separated from each other by nitrogen atoms, the organisms are in effect growing on one-carbon units. Their energy metabolism is thus of direct interest. There are also many cases of organisms growing on N-methyl derivatives such as choline, betaine, dimethylglycine and sarcosine where the methyl group is cleaved from the substrate and utilized along with the remainder of the molecule. Although the organisms are not growing on a one-carbon substrate, it is useful to compare the mechanism(s) of cleavage of the one-carbon unit from the nitrogen atom to those found in the case of the methylamines.

The microbial utilization of methylated amines is of special interest in marine microbiology due to the widespread occurrence of amines and amine oxides as excretory products of fish and as products encountered in fish decomposition (Shewan, 1951; Dyer, 1952). In this respect, attention should be drawn to a study of marine bacteria which utilize alkylamines as sole nitrogen source during growth on glucose (Budd and Spencer, 1968; Budd, 1969). The authors have concluded that these amines are broken down by stepwise demethylation, a process that needs concurrent utilization of endogenous or exogenous sources of energy. There may be an interesting parallel here to the co-oxidation observed in the case of microbial hydrocarbon utilization (Leadbetter and Foster, 1960).

METABOLISM OF ONE-CARBON COMPOUNDS BY MICRO-ORGANISMS   151

a. *Trimethylamine.* Kung and Wagner (1970b) isolated an organism called *Pseudomonas* MS, by enrichment culture on trimethylsulphonium chloride. The organism was also found to grow on trimethylamine, dimethylamine and methylamine but not on methanol, formaldehyde or formate. The organism when grown on any of the amines or a trimethyl-sulphonium salt oxidized formaldehyde and formate (but not methanol) without lag. This indicated that these growth substrates were oxidized *via* formaldehyde and formate, and this was supported by the presence of soluble formaldehyde and formate dehydrogenases in cell-free extracts of the organism when grown on any of the three amines or trimethyl-sulphonium salts. In cell-free extracts of *Pseudomonas* MS, oxidation of trimethylamine, dimethylamine or methylamine was carried out in a particulate fraction which catalysed a total oxidation of the substrate to carbon dioxide using oxygen as electron acceptor. The particulate fraction was capable of oxidizing formaldehyde and formate but at a slower rate than methylamine. This suggests that formaldehyde and formate are not free intermediates during the oxidation of methylamine. Attempts to solubilize the methylamine-oxidizing activity were not successful, nor could the oxidation be coupled to any artificial electron acceptors. It is not known whether one-carbon units are abstracted from this integrated oxidative system for biosynthetic purposes, or what the physiological role of the soluble formaldehyde- and formate dehydrogenases might be. It is possible that these two enzymes have been partly shaken out of the methylamine-oxidizing particle during cell breakage.

The inability to grow on methanol can be correlated with its non-involvement as an intermediary metabolite during one-carbon oxidation. Formaldehyde and formate may be unable to serve as growth substrates owing to difficulties of entry into the particulate oxidative assembly.

Colby and Zatman (1971) have isolated an organism, *Bacterium* 4B6, which is capable of growth on trimethylamine, dimethylamine or methylamine. Washed suspensions of the organism grown on either of the first two substrates will oxidize either of the first two substrates but not methylamine. Cell-free extracts of the trimethylamine-grown organism contain a soluble trimethylamine dehydrogenase which can be coupled to phenazine methosulphate:

$$(CH_3)_3NH^+ + H_2O + PMS \rightarrow (CH_3)_2NH_2^+ + HCHO + PMSH_2 \qquad (11)$$

a soluble $NAD(P)H_2$-linked dimethylamine mono-oxygenase (cf. Eady and Large, 1969):

$$(CH_3)_2NH_2^+ + O_2 + NAD(P)H_2 \rightarrow CH_3NH_3^+ + HCHO + H_2O + NAD(P) \qquad (12)$$

152   J. R. QUAYLE

and a glutathione-independent NAD-linked aldehyde dehydrogenase.
The trimethylamine dehydrogenase has been purified to homogeneity,
and has a molecular weight of about 116,000.

The first steps in oxidation of trimethylamine thus appear to be
successive oxidation of two of the methyl groups to formaldehyde. It
remains to be determined how the residual methylamine is further
oxidized.

b. *Trimethylamine-N-Oxide*. This compound is present in the tissues and
excreta of many marine invertebrates and vertebrates. An organism,
Bacillus PM6, has been isolated by Myers and Zatman (1971) which can
grow on trimethylamine-N-oxide, trimethylamine, dimethylamine or
methylamine as sole sources of carbon, nitrogen and energy. Washed
suspensions of the organism oxidize trimethylamine-N-oxide, dimethylamine and methylamine but not trimethylamine. An enzyme, trimethylamine-N-oxide demethylase, has been purified to homogeneity; the
enzyme catalyses the overall reaction:

$$(CH_3)_3NO \rightarrow (CH_3)_2NH + HCHO \qquad (13)$$

The initial step in the metabolism of trimethylamine-N-oxide by this
organism thus appears to be an intramolecular oxidation-reduction
resulting in dimethylamine formation rather than reduction to trimethylamine. This is consistent with the oxidative behaviour of washed cell
suspensions. The authors report that the activity of the enzyme is
strongly stimulated by ferrous iron, glutathione and L-ascorbate; the
mechanism of the reaction presents an interesting problem.

c. *Dimethylamine*. Mention has already been made of the ability of
Pseudomonas MS to grow on dimethylamine and of the total oxidation of
this substrate by a particulate enzyme system (Kung and Wagner,
1970b). Rather different behaviour has been found by Eady and Large
(1969) and Jarman *et al.* (1970) in the case of *Pseudomonas aminovorans*.
This organism, when grown on dimethylamine, oxidizes dimethylamine,
methylamine and formate, but not methanol or formaldehyde. An
enzyme has been partially purified which catalyses the reaction:

$$(CH_3)_2NH_2Cl + O_2 + NAD(P)H_2 \rightarrow CH_3NH_3Cl + HCHO + NAD(P) + H_2O \qquad (14)$$

Hence, as in the case of *Bacterium* 4B6 or *Bacillus* PM6 growing on
trimethylamine or trimethylamine-N-oxide respectively, the first step
in oxidation of the amine substrate is oxidative removal of one of the
methyl groups as formaldehyde. Eady and Large (1969) were not able
to detect any methylamine-oxidizing activity in extracts and hence, as
with *Bacterium* 4B6, the mechanism of further oxidation of the methylamine residue remains unknown.

The partially purified dimethylamine mono-oxygenase has an absorption maximum at 412 nm. Its spectral behaviour on reduction or on

METABOLISM OF ONE-CARBON COMPOUNDS BY MICRO-ORGANISMS   153

treatment with carbon monoxide or ethanol suggests that a P-450-type cytochrome may be involved as in the camphor methylene hydroxylase of *Ps. putida* (Katagiri *et al.*, 1968) and the octane hydroxylase of *Corynebacterium* 7E1C (Cardini and Jurtshuk, 1968).

d. *Methylamine*. *Pseudomonas* AM1 grows readily on methylamine, methanol or formate and a detailed study has been made of methylamine oxidation by this organism by Eady and Large (1968). Washed suspensions of the methylamine-grown organism oxidized methylamine, formaldehyde and formate. Methanol was only slightly oxidized. Cell-free extracts were found to contain an enzyme which oxidized methylamine to formaldehyde in the presence of phenazine methosulphate, cytochrome *c*, ferricyanide, brilliant cresyl blue or dichlorophenolindophenol as electron acceptors. Phenazine methosulphate was the most effective:

$$CH_3NH_3Cl + PMS + H_2O \rightarrow HCHO + PMSH_2 + NH_4Cl \qquad (15)$$

The enzyme was purified 20-fold and its properties studied. It is non-specific with respect to aliphatic primary amine substrate, but secondary and tertiary amines and quaternary ammonium salts are not oxidized. It is remarkably heat-stable as it will withstand heating for 15 minutes at 80° without loss of activity. Advantage has been taken of the substrate specificity of the dehydrogenase to develop an enzymic method for the micro-estimation of methylamine, ethylamine and *n*-propylamine in biological mixtures (Large *et al.*, 1969).

The essential role of methylamine dehydrogenase in methylamine metabolism is shown by its induced synthesis during the lag period of no growth when succinate-grown *Pseudomonas* AM1 adapts to growth on methylamine. The non-involvement of methanol as an intermediary metabolite in methylamine breakdown is shown by: (a) the inability of the methylamine-grown organism to oxidize methanol; (b) the absence of methanol dehydrogenase in cell-free extracts; (c) a mutant, M-15A, of *Pseudomonas* AM1 which lacks methanol dehydrogenase can oxidize and grow on methylamine (Heptinstall and Quayle, 1970).

Shaw *et al.* (1966) have studied the metabolism of methylamine by a *Pseudomonas* sp. which grows on methylamine but not on dimethylamine, trimethylamine, methanol, formaldehyde or formate. Methylamine-oxidizing activity could not be detected in cell-free extracts of the organism, neither could such extracts or indeed, whole cells, oxidize methanol or formaldehyde.

It is clear from the above survey that there is some diversity of mechanism for methylamine oxidation. *Pseudomonas* MS oxidizes it in a particulate system; *Ps. aminovorans* and *Bacterium* 4B6 oxidize it by an unknown mechanism; *Pseudomonas* AM1 oxidizes it by a soluble

dehydrogenase; a *Pseudomonas* sp. oxidizes it by an unknown mechanism which probably does not involve either methanol or formaldehyde (Shaw *et al.*, 1966).

e. *N-Methylurea*. N-Methylurea can function as a source of nitrogen for various bacteria (den Dooren de Jong, 1926) and Iyer and Kallio (1958) isolated an organism, *Bacillus sphaericus*, which would grow on this compound as combined carbon and nitrogen source. It was found that methylurea was converted to formaldehyde by cell-free extracts in a two-step reaction:

$$CH_3.NH.CO.NH_2 + H_2O \rightarrow CH_3NH_2 + CO_2 + NH_3 \qquad (16)$$

$$CH_3NH_2 + \tfrac{1}{2}O_2 \rightarrow HCHO + NH_3 \qquad (17)$$

The enzymes catalysing these steps were not characterized, nor could further oxidation of formaldehyde in cell extracts be detected.

f. *Methylsulphonium Salts*. It has been mentioned previously that oxidation of trimethylsulphonium salts by *Pseudomonas* MS grown on such compounds as sole carbon source probably involves formaldehyde and formate as intermediates (Kung and Wagner, 1970b). Wagner and his colleagues have partially purified an enzyme from trimethylsulphonium nitrate-grown *Pseudomonas* MS which catalyses the following reaction (Wagner *et al.*, 1966, 1967):

$$(CH_3)_3S^+ + \text{tetrahydrofolate} \rightarrow (CH_3)_2S + 5\text{-}CH_3 \text{ tetrahydrofolate} + H^+ \qquad (18)$$

It is likely that this reaction represents a major initial step in the utilization of the substrate because dimethylsulphide can be aspirated from growing cultures or washed suspensions of the organism incubated with a trimethylsulphonium salt.

Oxidation of 5-methyltetrahydrofolate to formaldehyde proceeds by way of dehydrogenation to 5,10-methylenetetrahydrofolate (Hornig and Wagner, 1968). The enzyme which catalyses this reaction has been partially purified and is thought to contain a flavin as cofactor (C. Wagner, personal communication).

g. *Choline, Betaine, Dimethylglycine and Sarcosine*. The ability of aerobic organisms from the soil to grow on choline has been shown to be a widespread property amongst organisms from several genera, particularly *Pseudomonas* and *Arthrobacter* (Kortstee, 1970). All the bacteria tested were found to be able to grow with betaine, N,N-dimethylglycine or sarcosine as sole carbon and nitrogen source. Suspensions of choline-grown cells oxidized these compounds at rates similar to that of choline. This indicates that the oxidation of choline proceeds by an initial oxidation to betaine:

$$(CH_3)_3\overset{+}{N}.CH_2.CH_2OH \rightarrow (CH_3)_3\overset{+}{N}.CH_2.CO_2H \qquad (19)$$

METABOLISM OF ONE-CARBON COMPOUNDS BY MICRO-ORGANISMS 155

followed by successive demethylations to sarcosine. This scheme is similar to that found for growth on choline by a strain of *Ps. fluorescens* (Goldstein, 1959; Fitch, 1963) and *Achromobacter cholinophagum* (Shieh, 1964, 1965, 1966a, b, 1968) except that, in the case of the last two organisms, sarcosine has been shown to be further demethylated to glycine. It is curious to find that *Ps. fluorescens* can grow on choline, betaine, dimethylglycine or sarcosine but is unable to grow on glycine plus formaldehyde or formate, though it can oxidize all three compounds (Fitch, 1963). It might be expected that glycine + a one-carbon compound would be equivalent to sarcosine as growth substrate.

Preliminary oxidation of choline to betaine seems to be the preferred route compared to the alternative oxidation to N,N-dimethylethanolamine and N-monomethylethanolamine. Evidence for operation of the latter pathway has been found in a minority of the choline-utilizing coryneform bacteria (Kortstee, 1970).

It will be of interest to see how the enzymic mechanism(s) responsible for oxidation of the choline methyl groups compares with those found for methylated amines. Little is known about this at present apart from the demonstration by Shieh that the one-carbon units can be trapped as formaldehyde.

3. *Methanol*

Studies of the oxidation of methanol by several methane-utilizing organisms have shown formaldehyde to be formed (see p. 143). The most detailed studies of microbial methanol oxidation have however been made with bacteria capable of growth on methanol but not on methane. Using whole-cell suspensions it has been shown that formaldehyde can be oxidized by *Pseudomonas* M27 (Anthony and Zatman, 1964a), "*Pseudomonas methanica*" (Iowa) (Harrington and Kallio, 1960), *Pseudomonas* AM1 (Peel and Quayle, 1961); in the first two organisms formaldehyde was also detected as a product of methanol oxidation.

The most detailed study of the enzymology of methanol oxidation has been made by Anthony and Zatman (1964a, b, 1965, 1967a, b) using *Pseudomonas* M27. No evidence could be found for the presence of an NAD-linked methanol dehydrogenase or for the involvement of catalase during the oxidation of methanol by *Pseudomonas* M27 (Anthony and Zatman, 1964a) or *Pseudomonas* AM1 (Johnson and Quayle, 1964). Instead, Anthony and Zatman (1964b, 1965) found an enzyme which catalyses the dehydrogenation of methanol in the presence of phenazine methosulphate:

$$CH_3OH + PMS \rightarrow HCHO + PMSH_2 \tag{20}$$

The enzyme catalyses the oxidation of *normal* alcohols but not secondary or tertiary alcohols and requires ammonia or methylamine as activator.

No hydrogen acceptor was found other than phenazine methosulphate. The methanol dehydrogenase has been purified by ammonium sulphate fractionation, followed by chromatography on DEAE-cellulose and Sephadex G-150 (Anthony and Zatman, 1967a). The resulting enzyme was more than 95% pure on the basis of ultracentrifugal and electrophoretic analysis. It contains insignificant amounts of any metal, has a molecular weight of 120,000–146,000, and its absorption spectrum shows peaks at 280 nm. and 350 nm. with little or no absorption at or above 450 nm. The enzyme accounts for approximately 10% of the total soluble protein of the organism.

There can be little doubt that this methanol dehydrogenase is a key enzyme in the metabolism of methanol by many methanol- (and methane-) utilizing organisms. It has been found in *Pseudomonas* AM1, *Protaminobacter ruber*, *Vibrio extorquens* and methanol-grown *Ps. methanica* by Johnson and Quayle (1964), methane-grown "*Methylosinus sporium*" (strain 5) by J. R. Quayle (unpublished results) and *Pseudomonas* PP by Ladner and Zatman (1969). The necessary involvement of the enzyme in methanol oxidation by *Pseudomonas* AM1 is evident from the isolation of the mutant, M-15A, which lacks methanol dehydrogenase (Heptinstall and Quayle, 1970); the mutant is unable either to oxidize or to grow on methanol. Growth of the mutant on methylamine or formate was however unimpaired, showing that its biosynthetic abilities with respect to a one-carbon substrate were intact. The oxidative role of methanol dehydrogenase may extend beyond that of oxidation of methanol to formaldehyde because there is circumstantial evidence suggesting that it may also catalyse the oxidation of formaldehyde (Ladner and Zatman, 1969). This is borne out by the finding that extracts of methanol-grown *Pseudomonas* AM1, under the conditions for assay of methanol dehydrogenase, will also oxidize formaldehyde at 26% of the rate for methanol. The fact that mutant M-15A lacks the ability to oxidize either of these substrates under these conditions suggests that one enzyme is involved (Heptinstall and Quayle, 1969, 1970).

The mechanism of action of methanol dehydrogenase is not known at present. The enzyme contains neither metal ions nor measurable quantities of such cofactors as riboflavin, a wide range of folate derivatives, vitamin B_{12}, nicotinic acid or its amide, pyridoxine and derivatives, biotin and derivatives, or thiamine (Anthony and Zatman, 1967b). These authors have adduced a variety of other evidence which indicates that the enzyme is not a flavoprotein. However, the enzyme does release a green fluorescent material, diffusible on dialysis, when it is treated with acid or alkali or when it is boiled. Kinetic studies showed a correlation (though not necessarily a causal relation) between loss of activity and appearance of the fluorescent material. Some purification of this material

METABOLISM OF ONE-CARBON COMPOUNDS BY MICRO-ORGANISMS 157

was achieved, the major component having a fluorescence maximum at 460 nm. with extinction maxima at 225 nm. and 365 nm. On the basis of the spectral behaviour, Anthony and Zatman (1967b) suggest that the compound may be a pteridine derivative, possibly related to pteroyl-glutamic acid. Its relation to the pteridine derivatives isolated from two species of methane-utilizing bacteria by Urushibara and Forrest (1970) remains to be determined.

The involvement of a pteridine as a prosthetic group would be of considerable interest. Such a cofactor might function as an electron acceptor as the standard electrode potential for the quinonoid dihydro-pterintetrahydropterin couple is appropriate ($E_0^1 = +0.15$ v) (Archer and Scrimgeour, 1970) for coupling with the dehydrogenation of methanol to formaldehyde ($E_0^1 = -0.18$ v). Alternatively, Anthony and Zatman (1967b) make the interesting suggestion that, if the prosthetic group were a pteroyl derivative, methanol might be oxidized to a 5,10-methylene derivative by way of a 5-methyl pteroate:

$$(21)$$

Pteroyl cofactor

$$(22)$$

Reaction (22) would be analagous to that proposed by Hornig and Wagner (1968) for *Pseudomonas* MS. Alternatively, the methanol might be bound as a 5-methyldihydro derivative, followed by intramolecular re-arrangement to a 5,10-methylene tetrahydro derivative:

$$(23)$$

This would be essentially the reverse of the reaction involved in the conversion of deoxyuridine to thymidine (Huennekens, 1963). The dihydro derivative might be regenerated by dehydrogenation either

158   J. R. QUAYLE

before or after transfer of the one-carbon-residue. In all cases there would first have to be a novel reaction (21) which effected the methylation of the N-5 atom of the pyrazine ring with methanol.

Mechanisms such as these would result in the formation of a coenzyme-bound one-carbon unit which might be channelled directly into the main carbon assimilation pathway of the cell (see p. 177). The elucidation of the nature of this prosthetic group is thus of importance not only to the understanding of methanol oxidation but, possibly, to the initial step in carbon assimilation as well.

4. *Formaldehyde*

It will be evident from Sections 1, 2 and 3 that formaldehyde is an intermediary metabolite in the bacterial oxidation of methyl groups arising from methane, methanol, methylated amines, trimethyl-sulphonium compounds and choline and its derivatives. Three enzymes which catalyse the dehydrogenation of formaldehyde have been characterized in organisms grown on one-carbon growth substrates, namely an NAD-linked dehydrogenase which requires reduced glutathione:

$$HCHO + NAD + H_2O \xrightarrow{\text{GSH}} HCO_2H + NADH_2 \qquad (24)$$

an NAD-linked dehydrogenase which does not require reduced gluta-thione, and a dehydrogenase which can be linked to dichlorophenol-indophenol as electron acceptor

$$HCHO + DCPIP + H_2O \rightarrow HCO_2H + DCPIPH_2 \qquad (25)$$

There is also a fourth possibility in the dual specificity of methanol dehydrogenase (see p. 156).

An enzyme which catalyses reaction (24) was found by Harrington and Kallio (1960) in methanol-grown *"Pseudomonas methanica"* (Iowa) and is similar to the formaldehyde dehydrogenase discovered in mammalian liver (Strittmatter and Ball, 1955) and yeast (Rose and Racker, 1962). These enzymes are strictly specific for NAD, require glutathione as cofactor, and are specific for formaldehyde (except for glyoxal and methylglyoxal in the case of the yeast enzyme).

The NAD-linked dehydrogenase which does not require glutathione was found by Kung and Wagner (1970b) and Colby and Zatman (1971) in extracts of *Pseudomonas* MS and *Bacterium* 4B6 respectively.

An enzyme which catalyses reaction (25) was discovered in methanol-grown *Pseudomonas* AM1 by Johnson and Quayle (1964) and purified. It is of broad specificity towards aliphatic aldehydes, but only dichloro-phenolindophenol and phenazine methosulphate serve as artificial electron acceptors. Despite its broad specificity the enzyme probably is involved in the oxidation of formaldehyde because its specific activity

METABOLISM OF ONE-CARBON COMPOUNDS BY MICRO-ORGANISMS   159

was twice as high in methanol-grown cells as in succinate-grown cells, Johnson and Quayle (1964) examined extracts of four species of bacteria grown on methanol and found that both enzymes (24) and (25) were present in *Pr. ruber*, *V. extorquens* and *Ps. methanica* but only enzyme (25) in *Pseudomonas* AM1.

The possible dual specificity towards methanol and formaldehyde of the phenazine methosulphate-linked methanol dehydrogenase (20) raises another possibility for formaldehyde oxidation. With more than one formaldehyde dehydrogenase present in a single organism, only mutant studies can show which of the enzymes are essential for the oxidation. Only one such mutant has been isolated and examined in this way, namely mutant M-15-A of *Pseudomonas* AM1 (Heptinstall and Quayle, 1970).

5. *Formate*

Enzymes catalysing the dehydrogenation of formate are widespread in animal, plant and microbial tissue; a variety of electron acceptors are involved. Apart from certain autotrophic organisms (see p. 160) the only formate dehydrogenases reported in aerobic bacteria growing on one-carbon compounds are NAD-linked.

Kaneda and Roxburgh (1959c) showed that extracts of methanol-grown *Pseudomonas* PRL-W4 catalysed the reduction of NAD in the presence of formate, but they did not further characterize the enzyme.

Pseudomonas oxalaticus contains high activities of an NAD-linked formate dehydrogenase when grown on oxalate or formate. The enzyme has been partially purified and, since it is specific with respect to formate, it forms the basis of a convenient assay for formate (Johnson *et al.*, 1964).

Johnson and Quayle (1964) found NAD-linked formate dehydrogenase activity in extracts of methanol-grown *Pr. ruber*, *V. extorquens*, *Ps. methanica* and *Pseudomonas* AM1. The same authors purified the formate dehydrogenase from *Pseudomonas* AM1 three-fold but found it to be rather unstable. The enzyme is specific for formate and is strongly inhibited by cyanide, ferrous and cupric ions.

Interesting studies have been made of the ability of some nitrifying bacteria to oxidize formate. It was observed by Silver (1960) that, of the many common organic substrates tested, formate stood out as the only one which was appreciably oxidized and which caused reduction of the cellular cytochromes of *Nitrobacter* species. Malavolta *et al.* (1962) prepared from nitrite-grown *Nitrobacter agilis* a particulate fraction which catalysed the oxidation of formate. The oxidation was stimulated by addition of cytochrome *c* or ATP but not by NAD, and it could be coupled to reduction of methylene blue. Since the bacteria can oxidize formate and can synthesize all of their cell constituents from carbon

160 J. R. QUAYLE

dioxide, it might be expected that the organism should be able to grow on formate. However, little or no formate carbon was assimilated by the cells (Delwiche and Finstein, 1965; Ida and Alexander, 1965) and cultures of *Nitrobacter winogradskyi* oxidizing both formate and nitrite together did not reduce more carbon dioxide than cultures oxidizing nitrite alone (Schön, 1965). This apparent paradox was further examined by van Gool and Laudelot (1966) who showed that in *N. winogradskyi* the rate of oxidation of formate was approximately one-fifth of the rate of nitrite oxidation, due partly to the very different pH optima for the two processes. Difference spectra of reduced *versus* oxidized cell suspensions were studied, the reduction of the pigments being effected with dithionite, nitrite or formate. It was found that the difference spectra after formate oxidation were identical to those after nitrite oxidation, showing that both compounds fed electrons into essentially the same respiratory chain from cytochrome *c* to oxygen. This finding emphasized the problem of why the free energy of formate oxidation should apparently not be utilized for autotrophic growth. The authors then examined the growth of *N. winogradskyi* in media containing as electron source either nitrite or formate or both, the pH value being maintained at the value optimal for *formate oxidation*. With nitrite only, the organism grew with a mean generation time of 18 hours; when formate was present in addition to nitrite, growth did not cease when the nitrite had all been oxidized but proceeded with a mean generation time of 144 hours. Measurements of short-term fixation of carbon dioxide indicated that one mole of carbon dioxide was fixed per 210 moles of formate oxidized as compared with one mole of cellular carbon being formed per 50–90 moles of nitrite oxidized during growth with nitrite. It appears from this that formate can indeed be used at certain pH values as carbon and energy source for growth of *N. winogradskyi*, albeit slowly and with a very low efficiency of free energy utilization.

Detailed studies have also been made of the particulate nitrite oxidase-formate oxidase system of *N. agilis* (O'Kelley *et al.*, 1970; O'Kelley and Nason, 1970). A pellet fraction prepared by these authors was able to fix inorganic phosphate into organic phosphate upon oxidation of $NADH_2$, nitrite or formate. However, attempts to grow the organism on formate as carbon and energy source have not, as yet, been successful. The authors suggest that the obstacle may be the large disparity (two pH units) between the pH optima for formate oxidation and for growth on nitrite.

6. *Carbon Monoxide*

Little is known about the mechanism of oxidation of carbon monoxide by organisms reputed to grow on this substrate as carbon source although

METABOLISM OF ONE-CARBON COMPOUNDS BY MICRO-ORGANISMS 161

some information has been obtained with other organisms. Yagi (1958, 1959) and Yagi and Tamiya (1962) have studied the oxidation of carbon monoxide by lactate-grown *Desulfovibrio desulfuricans*. Cell-free extracts catalysed the oxidation of carbon monoxide to carbon dioxide in the presence of substrate amounts of benzyl viologen:

$$CO + \text{oxidized benzyl viologen} + H_2O \rightarrow CO_2 + \text{reduced benzyl viologen} + 2H^+ \quad (26)$$

In the presence of catalytic quantities of benzyl viologen, hydrogen was produced, probably due to involvement of hydrogenase.

Chappelle (1962) has observed in several species of algae an oxygen-dependent oxidation of carbon monoxide which is stimulated by light. The product of the reaction was presumed to be carbon dioxide on the basis of its fate in the intact organisms.

7. *Involvement of Cytochromes*

Although a considerable amount of work has been reported on the enzymology of oxidation of organic one-carbon compounds, very little has appeared on the electron-transport chain involved. Whittenbury (1969) states that preliminary studies have shown the presence of cytochromes of the c, b and a type in methane-utilizing organisms. Anthony (1970) has reported that insoluble cytochromes of the b and a type and a soluble cytochrome c are present in *Pseudomonas* AM1. The extent of reduction of cytochrome c by various substrates was measured from difference spectra observed with whole cells. It was concluded that cytochrome c is an essential intermediate in the oxidation of methanol, methylamine, formaldehyde, formate, succinate and malate by *Pseudomonas* AM1.

B. ANAEROBIC DISMUTATION

Formidable problems of energy metabolism are posed by the methane fermentations of one-carbon substrates. The bacteria involved are able to derive the energy necessary for growth from the overall reactions:

$$4CH_3OH \rightarrow CO_2 + 2H_2O + 3CH_4 \qquad \Delta G^\circ = -74{\cdot}59 \text{ kcal} \qquad (27)$$
$$4HCO_2H \rightarrow 3CO_2 + 2H_2O + CH_4 \qquad \Delta G^\circ = -62{\cdot}05 \text{ kcal} \qquad (28)$$
$$4H_2 + CO_2 \rightarrow 2H_2O + CH_4 \qquad \Delta G^\circ = -33{\cdot}21 \text{ kcal} \qquad (29)$$

and possibly:

$$4CO + 2H_2O \rightarrow 3CO_2 + CH_4 \qquad (30)$$

In this section, all free energies are calculated from the data given by Burton (1961) for methanol, formic acid and carbon dioxide in aqueous solution, methane in gaseous state. Two major problems of the energy metabolism are: (1) the path of carbon in which the substrate is reduced

162 J. R. QUAYLE

to methane; and (2) the generation of ATP and utilizable reducing power from the overall reactions. These two aspects will be treated separately. Only a brief outline will be given; further details can be found in a full review by Stadtman (1967) and in the original papers discussed in the present review.

1. *Path of Carbon*

It has long been apparent that carbon dioxide is not reduced by way of the free intermediates, formate, formaldehyde and methanol during the methane fermentation. Also, substrates such as methanol or formate are not totally oxidized to carbon dioxide before being reduced to methane. These ideas were expressed by Barker (1956) in a scheme which has formed the model for subsequent investigations:

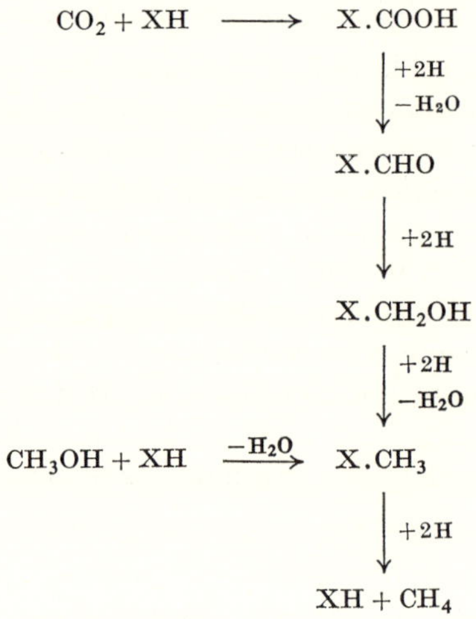

$$CO_2 + XH \longrightarrow X.COOH$$

$$\downarrow \begin{array}{l} +2H \\ -H_2O \end{array}$$

$$X.CHO$$

$$\downarrow +2H$$

$$X.CH_2OH$$

$$\downarrow \begin{array}{l} +2H \\ -H_2O \end{array}$$

$$CH_3OH + XH \xrightarrow{-H_2O} X.CH_3$$

$$\downarrow +2H$$

$$XH + CH_4$$

FIG. 8. Reduction of one-carbon compounds to methane. Based on Barker (1956).

In this scheme (Fig. 8) XH was a hypothetical compound which Barker suggested might be a metabolite as in photosynthesis, a coenzyme as in fatty acid oxidation, or a series of one-carbon carriers.

Investigation at the enzymological level commenced with the discoveries of Wolin *et al.* (1963a) and Blaylock and Stadtman (1963) of methane formation in cell-free extracts of "*Methanobacterium omelianskii*" and *Methanosarcina barkerii*, respectively. Most of the subsequent work has been performed with these two systems. It may be recalled that in 1967 Bryant *et al.* showed that "*Mtb. omelianskii*" was a symbiotic mixture of two organisms—a methanogenic organism, *Methanobacterium* strain M.o.H., fermenting hydrogen plus carbon dioxide to methane, and a non-methanogenic organism "S" fermenting ethanol to hydrogen

METABOLISM OF ONE-CARBON COMPOUNDS BY MICRO-ORGANISMS   163

plus acetate. Following this discovery it was important to establish that extracts of *Methanobacterium* strain M.o.H. alone were responsible for the catalysis of methane formation from one-carbon donors observed in extracts of "*Mtb. omelianskii*". This was confirmed by Bryant *et al.* (1968). However, all detailed work, both at the enzymological and whole cell level, done by Wolfe and his coworkers on "*Mtb. omelianskii*" up to 1968 relates to a mixed culture. Work subsequent to 1968 by this group has centred on the pure culture of *Methanobacterium* strain M.o.H.

a. *Reduction of One-Carbon Units to the Level of Methanol.* A variety of free or coenzyme-bound one-carbon components can be reduced to methane in cell-free extracts of methanogenic bacteria. Thus, extracts of "*Mtb. omelianskii*" catalysed methane formation from methyl-cobalamin, 5-methyltetrahydrofolate, 5,10-methylenetetrahydrofolate, the β-carbon of serine or pyruvate and carbon dioxide (Wolin *et al.*, 1963a, b; Wood and Wolfe, 1965; Wood *et al.*, 1965). In all cases ATP was necessary, and in all cases except that of methylcobalamin a source of reducing power was necessary. Extracts of *Mts. barkerii* catalysed methane formation from methylcobalamin, 5-methyltetrahydrofolate, methanol, formaldehyde, formate, the carboxyl carbon of pyruvate and carbon dioxide (Blaylock and Stadtman, 1966). Molecular hydrogen or pyruvate could be utilized as sources of reducing power; in all cases ATP was needed, and for pyruvate, coenzyme A as well.

The activity of the one-carbon folate derivatives suggests that such components may be the intermediary metabolites between at least formate and 5-methyltetrahydrofolate during methanogenesis. There is enzymic evidence which is consistent with this possibility. Extracts of *Mts. barkerii* have been shown (Blaylock and Stadtman, 1966) to possess low levels of tetrahydrofolate formylase activity:

$$H.COOH + \text{tetrahydrofolate} + ATP \rightleftharpoons 10\text{-formyltetrahydrofolate} + ADP + Pi \qquad (31)$$

Extracts of "*Mtb. omelianskii*" have been shown (Wood *et al.*, 1965) to catalyse the transformations:

$$\text{Serine} + \text{tetrahydrofolate} \rightleftharpoons \text{glycine} + 5,10\text{-methylenetetrahydrofolate} \qquad (32)$$

$$5,10\text{-methylenetetrahydrofolate} + NADH_2 \rightleftharpoons 5\text{-methyltetrahydrofolate} + NAD \qquad (33)$$

Although these enzymes are present this does not prove that they are obligatory for methanogenesis. They are ubiquitous enzymes in many kinds of living tissue. There are also uncertainties relating to the lack of specificity for optical isomers of 5-methyltetrahydrofolate and to the action of the folate antagonist, aminopterin (Stadtman, 1967). The natural cofactors could of course be derived forms such as polyglutamates which may differ from the monoglutamates in their behaviour in these transformations.

164   J. R. QUAYLE

If the one-carbon units are carried as folates it is not known how carbon dioxide or carbon monoxide could enter the metabolic scheme. Extracts of *"Mtb. omelianskii"* possess a ferredoxin-linked formate dehydrogenase which catalyses the interconversion of formate and carbon dioxide (Brill *et al.*, 1964). However, the same extracts do not convert formate to methane, nor are there reports of the presence in these extracts of an enzyme catalysing reaction (31). The requirement for only catalytic quantities of ATP also raises further problems (see p. 167). It is clear that much further work is needed before the role of folates in methanogenesis is understood completely.

b. *Reduction of Methyl Groups to Methane.* The discovery that the methyl group of methylcobalamin could be converted into methane by extracts of *Mts. barkerii* (Blaylock and Stadtman, 1963) and *"Mtb. omelianskii"* (Wolin *et al.*, 1963b), coupled with the knowledge that methanogenic bacteria are rich sources of corrinoids (see Stadtman, 1967) focused attention on methyl cobamides as possible substrates for the final reduction step to methane. Furthermore, transfer of the methyl group from 5-methyltetrahydrofolate to a cobamide which occurs during cobalamin-dependent methylation of homocysteine to methionine in *Escherichia coli* (see Blakley, 1969) could provide a model for linking the reduction of folate-bound one-carbon units to the methyl level to a terminal reduction step from methyl cobamide to methane.

Inhibitor studies indicate that corrinoids are involved in methane formation from one-carbon compounds in extracts of *Mts. barkerii* and *"Mtb. omelianskii"*. "Intrinsic factor" (a glycoprotein which binds vitamin B_{12}-like compounds) was found to be an effective inhibitor of methane formation in extracts of *Mts. barkerii* from all methanogenic one-carbon substrates (Blaylock and Stadtman, 1966). Treatment of extracts of *"Mtb. omelianskii"* with propyl iodide in the dark under reducing conditions resulted in total loss of activity in catalysing methane formation from methyl cobalamin, 5-methyltetrahydrofolate or pyruvate (Wood and Wolfe, 1966a). Subsequent exposure to light resulted in partial recovery (50–60%) of activity. This behaviour is similar to that previously observed by Brot and Weissbach (1965) in the methyl transfer system involved in methionine biosynthesis in *E. coli* and strongly suggests the involvement of a cobamide derivative. It is a general property of the cobalt ion bound by a corrin ligand that when reduced to the univalent state (Co^+) it may be alkylated with an alkyl halide to a stable alkyl derivative. The Co-alkyl bond is cleaved on exposure to light, regenerating an alkane and the non-alkylated Co-corrin grouping. For an account of the chemistry of organocobalt compounds, see Schrauzer (1968).

This technique of reversible alkylation of Co-corrinoids was success-

METABOLISM OF ONE-CARBON COMPOUNDS BY MICRO-ORGANISMS 165

fully used by Wolfe and his coworkers in fractionation of extracts of "*Mtb. omelianskii*" (Wood and Wolfe, 1966b). Such extracts were reductively alkylated with [^{14}C]propyl iodide and then fractionated on columns of TEAE-cellulose, hydroxylapatite and Sephadex G-200. The preliminary alkylation served not only to stabilize any corrinoids but also to mark them with tracer for following through the fractionation procedure. After purification in this way, two radioactive fractions were obtained, photolysed to remove the [^{14}C]propyl group, and then tested for ability to stimulate methane formation in the following assay system. It was already known that extracts of "*Mtb. omelianskii*" could catalyse methane formation from methylcobalamin in the presence of ATP and a source of reducing power. Assuming that the methyl group of methylcobalamin is transferred by the extract to the actual corrinoid carrier involved in the terminal reduction step, an assay could be set up in which the concentration of the physiological carrier imposed a rate-determining step in the overall rate of methane production from methylcobalamin in the presence of crude extract:

Methyl cobalamin + terminal carrier $\longrightarrow$

cobalamin + methylated terminal carrier (34)

Methylated terminal carrier $\xrightarrow{\text{ATP} + \text{H}_2}$ terminal carrier + methane (35)

The amount of crude extract used had to be chosen such that it could, by itself, catalyse only a low rate of methane formation; stimulation by addition of the putative terminal carrier could then be followed. One of the two radioactive fractions obtained from the column chromatography stimulated methane formation in this assay system. Its catalytic specific activity was 100-fold greater than that in crude extracts. This enzyme fraction, obtained in 5% overall yield, contained a prosthetic group identified as a derivative of 5-hydroxybenzimidazolylcobamide. The activity of this cobamide enzyme was dependent on the presence of ATP, reducing power, and small amounts of crude extract showing that further component(s) present in the crude extract were necessary for the overall reaction from methyl cobalamin to methane. It is important to note that 5-methyltetrahydrofolate was as effective as methylcobalamin in this system.

In further studies on the detailed mechanism of the final reduction step to methane, Blaylock (1968) used a quite different approach from that used by Wood and Wolfe (1966b). It was known that extracts of *Mts. barkerii* could catalyse an ATP-dependent generation of methylcobalamin from methanol and substrate amounts of cob(I)alamin (Vitamin B$_{12s}$) (Blaylock and Stadtman, 1966). This was termed "methyl transferase activity". Since the extracts could also catalyse an

ATP-dependent reduction of methanol to methane it was possible that methanol was converted to a methylated intermediate common to both transmethylation and methanogenesis. Thus, identification of any enzymes or cofactors involved in the transmethylation of cob(I)alamin by methanol might lead also towards the terminal substrate for reduction to methane. Extracts of *Mts. barkerii* were then fractionated with respect to the methyl transferase activity, the product (methylcobalamin) being assayed directly. Chromatography on DEAE-cellulose split the extract into four principal fractions, called A, R, F and "acid-stable factor".

Fraction A was further split into two components A_{1-2} and A_{3-4}. It was concluded that A_{1-2} and A_{3-4} both contained the same enzyme protein but that A_{1-2} contained in addition a ferredoxin.

Fraction R was red in colour and was further purified by chromatography on Bio-gel P-100 and TEAE-cellulose. Its absorption spectrum indicated that it contained a cobamide derivative as prosthetic group; the molecular weight of the cobamide enzyme was estimated to be 100,000–200,000.

Fraction F contained ferredoxin but its relationship to fraction A_{1-2} was not clear. The "acid-stable factor" was stable in both 2 N-acid and 2 N-base at 100° and was anionic in character; it was not identified.

Reconstitution of methyltransferase activity required A_{1-2} (or A_{3-4} + ferredoxin) + R + "acid-stable factor" in the presence of hydrogen and ATP. Low rates of methane synthesis from methanol in crude extracts of *Mts. barkerii* could be stimulated by each of the above factors, and hence it seems likely that the methyl transferase system is indeed a portion of the overall system which catalyses the reduction of methanol to methane.

All the evidence points towards a methyl-cobamide being very close to, or identical with, the terminal substrate for the last reduction step to methane. The uncertainty resides in the continued requirement for crude extract in the reduction of the methylated intermediates isolated in the "*Mtb. omelianskii*" system and the *Mts. barkerii* system. It is not known whether the crude extract is supplying yet further carriers for the methyl groups or some other cofactor or enzyme.

c. *ATP Requirement During Methanogenesis.* Previous work by Wood and Wolfe (1966c) with extracts of "*Mtb. omelianskii*" indicated that one mole of ATP was required for generation of one mole of methane from methylcobalamin. A re-examination of the ATP requirement, using extracts of *Methanobacterium* strain M.o.H., showed that only catalytic quantities of ATP were required for methane formation from either methylcobalamin or carbon dioxide (Roberton and Wolfe, 1969). The apparent 1:1 stoicheiometry obtained in the previous measurements was

METABOLISM OF ONE-CARBON COMPOUNDS BY MICRO-ORGANISMS   167

probably fortuitous due to the fact that "*Mtb. omelianskii*" was a mixed culture.

Roberton and Wolfe (1969) were able to show that the lack of stoicheio-metry with respect to ATP was not due to partial ATP resynthesis during methanogenesis; the rate of ATP synthesis was too slow. Further-more, depletion of the ATP by a hexokinase trap did not stop methane synthesis. The role of the ATP remains unknown, but the fact that it is only needed in catalytic quantities for methanogenesis from carbon dioxide and hydrogen would appear to eliminate routes involving stoicheiometric amounts of ATP as a cofactor for the initial fixation of the carbon dioxide, e.g. reduction to formate followed by formation of 10-methylenetetrahydrofolate by reaction (31).

2. *Energy Generation*

Stadtman (1967) has pointed out that in the overall reduction of carbon dioxide to methane (reaction 29) the principal site of the release of free energy probably lies in the terminal reduction step from the level of methanol. This can clearly be seen by splitting reaction (29) into two parts:

$$3H_2 + CO_2 \rightarrow CH_3OH + H_2O \qquad \Delta G° = -6·26 \text{ kcal} \qquad (36)$$

$$H_2 + CH_3OH \rightarrow CH_4 + H_2O \qquad \Delta G° = -26·95 \text{ kcal} \qquad (37)$$

Fermentation of methanol (reaction 27) may be considered in terms of the reducing power from one molecule of methanol being used to reduce three molecules of methanol to methane:

$$CH_3OH + H_2O \rightarrow CO_2 + 3H_2 \qquad \Delta G° = +6·26 \text{ kcal} \qquad (38)$$

$$3H_2 + 3CH_3OH \rightarrow 3CH_4 + 3H_2O \qquad \Delta G° = -80·85 \text{ kcal} \qquad (39)$$

It is most unlikely that steps (36) or (38) could yield more than one mole of ATP. Hence, if growth yield measurements showed unequivo-cally that more than one mole of ATP were being made per four moles of methanol or per mole of carbon dioxide fermented, then it would indicate that some energy coupling must be taking place in the terminal reduction step from the level of methanol to methane (reactions (37) and (39)). Stadtman (1967) quotes figures for growth of *Mts. barkerii* on methanol giving a cell yield of 3·3 g. dry weight/mole of methanol fermented. Roberton and Wolfe (1970) quote figures for growth of *Methano-bacterium* strain M.o.H. on hydrogen plus carbon dioxide giving a cell yield of 2·32 g. dry weight/mole of carbon dioxide fermented. Assuming a yield of 10 g. dry weight for each mole of ATP produced (Stouthamer, 1969), ATP yields of 1·2 per 4 moles of methanol and 0·23 per mole of carbon dioxide fermented may be calculated for *Mts. barkerii* and *Methanobacterium* strain M.o.H., respectively. In the case of *Methano-bacterium* strain M.o.H. the growth yield is obviously too small to have

168   J. R. QUAYLE

to invoke energy coupling in reaction (37). The figures for *Mts. barkerii* are so close to the critical value of one that more accurate measurements will be needed before a deduction can safely be made. At present, therefore, there is little evidence in favour of the terminal reduction step being coupled to ATP synthesis, despite the attractive thermodynamic possibilities.

It is unfortunate that knowledge of the energy coupling mechanism at the biochemical level lags far behind pencil and paper thermodynamics. Roberton and Wolfe (1970) have shown that substances which uncouple oxidative phosphorylation in mitochondria, e.g. 2,4-dinitrophenol, carbonylcyanide-*m*-chlorophenylhydrazone and pentachlorophenol inhibited, methane production from $H_2 + CO_2$ by whole cells of *Methanobacterium* strain M.o.H. and caused the intracellular ATP and AMP pools to decrease and increase respectively. Thus, in contrast to oxidative phosphorylation, these inhibitors affected electron transport as well as phosphorylation. ATP synthesis could not be detected in cell-free extracts, but the situation is complicated by the simultaneous catalytic *requirement* for ATP during methanogenesis in such extracts.

Little is known of the mechanism of generation of utilizable reducing power during fermentation of methanol by *Mts. barkerii* (reaction (27)). It would be interesting to know whether the methanol is oxidized to carbon dioxide by way of free intermediates or by way of one-carbon folate intermediates. Wood *et al.* (1965) showed that extracts of "*Mtb. omelianskii*" oxidized the methylene group of 5,10-methylenetetrahydrofolate to carbon dioxide and suggested that the oxidation might take place by way of 10-formyltetrahydrofolate and formate. It may be recalled that extracts of this mixed culture contained a ferredoxin-linked formate dehydrogenase (Brill *et al.*, 1964).

C. PHOTOMETABOLISM

Consideration of the energy metabolism of photosynthetic bacteria lies outside the scope of this review. The role of the organic substrate with respect to energy transduction during photosynthetic growth on one-carbon compounds is probably limited to provision of electrons for photoactivation. In this regard, mention may be made of two cases where enzyme systems catalysing the dehydrogenation of a one-carbon substrate have been studied.

Qadri and Hoare (1968) showed that a strain of *Rhodopseudomonas palustris* which was able to grow anaerobically on formate in the light contained a soluble formate dehydrogenase and a particulate hydrogenase. The formate dehydrogenase could be linked to phenazine methosulphate, dichlorophenolindophenol, methylene blue, ferricyanide,

METABOLISM OF ONE-CARBON COMPOUNDS BY MICRO-ORGANISMS   169

methyl- or benzyl viologen, triphenyltetrazolium and FAD. Activity was not observed with NAD or NADP. The pH optimum, using ferricyanide as electron acceptor, was 6·3.

Yoch and Lindstrom (1969) studied the formate dehydrogenase(s) from *Rh. palustris* (ATCC 11168) grown photosynthetically in a medium containing formate. They showed the enzyme(s) to be soluble and could be linked to NAD (pH optimum, 8·0) or cytochrome *c* or dichlorophenolindophenol (pH optimum, 6·8). In both cases, FAD stimulated the rate of dehydrogenation. It is not known whether two different dehydrogenases are present in the crude extracts, one NAD-linked and the other cytochrome *c* or dichlorophenolindophenol-linked, or whether the pH value required for a single enzyme to bind NAD is different from that for cytochrome *c* and dichlorophenolindophenol.

IV. Carbon Assimilation

The unique biosynthetic capabilities of a one-carbon-utilizing organism can be narrowed down to its ability to synthesize, in net fashion a C_3 skeleton from one-carbon compounds. There is no reason to expect in these organisms a fundamental difference in the way a C_3 compound such as pyruvate, phosphopyruvate or triose phosphate thereafter serves as precursor of other cell constituents.

The first sequence of reactions to be discovered which effects the net synthesis of a C_3 compound from a one-carbon compound was the ribulose diphosphate cycle of carbon dioxide fixation (Bassham *et al.*, 1954). This has been found to be a ubiquitous cycle amongst autotrophic organisms but apparently rarely operates in organisms growing on reduced one-carbon compounds. The relative wastage of energy consequent on oxidation of reduced carbon to carbon dioxide followed by its re-reduction to the level of $[CH_2O]$, as compared to direct assimilation as a reduced one-carbon unit, doubtless explains the rarity of this process amongst heterotrophic organisms. Instead, other biosynthetic pathways exist in which the reduction level of the entering one-carbon unit is conserved.

The following sections will outline those mechanisms which offer a solution to the problem of net biosynthesis from one-carbon units at reduction levels ranging from carbon dioxide to methane.

A. ASSIMILATION OF CARBON DIOXIDE

1. *Ribulose Diphosphate Cycle*

A detailed description of this well-known cycle will not be given in this review because it has been adequately covered elsewhere as part of the

170   J. R. QUAYLE

general fields of photosynthesis and autotrophy. Appropriate references
have been given in the Introduction. Attention will here be drawn to the
few instances where this cycle has been implicated in growth on reduced
one-carbon compounds, and to one instance where it probably will be
implicated when the necessary tests are done. The temptation will be
resisted to discuss whether such organisms should be called heterotrophs
or autotrophs, even though this question affords all the grave excitement
of a mediaeval theological disputation.

a. *Growth of* Pseudomonas oxalaticus *on Formate*. Evidence for the
operation of the ribulose diphosphate cycle during growth of *Pseudomonas oxalaticus* on formate has been given previously (Quayle and
Keech, 1959a, b; Quayle, 1961; Blackmore *et al.*, 1968) and will not be
discussed further here.

b. *Growth of Photosynthetic Bacteria on Formate*. Stokes and Hoare
(1969) studied the carbon assimilation of a strain of *Rh. palustris* when
grown anaerobically in the light on formate. They showed that during
photoassimilation of the formate virtually all of the formate carbon
passed through the stage of carbon dioxide before being incorporated
into cell material. Pulse-labelling experiments with [^{14}C]formate and
$^{14}CO_2$ showed a pattern of uptake of tracer that was characteristic of
the ribulose diphosphate cycle, *viz.* predominant labelling of phosphorylated compounds at the earliest times. Cell-free extracts of the organism
grown photosynthetically on formate contained ribulose diphosphate
carboxylase, phosphoribulokinase and phosphoriboisomerase.

It will be seen later (see p. 175) that Yoch and Lindstrom (1967)
obtained very different results with *Rh. palustris* strain ATCC 11168
grown photosynthetically on formate. This organism appeared to
incorporate carbon dioxide predominantly into glutamate and other
amino acids rather than phosphorylated compounds. The reason for the
different results may lie in strain differences and calls for a direct
comparison of the two organisms.

It will be interesting to see what type of metabolism occurs in species of
Chloropseudomonas isolated by Kondrat'eva and Trotsenko (1969). The
authors state that formate is photometabolized to carbon dioxide but it
is not known to what relative extent formate and carbon dioxide are
assimilated.

c. *Growth of* Bacterium formoxidans *on Formate*. Sorokin (1961) has
shown by isotopic methods that, when this organism grows aerobically
on formate, 95–97% of the cell carbon is synthesized from carbon dioxide.
This may be compared with figures of 4–11% when the organism is
grown on glucose, lactate or acetate. No enzymic data are given but it
seems likely that the ribulose diphosphate cycle operates during growth
on formate in a fashion similar to that in *Ps. oxalaticus*.

2. Acetyl-CoA Cycle (Reductive Carboxylic Acid Cycle)

The discovery in photosynthetic bacteria and other anaerobic bacteria of two new carboxylation reactions, pyruvate synthase (40) (Bachofen et al., 1964; Buchanan et al., 1964; Evans and Buchanan, 1965) and α-oxoglutarate synthase (41) Buchanan and Evans, 1965):

Acetyl-CoA + CO_2 + reduced ferredoxin → pyruvate + CoA + oxidized ferredoxin (40)

Succinyl-CoA + CO_2 + reduced ferredoxin →
$$\alpha\text{-oxoglutarate} + CoA + \text{oxidized ferredoxin} \qquad (41)$$

led Evans et al. (1966) to propose the operation of a new cycle of net carbon dioxide fixation, which they called the reductive carboxylic acid cycle, during growth of Chlorobium thiosulphatophilum anaerobically in the light on carbon dioxide as sole carbon source. The cycle (Fig. 9), which involves the reversal of several reactions of the tricarboxylic acid cycle, results in a net synthesis of one molecule of oxaloacetate from four molecules of carbon dioxide and a regeneration of the first carbon dioxide

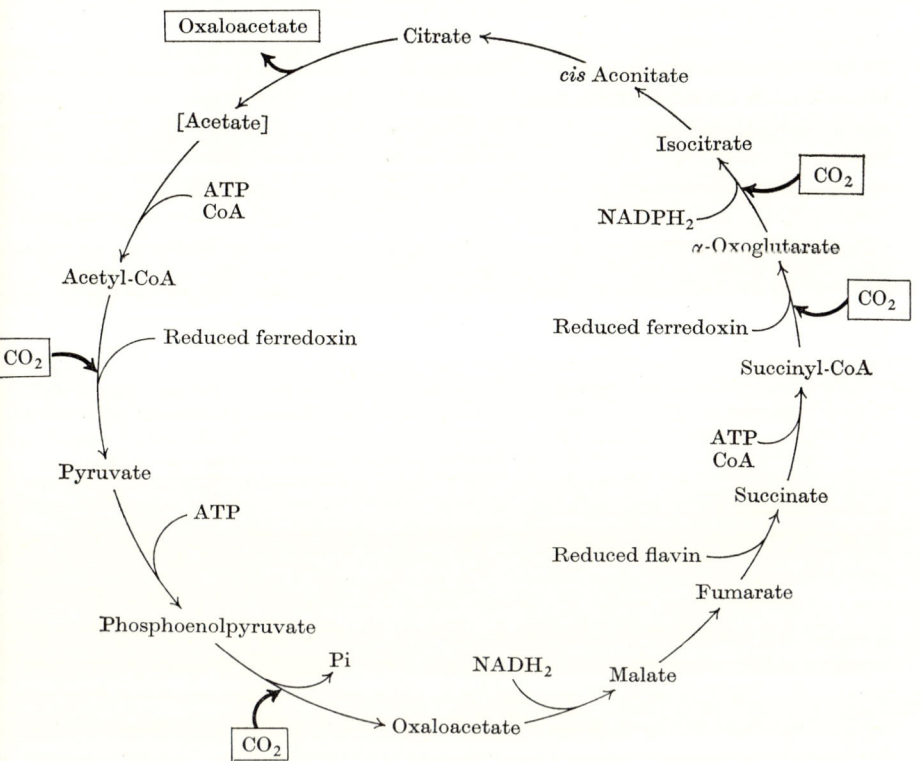

FIG. 9. Acetyl-CoA cycle of carbon dioxide fixation. Based on Evans et al. (1966).

172 J. R. QUAYLE

acceptor, acetyl-CoA. It is thus more accurately described as the acetyl-CoA cycle of carbon dioxide fixation.

Evidence for operation of the cycle is based on the rapid incorporation of $^{14}CO_2$ into amino acids, particularly glutamate, and the presence of the necessary enzymes in cell-free extracts. The specific activities of the

TABLE 5. Enzymes Implicated in the Acetyl-CoA Cycle (Reductive Carboxylic Acid Cycle)

Enzyme	Chlorobium thiosulphatophilum (data from Evans et al., 1966)		Rhodospirillum rubrum (data from Buchanan et al., 1967)	
	Enzyme activity[a]	Rate of carbon dioxide fixation[b]	Enzyme activity[a]	Rate of carbon dioxide fixation[b]
Acetyl-CoA synthetase	0·8	1·8	24	2·4
Pyruvate synthetase	0·2		0·06	
Phosphoenolpyruvate synthase	2·3		0·7	
Phosphoenolpyruvate carboxylase	4·8		6·0	
Malate dehydrogenase	37		159	
Fumarate hydratase	118		128	
Succinate dehydrogenase	0·85		1·2	
Succinyl-CoA synthetase	1·6		3·3	
α-Oxoglutarate synthase	0·4		0·012	
Isocitrate dehydrogenase	102		70	
Aconitate hydratase	3·1		7·3	
Citrate lyase	0·15		0 (0·17)[c]	

The organisms were grown anaerobically in the light on sodium thiosulphate plus carbon dioxide (*Chl. thiosulphatophilum*) or succinate (9 mM) plus hydrogen plus carbon dioxide (*Rh. rubrum*) except where indicated otherwise.

[a] Specific activities of enzymes expressed as μmoles/hr./mg. protein.

[b] Rates of carbon dioxide fixation by whole cells expressed as μmoles/hr./mg. of total soluble protein released by sonication.

[c] Grown anaerobically in the light on hydrogen + carbon dioxide.

enzymes concerned may be compared with the rate of carbon-dioxide fixation by whole cells (Table 5). It will be seen that the activities of several of the enzymes are lower than that required to match the performance of whole cells. However, the authors did not search for optimum assay conditions, and pointed out that there are deficiencies in the activities of some enzymes of the ribulose diphosphate cycle in organisms which undoubtedly use that cycle for growth (Peterkofsky and Racker,

METABOLISM OF ONE-CARBON COMPOUNDS BY MICRO-ORGANISMS 173

1961; Latzko and Gibbs, 1969). Nevertheless it would still seem important to examine closely the activities of deficient enzymes, especially those which are of key importance and/or which catalyse reactions in a reverse direction to that normally encountered, e.g. pyruvate synthase (40), α-oxoglutarate synthase (41), fumarate reductase and citrate lyase:

$$\text{Citrate} \rightarrow \text{oxaloacetate} + \text{acetate} \qquad (42)$$

There is not sufficient detail in the paper to enable a proper assessment to be made of the assay procedures that were used. This applies particularly to citrate lyase, which is vital to the whole cycle.

Chlorobium thiosulphatophilum contains enzymes of the ribulose diphosphate cycle (Smillie *et al.*, 1962) and therefore has the potentiality of assimilating carbon dioxide *via* ribulose diphosphate carboxylase. At the earliest time of sampling, 75% of the total $^{14}CO_2$ fixed was in glutamate and 10% in phosphate esters. In these circumstances it must be asked whether the organism does actually use the acetyl-CoA cycle *as a full cycle* of net carbon dioxide fixation, either singly or concurrently with the ribulose diphosphate cycle. It could be argued that possession of enzymes (40) and (41) in a cell growing under reducing conditions in the presence of carbon dioxide would inevitably result in rapid labelling of alanine and glutamate without this necessarily meaning that the acetyl-CoA cycle as a whole operates. However, experiments involving the use of fluoroacetate as inhibitor have been interpreted in favour of operation of the complete cycle in *Chl. thiosulphatophilum* (Sirevåg and Ormerod, 1970). These authors have shown that anaerobic fixation of carbon dioxide in the light in the presence of thiosulphate is inhibited by 50% on addition of 1 mM-fluoroacetate. This inhibition is accompanied by a 50% reduction in the polysaccharide content of the cells and a doubling in the amount of α-oxoglutarate which is excreted by the organism into the supernatant: the incubation time of the experiment was 90 minutes. If accumulation of α-oxoglutarate is taken as a reflection of initial accumulation of isocitrate, these results do indicate operation of the complete acetyl-CoA cycle, provided that the fluoroacetate specifically inhibited aconitate hydratase and nothing else. Confirmation of this point by testing the action of fluoroacetate on other autotrophic organisms would be very valuable.

Enzymes implicated in the acetyl-CoA cycle have been found in extracts of *Rhodospirillum rubrum* grown anaerobically in the light in a medium containing carbon dioxide plus 9 mM-succinate as carbon source, using hydrogen as reductant (Buchanan *et al.*, 1967). This has been taken as evidence that the acetyl-CoA cycle may also function in this organism, although why it should need to function in a cell growing photoheterotrophically on succinate, other than perhaps as an electron sink, is not

174   J. R. QUAYLE

clear. The specific activities of the enzymes measured in cell-free extracts
may be compared with the rate of carbon-dioxide fixation measured in
whole cells (Table 5). Citrate lyase activity was not detected in cells grown
on succinate plus carbon dioxide, but activity was detected in photoautotrophically-grown cells (i.e. grown on hydrogen plus carbon dioxide
anaerobically in the light). As with *Chl. thiosulphatophilum*, conditions
for optimum assay of the various enzymes were not sought and there is
insufficient detail to allow assessment of some of the procedures, particularly with respect to citrate lyase. *Rhodospirillum rubrum* can synthesize
enzymes of the ribulose diphosphate cycle (Fuller and Gibbs, 1959) and
this raises the same question as was raised in the case of *Chl. thiosulphatophilum*, namely does the scheme of reactions shown in Fig. 9 (p. 117)
actually function *in vivo* as a cycle of net carbon dioxide fixation? The
pattern of uptake of $^{14}CO_2$ in the autotrophically-grown organism under
autotrophic conditions is needed to answer this question but the significance of some of the experiments which have been performed is obscured
by having used photoheterotrophically-grown cells for studying $^{14}CO_2$
uptake in the presence of hydrogen. Thus, Glover *et al.* (1952) used cells
grown on malate or acetate; suspensions of such cells incubated with
$^{14}CO_2$ under hydrogen incorporated the tracer predominantly into
phosphoglycerate. Hoare (1963) used cells grown on malate. When these
were incubated with $^{14}CO_2$ and hydrogen the isotope appeared primarily
in phosphate esters with somewhat lower incorporation into malate,
glutamate and aspartate. In the one case where the autotrophically
grown organism has been used to study $^{14}CO_2$ incorporation under
hydrogen the pattern of isotope incorporation corresponds unequivocally
with a major operation of the ribulose diphosphate cycle; after one
second incubation 75% of the fixed radioactivity was in phosphoglycerate, this proportion dropping rapidly with increasing time of
incubation (Anderson and Fuller, 1967a). In cells grown on malate,
isotope experiments indicated that carboxylation reactions played a
minor role as compared with direct utilization of the malate carbon
(Anderson and Fuller, 1967b). These whole-cell data are in full accord
with the levels of ribulose diphosphate carboxylase and phosphoribulokinase in *Rh. rubrum* under the different growth conditions (Anderson
and Fuller, 1967c). The results clearly demonstrate the profound changes
in metabolism which may result from the presence of organic substrates
in the growth medium.

    The combined evidence certainly does not support the operation of the
acetyl-CoA cycle as a full cycle in *Rh. rubrum*, despite statements to the
contrary (Evans *et al.*, 1966; Buchanan *et al.*, 1967; Evans and Whatley,
1970). It can also be misleading to quote the experiments of Fuller *et al.*
(1961) with *Chromatium* as showing another example of amino acids

METABOLISM OF ONE-CARBON COMPOUNDS BY MICRO-ORGANISMS 175

being the main early products of $^{14}CO_2$ fixation (Evans et al., 1966; Buchanan et al., 1967). Under photoautotrophic conditions, using the photoautotrophically-grown organism, $^{14}CO_2$ was incorporated most rapidly into aspartic acid and phosphorylated compounds. Fuller et al. concluded that the ribulose diphosphate cycle operated under these conditions, the aspartic acid arising from carboxylation of pyruvate or phosphopyruvate.

The acetyl-CoA cycle has been implicated by Yoch and Lindstrom (1967) in the growth of Rh. palustris (strain ATCC 11168) on formate anaerobically in the light. They showed that, when the organism was incubated with [^{14}C]formate for 30 seconds, 90% of the radioactivity fixed into the ethanol-soluble fraction of the cells was present in glutamate. The C-1 atom of glutamate contained 70% of the radioactivity. Over a period of incubation with [^{14}C]formate of 0·25–8 minutes, the percentage of fixed radioactivity which was present in glutamate dropped and at no time did phosphate esters contain more than 7% of the fixed radioactivity. Similar results were obtained when the organism was incubated with $^{14}CO_2$ in the presence of thiosulphate as electron donor. The results contrasted with those obtained with malate-grown cells which incorporated $^{14}CO_2$ predominantly into sugar phosphates. Yoch and Lindstrom interpreted these results as meaning that carbon was assimilated by the acetyl-CoA cycle of carbon dioxide fixation during photosynthetic growth either on formate or on carbon dioxide plus thiosulphate. During photosynthetic growth on malate, excess reducing power was disposed of by carbon dioxide fixation via the ribulose diphosphate cycle. These interesting results clearly need further work at the enzyme level. By contrast, it may be recalled that Stokes and Hoare (1969) concluded that a strain of Rh. palustris isolated by Qadri and Hoare (1968) incorporated carbon via the ribulose diphosphate cycle during photosynthetic growth on formate.

3. Net Fixation of Carbon Dioxide by Non-Photosynthetic Anaerobes

Nothing is known of the pathway of net carbon assimilation which operates during growth of methanogenic bacteria on hydrogen plus carbon dioxide, nor is it known what type of metabolism was involved during growth of Clostridium aceticum on hydrogen plus carbon dioxide in the presence of mud extract (Wieringa, 1940). However, there have been important developments in the study of the homoacetate fermentation carried out by Clostridium thermoaceticum which may be very relevant to the problem of growth on one-carbon compounds and carbon dioxide as sole carbon source. A full review of these extensive investigations has been written by Ljungdahl and Wood (1969).

176   J. R. QUAYLE

Clostridium thermo-aceticum is one of several species of strictly anaerobic bacteria which carry out a homoacetate fermentation of glucose (Ljungdahl and Wood, 1969; Andreesen *et al.*, 1970). The fermentation may be represented by:

$$C_6H_{12}O_6 + 2H_2O \rightarrow 2CH_3.CO_2H + 8H + 2CO_2 \tag{43}$$

$$2CO_2 + 8H \rightarrow CH_3.CO_2H + 2H_2O \tag{44}$$

$$\text{Sum: } C_6H_{12}O_6 \rightarrow 3CH_3.CO_2H \tag{45}$$

Isotopic and mass spectrometric experiments by Wood (1952), using *Cl. thermo-aceticum*, showed conclusively that one out of three molecules of acetate is formed from two molecules of carbon dioxide, as represented by reaction (44). In an attempt to trace the path of carbon from carbon dioxide to acetate, Ljungdahl and Wood (1965) carried out a series of five-second incubations of suspensions of *Cl. thermo-aceticum* with $^{14}CO_2$, using approximately 100 g. (wet weight) of organism each time. The cells were killed with perchloric acid and the radioactive cell constituents were analysed and, in some cases, degraded. Sugar phosphates, carboxylic acids and amino acids were labelled but their pattern of labelling gave no clue as to the mechanism of acetate biosynthesis from carbon dioxide. However, a clue emerged from the finding that synthesis of acetate from carbon dioxide in cell-free extracts of *Cl. thermo-aceticum* was inhibited by "intrinsic factor" and that the methyl group of methylcobalamin was converted to the methyl group of acetate (Poston *et al.*, 1964). Forewarned that the one-carbon units involved in the transformation might be corrin-bound, Ljungdahl *et al.* (1965) performed a 15-second incubation of 350–400 g. (wet weight) of *Cl. thermo-aceticum* with 2 mCi of $^{14}CO_2$. The incubation was stopped with acetone and the cells analysed for corrinoid derivatives using appropriate special techniques. Two radioactive methyl-corrinoids were isolated, namely [^{14}C]methylcobyric acid and [^{14}C]methyl-α-(5-methoxybenzimidazolyl)cobamide. Their isotopic specific activities were higher than that of acetate showing that they could be precursors of the acetate. Further work at the enzyme level (see Ljungdahl and Wood, 1969; Sun *et al.*, 1969) has shown that reaction (44) probably proceeds as follows:

$$CO_2 + NADPH_2 \rightarrow H.COOH + NADP \tag{46}$$

$$HCO_2H + \text{tetrahydrofolate} + ATP \rightarrow \text{10-formyltetrahydrofolate} + ADP + Pi \tag{31}$$

$$\text{10-formyltetrahydrofolate} + 2NADPH_2 \rightarrow \rightarrow$$
$$\text{5-methyltetrahydrofolate} + 2NADP \tag{47}$$

$$\text{5-methyltetrahydrofolate} + [\text{corrinoid.E}] \rightarrow$$
$$\text{tetrahydrofolate} + [CH_3.\text{corrinoid.E}] \tag{48}$$

$$CO_2 + [CH_3.\text{corrinoid.E}] + ATP \rightarrow \begin{bmatrix} HO_2C.CH_2.\text{corrinoid.E} \\ \text{or} \\ CH_3.COO.\text{corrinoid.E} \end{bmatrix} + ADP + Pi \tag{49}$$

$$[\text{HOOC.CH}_2\text{.corrinoid.E}] + \text{NADPH}_2 \rightarrow \text{CH}_3\text{.COOH} + [\text{corrinoid.E}] + \text{NADP} \quad (50)$$

$$or \; [\text{CH}_3\text{.COO.corrinoid.E}] + \text{H}_2\text{O} \rightarrow \text{CH}_3\text{.COOH} + [\text{oxidized corrinoid.E}] \quad (51)$$

The corrinoid derivatives involved in reactions (48–51) remain enzyme-bound throughout. The mechanism of the final reaction is not yet fully understood.

Two important points emerge from this work with *Cl. thermo-aceticum*. Firstly, although the formation of acetate from carbon dioxide probably functions primarily as a sink for electrons in this particular case, it nevertheless provides a clear model system in which a C_2 skeleton is constructed in net fashion from two one-carbon units. If such a pathway were coupled to a reductive carboxylation of acetyl-CoA (reaction (40)), as has been observed by Linke (1969) in *Cl. formico-aceticum*, or to the reactions of the glyoxylate cycle, new mechanisms of net biosynthesis of cell constituents would be possible. The fact that one-carbon derivatives of folates and corrinoids are probably involved in the reduction of carbon dioxide to methane by methanogenic bacteria makes it an attractive possibility in such organisms. Secondly, the results demonstrate the difficulties that can arise when crucial biosynthetic reactions are carried out by way of enzyme-bound intermediates. In such cases, standard conditions for short-term isotope experiments in which 20–100 mg. (dry weight) of bacteria may be incubated with isotope (Quayle, 1971) would not reveal the intermediate steps. It was only by performing a special experiment on a very large scale and by using special techniques for isolation of corrinoid derivatives that Ljungdahl *et al.* (1965) were able to demonstrate the involvement of methylated corrinoids. In other words, the short-term experiment was of no use as a predictive tool, as in photosynthesis, but only of use in confirmation of a possibility suggested by other approaches. This may be of importance in other areas where key biosynthetic intermediates have not been observed in isotopic experiments. Ljungdahl and Wood (1969) have emphasized this point and indeed conjecture that, if NADPH_2 were replaced by H_2O in equation (50), glycollate might be formed as a product. If this were connected, for example, with the early appearance of labelled glycollate from labelled carbon dioxide during photosynthesis, it would require special techniques for any enzyme-bound intermediates to be detected.

B. ASSIMILATION OF REDUCED ONE-CARBON COMPOUNDS

1. *Serine Pathway*

There is now a considerable amount of evidence which indicates that a pathway of net carbon assimilation involving hydroxymethylation of glycine to serine operates in bacteria during growth on a variety of one-carbon compounds varying in reduction level between methane and

formate. Since some of the bacteria involved belong to the widespread group of one-carbon utilizers typified by *Vibrio extorquens*, this pathway, termed the "serine pathway", may turn out to be as widely distributed amongst this type of organism as is the ribulose diphosphate cycle amongst autotrophs.

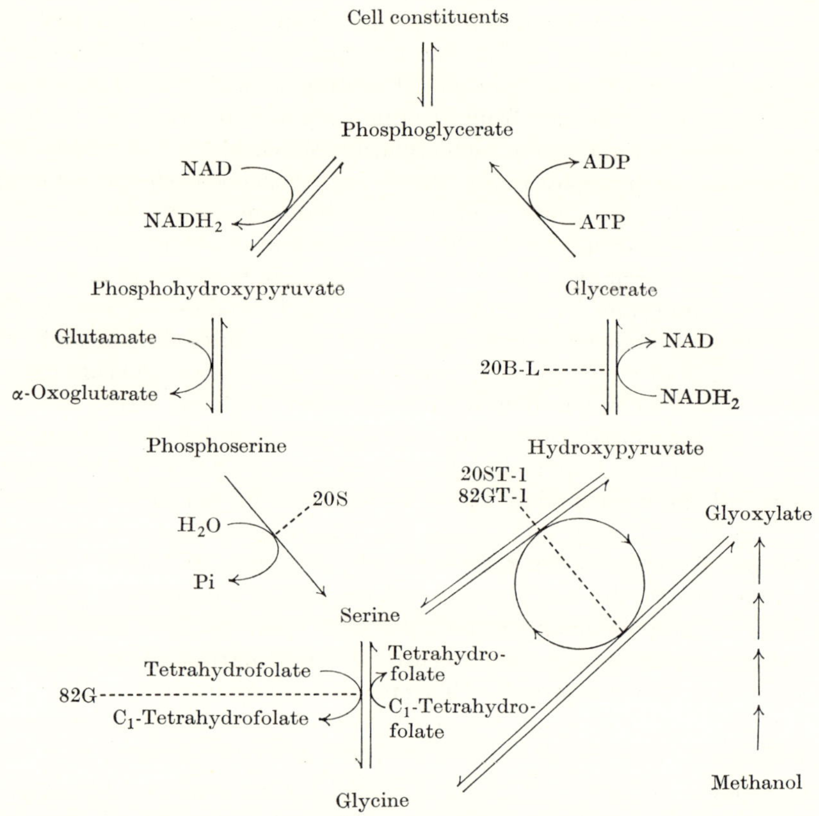

Cell constituents

Phosphoglycerate

NAD

NADH$_2$

ADP

ATP

Phosphohydroxypyruvate

Glycerate

Glutamate

α-Oxoglutarate

20B-L

NAD

NADH$_2$

Phosphoserine

Hydroxypyruvate

20ST-1
82GT-1

Glyoxylate

20S

H$_2$O

Pi

Serine

Tetrahydrofolate

Tetrahydro-folate

82G

C$_1$-Tetrahydrofolate

C$_1$-Tetrahydro-folate

Methanol

Glycine

FIG. 10. Metabolic interrelationship of reactions of the serine pathway. The dotted lines identify the lesions present in the designated mutants.

The serine pathway accomplishes the conversion of a one-carbon unit and a molecule of glycine to phosphoglycerate by the reactions shown on the right hand side of the hexagonal scheme in Fig. 10. The left hand side depicts a scheme of reactions used by the same bacteria to convert phosphoglycerate to glycine plus a one-carbon unit during growth on compounds such as lactate or succinate.

The evidence for these pathways will now be summarized in sections which relate to the different types of experimental approach which have

METABOLISM OF ONE-CARBON COMPOUNDS BY MICRO-ORGANISMS   179

been used. The question of the net synthesis of the glycine skeleton from one-carbon units will be discussed in the last section.

1. *Isotope Studies with Intact Cells.* Five species of bacteria growing on reduced one-carbon compounds have shown a closely similar pattern of ^{14}C uptake under the conditions of short-term isotope incubation; these are: methanol-grown *Pseudomonas* PRL-W4 (Kaneda and Roxburgh, 1959b); methanol- and formate-grown *Pseudomonas* AM1 (Large *et al.*, 1961); methanol-grown *Hyphomicrobium vulgare* (Large *et al.*, 1961); methylamine-grown *Diplococcus* PAR (Leadbetter and Gottlieb, 1967); methane-grown *Mtn. methano-oxidans* (Lawrence *et al.*, 1970). In all cases except the last, incorporation of both labelled substrate and labelled carbon dioxide was followed; in the last case, only the incorporation of [^{14}C]methanol was followed. With all tracers and organisms the patterns obtained were complex but the following features stand out: (a) radioactivity mainly appeared at early times in malate/aspartate and other amino acids; labelled phosphates appeared later; (b) the radioactivity incorporated from [^{14}C]methanol or [^{14}C]formate into amino acids other than aspartate was mainly in serine and to a lesser extent in glycine; (c) glycine contained more radioactivity at early times than serine when [^{14}C]bicarbonate was used as tracer; (d) radioactivity rapidly appeared in glycollate from [^{14}C]methanol in the case of *Hyphomicrobium vulgare* and in unknown compound(s) tentatively assumed to be folates in *Mtn. methano-oxidans*; and (e) the specific radioactivity of cellular material obtained from *Pseudomonas* AM1 grown on [^{14}C]methanol in air was decreased by one-half on bubbling air-carbon dioxide (99:1, v/v) mixture through the growing culture. This indicates that at least 50% of the carbon incorporated must have passed through the stage of carbon dioxide or a compound in ready equilibrium with it.

Studies by Large *et al.* (1962a) of the distribution of isotope in metabolites isolated from methanol-grown *Pseudomonas* AM1 after incubation with [^{14}C]methanol or [^{14}C]bicarbonate indicated that:

(a) the carboxyl group of glycine is mainly derived from carbon dioxide, and the methylene carbon from methanol;

(b) the hydroxymethyl group of serine is derived from methanol, whereas the distribution of radioactivity between C-1 and C-2 of serine is the same under all conditions as that between C-1 and C-2 of glycine; and

(c) the labelling patterns of serine and malate are consistent with the formation of malate by carboxylation of a C_3 fragment derived from serine.

These data are consistent with the right-hand scheme of reactions in Fig. 10.

b. *Enzyme Studies with Cell-free Extracts.* Evidence has been obtained (Large and Quayle, 1963; Heptinstall and Quayle, 1970; Blackmore and Quayle, 1970; Harder and Quayle, 1971a) for the presence of serine hydroxymethylase (32) and enzymes catalysing the following reactions in cell-free extracts of *Pseudomonas* AM1:

$$\text{serine} + \text{glyoxylate} \; \rightleftharpoons \; \text{hydroxypyruvate} + \text{glycine} \tag{52}$$

$$\text{hydroxypyruvate} + \text{NADH}_2 \; \rightleftharpoons \; \text{glycerate} + \text{NAD} \tag{53}$$

$$\text{glycerate} + \text{ATP} \; \rightarrow \; \text{3-phosphoglycerate} + \text{ADP} \tag{54}$$

Each of these enzymes is present at a higher specific activity in methanol-*Pseudomonas* AM1 than in succinate-grown *Pseudomonas* AM1, consistent with their playing a special role during growth on a one-carbon substrate.

The following reactions have also been demonstrated using cell-free extracts of *Pseudomonas* AM1 (Heptinstall and Quayle, 1970):

$$\text{3-phosphoglycerate} + \text{NADH}_2 \; \rightarrow \; \text{3-phosphohydroxypyruvate} + \text{NAD} \tag{55}$$

$$\text{3-phosphohydroxypyruvate} + \text{glutamate} \; \rightleftharpoons \; \text{3-phosphoserine} + \alpha\text{-oxoglutarate} \tag{56}$$

$$\text{3-phospherine} + \text{H}_2\text{O} \; \rightarrow \; \text{serine} + \text{Pi} \tag{57}$$

The specific activities of each of the enzymes catalysing these reactions were similar in both the succinate- and methanol-grown organisms.

These results show that *Pseudomonas* AM1 possesses the enzymic capabilities of carrying out the transformations shown in Fig. 10. In addition, Large *et al.* (1962b) have purified a phosphoenolpyruvate carboxylase from *Pseudomonas* AM1 which catalyses the reaction:

$$\text{phosphoenolpyruvate} + \text{CO}_2 + \text{H}_2\text{O} \; \rightarrow \; \text{oxaloacetate} + \text{Pi} \tag{58}$$

An integral step in the serine pathway is the deamination of serine to hydroxypyruvate. Large and Quayle (1963) found that extracts of *Pseudomonas* AM1 catalysed a transamination between serine and α-oxoglutarate or pyruvate, but the specific activities were low. It now seems likely that the enzyme used *in vivo* is a serine-glyoxylate aminotransferase (52) (Blackmore and Quayle, 1970; Harder and Quayle, 1971a). This enzyme is present in methanol-grown cells at specific activities which are 7–10 fold higher than those in succinate-grown cells. As will be seen in the succeeding section, the behaviour of mutants of *Pseudomonas* AM1 indicates a key role for this enzyme during growth on one-carbon compounds.

Extracts of *Pseudomonas* AM1 contain tetrahydrofolate formylase (31) and methylenetetrahydrofolate dehydrogenase (Large and Quayle, 1963):

$$\text{5,10-methylenetetrahydrofolate} + \text{NADP} \; \rightleftharpoons \; \text{5,10-CH}{=}\text{H}_4\text{folate} + \text{NADPH}_2 \tag{59}$$

METABOLISM OF ONE-CARBON COMPOUNDS BY MICRO-ORGANISMS 181

The methylenetetrahydrofolate necessary for reaction (32) could be made from formate by the concerted action of enzyme (31), methenyltetrahydrofolate cyclohydrolase (see Blakley, 1969), and enzyme (59) working from right to left. Methylenetetrahydrofolate could also be made, in theory, from methanol by the action of enzyme (20) followed by non-enzymic (or perhaps enzymic) condensation of formaldehyde with tetrahydrofolate. It is interesting to note, however, that the specific activities of enzymes (59) and (31) are respectively nine-fold and two-fold higher in both methanol- and formate-grown cells as compared to succinate-grown cells (Large and Quayle, 1963). This could mean that the two enzymes are co-ordinately induced or derepressed along with other enzymes of one-carbon metabolism, but are not actually used to any greater extent in growth on methanol than in growth on succinate, or that the formaldehyde resulting from oxidation of methanol is oxidized to formate which is then converted to methylene tetrahydrofolate via reactions (31) and (59). In other words, the reduced one-carbon substrate would be oxidized through to formate in an oxidative assembly as in $Pseudomonas$ MS, the carbon in the intermediate oxidation levels not being available for assimilation.

The sequence of reactions (55–57), followed by reaction (32), constitutes the well-known phosphorylated pathway of serine/glycine biosynthesis found in $E. coli$ (Pizer, 1963, 1965; Pizer and Potochny, 1964; Umbarger et al., 1963), $Haemophilus influenzae$ (Pizer et al., 1969) and $Salmonella typhimurium$ (Umbarger and Umbarger, 1962; Umbarger et al., 1963).

c. $Effect$ of $Folate$ $Antagonists$ on $Growth$ of Pseudomonas $AM1$. The involvement of one-carbon-folate derivatives in the main carbon assimilatory pathway of an organism growing on a one-carbon compound might be expected to render the organism more sensitive to folate antagonists during growth on one-carbon substrates than on multicarbon compounds such as succinate. This prediction has been clearly verified by Hollinshead (1966) who showed that 1 mM-sulphanilamide caused almost complete inhibition of growth of $Pseudomonas$ AM1 on methanol but had little effect on growth on succinate. The inhibition could be reversed by p-aminobenzoic acid. Similar results were obtained with $Protaminobacter$ $ruber$ and $Vibrio$ $extorquens$ (Bassalik) and contrasted with those obtained with $Ps.$ $oxalaticus$. Sulphanilamide inhibited growth of the last organism to the same extent on either formate or succinate as would be expected from operation of the ribulose diphosphate cycle, and not the serine pathway, during growth on formate.

d. $Studies$ $with$ $Mutants$. A series of mutants of $Pseudomonas$ AM1, lacking enzymes of the metabolic scheme shown in Fig. 10, have been prepared by use of N-methyl-N′-nitro-N-nitrosoguanidine as mutagen followed by one or more cycles of penicillin enrichment (Heptinstall and

TABLE 6. Mutants of Pseudomonas AM1

Mutant	Derivation	Growth on unsupplemented carbon source		Growth on succinate supplemented with:				Enzyme loss or recovery
		Succinate	Methanol	Serine (1 mM)	Glycine (1 mM)	Glyoxylate (1 mM)	Methanol (2 mM)	
20B-L		+	—	—	+	+	+	Hydroxypyruvate reductase⁻
20S		—	+	+	+	+	+	Phosphoserine phosphohydrolase⁻
82G		—	+	—	+	+	—	Serine hydroxymethylase⁻ when grown on supplemented succinate.
20ST-1	Double mutant derived from 20S	—	—	+	+	—	—	Phosphoserine phosphohydrolase⁻. Serine-glyoxylate aminotransferase⁻.
82GT-1	Double mutant derived from 82G	—	+	—	+	—	—	Serine hydroxymethylase⁻. Serine-glyoxylate aminotransferase⁻.
20STR$_p$	Partial revertant of 20ST-1	—	+	+	+	+	+	Phosphoserine phosphohydrolase⁻. Serine-glyoxylate aminotransferase⁺.
82GTR$_p$	Partial revertant of 82GT-1	—	+	—	+	+	+	Serine hydroxymethylase⁻. Serine-glyoxylate aminotransferase⁺.

METABOLISM OF ONE-CARBON COMPOUNDS BY MICRO-ORGANISMS 183

Quayle, 1970; Harder and Quayle, 1971a, b). The nutritional character-istics and enzyme lesions of these mutants have been summarized in Table 6.

Mutant 20B-L was isolated on the basis of its ability to grow on succinate but inability to grow on methanol (Heptinstall and Quayle, 1970); it was also unable to grow on methylamine or formate. The mutant lacked hydroxypyruvate reductase and revertants that regained the ability to grow on methanol, methylamine and formate had regained the ability to synthesize hydroxypyruvate reductase. This shows that hydroxypyruvate reductase is necessary for growth on one-carbon compounds.

Mutant 20S was isolated through its inability to grow on succinate unless supplemented with serine; it was also unable to grow on lactate or ethanol but was able to grow on methanol (Harder and Quayle, 1971a). The mutant lacked phosphoserine phosphohydrolase and revertants which regained the ability to grow on succinate, lactate and ethanol regained phosphoserine phosphohydrolase activity. This shows that the phosphorylated pathway of serine biosynthesis is necessary for growth on these three substrates but not on methanol.

Mutant 82G was isolated on the basis of its inability to grow on succinate unless supplemented with 1 mM-glycine plus 1 mM-formate; in fact, formate was later found to be unnecessary. The mutant was also unable to grow on lactate or ethanol. When grown on succinate supple-mented with glycine and formate, the mutant lacked serine hydroxy-methylase. Revertants which regained the ability to grow on succinate, lactate or ethanol, regained serine hydroxymethylase activity. This shows that during growth on these substrates, glycine is made from serine *via* serine hydroxymethylase. The fact that supplemental formate was unnecessary showed that the organism can obtain one-carbon units from glycine.

Mutant 82G, which lacked serine hydroxymethylase when grown on supplemented succinate medium was, nevertheless, able to grow on one-carbon compounds. Serine hydroxymethylase activity appeared in the mutant when washed suspensions, grown on supplemented succinate medium, were incubated with methanol or formate. This shows that serine hydroxymethylase is necessary for growth on one-carbon com-pounds and suggests that *Pseudomonas* AM1 may synthesize two such enzymes, namely a constitutive enzyme, the function of which is to supply glycine during growth of the organism on succinate or similar compounds, and an inducible enzyme, the function of which is to convert glycine to serine as part of the pathway involved in assimilation of one-carbon compounds. Mutant 82G might lack the constitutive enzyme, thus explaining the requirement for glycine during growth on succinate.

7

184   J. R. QUAYLE

e. *Net Synthesis of Glycine Skeleton*. Operation of the serine pathway during synthesis of cell constituents from one-carbon compounds requires the synthesis of one molecule of glycine for each molecule of phosphoglycerate formed. Evidence has now been obtained by Harder and Quayle (1971b) which indicates that glyoxylate is the immediate precursor of glycine. Glyoxylate can substitute for glycine or methanol as supplement for growth of mutants 20S and 82G on succinate; it will be recalled that these mutants are unable to synthesize glycine from succinate. Hence if glyoxylate is an intermediary metabolite between methanol and glycine, it should be possible to isolate double mutants from 20S and 82G in which the second lesion blocks the conversion of glyoxylate to glycine. Such double mutants should still be able to grow on succinate plus supplemental glycine, but no longer be able to grow on succinate plus supplemental glyoxylate, succinate plus supplemental methanol, or methanol itself. Several double mutants of this type have been isolated, two of which, 20ST-1 and 82GT-1, are listed in Table 6. Both the double mutants lacked serine-glyoxylate aminotransferase activity in addition to the original lesions present in the respective parent mutants. Partial revertants $20STR_p$ and $82GTR_p$ were obtained which regained the ability to grow on succinate plus 1 mM-glyoxylate, succinate plus 2 mM-methanol and methanol alone. The revertants had regained the ability to synthesize serine-glyoxylate aminotransferase. These results therefore show that in *Pseudomonas* AM1: (a) serine-glyoxylate aminotransferase is necessary for growth on one-carbon compounds; (b) serine-glyoxylate aminotransferase is involved in the conversion of methanol into glycine; and (c) glyoxylate is the precursor of glycine during growth on one-carbon compounds.

These conclusions are consistent with the finding that the specific activity of serine-glyoxylate aminotransferase is 6–10-fold higher in methanol-grown *Pseudomonas* AM1 than in the succinate-grown organism.

The mechanism of the net synthesis of glyoxylate from one-carbon units remains a mystery. It must happen by one of two general ways: (a) a direct condensation of two one-carbon units, which the isotopic data would suggest to be a reduced one-carbon compound and carbon dioxide respectively, or (b) extension of the carbon skeleton of phosphoglycerate by at least one carbon atom followed by cleavage of the molecule to give glyoxylate and a residual multicarbon molecule. The multicarbon molecule would represent the net biosynthetic product of the overall cycle of one-carbon fixation.

These two possibilities will now be examined separately.

i. *Direct synthesis*. A direct condensation of two one-carbon units to give a C_2 compound has not, as yet, been reported to occur in cell-free

METABOLISM OF ONE-CARBON COMPOUNDS BY MICRO-ORGANISMS   185

extracts of an organism which utilizes the serine pathway during growth on one-carbon compounds. The properties of mutants 20S and 82G, isolated by Harder and Quayle (1971a), suggested that such a condensation might be occurring in *Pseudomonas* AM1. It was observed that these mutants, both of which were blocked in glycine synthesis from succinate, would grow on succinate if supplemented with glycine or glyoxylate or with a one-carbon compound such as methanol or formate. It is known that in *E. coli*, serine furnishes the one-carbon units necessary for synthesis of purines and methionine by way of serine hydroxymethylase (Pizer and Potochny, 1964). Hence mutants 20S and 82G must be able to cleave glycine or glyoxylate to one-carbon units, and also to make glycine or glyoxylate from methanol, during growth on supplemental succinate. It is tempting to think that these two metabolic capabilities are manifestations of a reversible cleavage reaction which could offer a solution to growth on one-carbon compounds. Cleavage of the glycine skeleton according to the following overall reaction:

glycine + tetrahydrofolate →

$$\text{5,10-methylenetetrahydrofolate} + CO_2 + NH_3 + 2H \qquad (60)$$

has been demonstrated in cell-free extracts of *Peptococcus glycinophilus* (Klein and Sagers, 1966) and *Arthrobacter globiformis* (Jones and Bridgeland, 1966; Kochi and Kikuchi, 1969). In the latter organism, this reaction is to a minor extent, reversible. All attempts to demonstrate cleavage of $[1^{14}C]$glycine, $[1^{14}C]$glyoxylate or $[1^{14}C]$glycollate in anaerobic reaction mixtures as described by Klein and Sagers (1966) or Kochi and Kikuchi (1969), using cell-free extracts of methanol-grown *Pseudomonas* AM1 or mutant 20S grown on succinate plus 1 mM-glycine, have failed. Under aerobic conditions a thiamine pyrophosphate-dependent cleavage of $[1^{14}C]$glyoxylate was found. The rate of this cleavage reaction was low but could account for the growth rate of mutant 20S on succinate plus 1 mM-glycine on the basis of generation of the necessary one-carbon units. It is probable that this cleavage of glyoxylate is catalysed by pyruvate oxidase as the mammalian enzyme is known to decarboxylate glyoxylate at half the rate that it decarboxylates pyruvate (Kohlhaw *et al.*, 1965). No conditions could be found for reversing this cleavage reaction in extracts of methanol-grown *Pseudomonas* AM1.

ii. *Indirect synthesis.* An obvious possibility for extension of the phosphoglycerate skeleton prior to cleavage is by carboxylation of the derived phosphoenolpyruvate *via* phosphoenolpyruvate carboxylase (reaction 58). If the resulting oxaloacetate or a derivative were cleaved to give two C_2 fragments, these might provide one C_2 skeleton for glyoxylate

186   J. R. QUAYLE

regeneration and another C_2 compound for biosynthesis of cell constituents. Such a scheme would yield glycine with the right labelling pattern from [^{14}C]methanol or [^{14}C]carbon dioxide. If the C_2 compound available for net biosynthesis could not itself be converted to glyoxylate, it would pose the problem as to how cell constituents could be synthesized in net fashion from it. Neither the glyoxylate cycle (Kornberg and Krebs, 1957) nor the glycerate pathway (Kornberg and Gotto, 1961) could be invoked since *Pseudomonas* AM1 does not synthesize the necessary key enzymes, isocitrate lyase and glyoxylate carboligase, during growth on one-carbon compounds. Furthermore, cleavage of oxaloacetate, malate, fumarate, succinate, aspartate or hydroxyaspartate into two C_2-compounds has not yet been observed in cell-free extracts of *Pseudomonas* AM1. It may be, of course, that further extension of the C_4 skeleton of oxaloacetate is carried out before cleavage, for example, if oxaloacetate were hydroxymethylated according to the reaction discovered by Ruffo *et al.* (1962):

$$\begin{array}{ccc}
\begin{array}{l} CO_2Na \\ | \\ CO \\ | \\ CH_2 \\ | \\ CO_2Na \end{array} & + \; HCHO \; \longrightarrow & \begin{array}{l} CO_2Na \\ | \\ CO \\ | \\ CH.CH_2OH \\ | \\ CO_2Na \end{array}
\end{array} \qquad (61)$$

cleavage of the α-oxo-β-hydroxymethylsuccinate between α and β carbon atoms might yield glyoxylate and glycerate. There is, however, no evidence for occurrence of such a reaction in cell-free extracts of *Pseudomonas* AM1 (W. Harder and J. R. Quayle, unpublished data).

Conversion of phosphoglycerate to sugar phosphates might provide alternative substrate(s) for the required cleavage reaction. Again, there is no evidence for occurrence of such a cleavage reaction in cell-free extracts of *Pseudomonas* AM1, nor is it obvious how the right labelling pattern for glycine could emerge without some rather bizarre biochemistry.

The best approach, at this stage, to finding the source of the glyoxylate would seem to be in isolating the reaction by means of mutants. It should now be possible to obtain the right mutants to show unequivocally whether glyoxylate is formed by direct synthesis or indirect cleavage. If this fact were established, it could orientate further work at the enzymic level with cell-free extracts.

2. *Involvement of N-Methylated Amino Acids*

N-Methylated amino acids have been implicated as intermediary metabolites during growth of two species of bacteria on methylamine as

carbon source. Shaw *et al.* (1966) found an enzyme in a methylamine-grown *Pseudomonas* sp. which catalysed the reaction:

$$
\begin{array}{ccc}
\begin{array}{l}
\text{CO}_2\text{H} \\
| \\
\text{CHNH}_2 + \overset{*}{\text{CH}_3}\text{NH}_2 \\
| \\
(\text{CH}_2)_2 \\
| \\
\text{CO}_2\text{H}
\end{array}
& \longrightarrow &
\begin{array}{l}
\text{CO}_2\text{H} \\
| \\
\text{CH}.\overset{*}{\text{NH}}.\text{CH}_3 + \text{NH}_3 \\
| \\
(\text{CH}_2)_2 \\
| \\
\text{CO}_2\text{H}
\end{array}
\end{array}
\tag{62}
$$

Isotopic experiments indicated that the entire methylamine molecule is incorporated into the N-methylglutamate by a direct displacement reaction. The involvement of the enzyme in growth on methylamine was suggested by the absence of the enzyme when the organism was grown on glycerol as carbon source and ammonia as nitrogen source. Short-term isotope experiments with whole cells showed that isotope from [^{14}C] methylamine appeared most rapidly in N-methylglutamate and less rapidly in serine, aspartate and alanine. It was thought that sarcosine was also an early-labelled product but later work by Kung (1969) makes it likely that this compound was, in fact, γ-glutamylmethylamide (see below).

The second bacterium in which N-methylated amino acids have been implicated in one-carbon metabolism is *Pseudomonas* MS, studied in detail by Wagner and his colleagues. Systematic isotope studies of the uptake of [^{14}C]methylamine by the methylamine-grown organism have shown (Kung, 1969) that 98% of the radioactivity which was incorporated within three seconds was present in amino acids; 85% of the radioactivity in the amino acids was contained in N-methylglutamate, γ glutamyl-methylamide and alanine. Kinetic studies were made of the isotope incorporation during 90-seconds incubation and from these studies curves were plotted of the percentage of total radioactivity fixed against time for each of the above three compounds. γ-Glutamylmethylamide and N-methylglutamate contained the majority of the radioactivity at early times; the only curve which showed a clear negative slope immediately after the introduction of isotope was that corresponding to γ-glutamylmethylamide. Labelled serine and glycine appeared later in small amounts. Parallel experiments performed with methylamine-grown *Pseudomonas* sp. (Shaw *et al.*, 1966) gave identical results. In contrast to Shaw *et al.* (1966) sarcosine was not detected as an early-labelled product, but the finding that γ-glutamylmethylamide co-chromatographs with sarcosine in the solvent systems used by Shaw *et al.* (1966) probably explains this apparent discrepancy.

Enzymes have been found in extracts of *Pseudomonas* MS which catalyse the incorporation of methylamine into N-methylglutamate and γ-glutamylmethylamide. The first enzyme is N-methylglutamate

synthetase, catalysing reaction (62); the second enzyme catalyses a similar reaction to that catalysed by glutamine synthetase:

$$\text{glutamate} + \text{ATP} + \text{methylamine} \rightarrow \gamma\text{-glutamylmethylamide} + \text{ADP} + \text{Pi} \qquad (63)$$

Although glutamine synthetase has been reported to use methylamine to a limited extent in place of ammonia (Levintow and Meister, 1954; Woolfolk *et al.*, 1966; Speck, 1949), Kung and Wagner (1969) present evidence which suggests that in *Pseudomonas* MS synthesis of glutamine and γ-glutamylmethylamide is carried out by separate enzymes.

A further interesting enzyme system has been discovered by Kung and Wagner (1970a) in extracts of *Pseudomonas* MS which catalyses the overall conversion:

$$\text{serine or pyruvate} + \text{methylamine} \rightarrow \text{N-methylalanine} \qquad (64)$$

The reaction is dependent on ATP, Mg^{2+}, tetrahydrofolate and $NAD(P)H_2$ but the mechanism has not yet been determined.

The results with *Pseudomonas* sp. (Shaw *et al.*, 1966) and *Pseudomonas* MS suggest that in these two organisms methylamine is incorporated into cell constituents by a pathway which involves initial fixation into γ-glutamylmethylamide and N-methylglutamate; the further steps in the assimilation of the N-methyl groups are not yet known. The relationship between the pathway of assimilation of methylamine by these organisms and the serine pathway in *Pseudomonas* AM1 remains to be determined, it could be that the N-methylated amino acids merely function as intermediary one-carbon carriers, the crucial steps of net biosynthesis being similar to those in *Pseudomonas* AM1. At present, however, it seems more likely that entirely different pathways are involved.

Growth of *Pseudomonas* MS on various one-carbon compounds may present further interesting and complex problems. Kung (1969) has shown that preliminary short-term isotope incubation experiments with dimethylamine- and trimethylamine-grown cells, using the appropriate radioactive amine, give a rather different distribution of radioactivity as compared with the methylamine-grown organisms, and he suggests that a different pathway of initial incorporation may be operating. Furthermore, Hornig and Wagner (1968) have observed a totally different pattern of labelled substrate uptake in trimethylsulphonium chloride-grown *Pseudomonas* MS. The radioactivity appeared most rapidly in serine, glycine and alanine. The activity of γ-glutamylmethylamide synthetase (63) in trimethylsulphonium-grown *Pseudomonas* MS was only 20% of that in the methylamine-grown organism. Hence growth

METABOLISM OF ONE-CARBON COMPOUNDS BY MICRO-ORGANISMS   189

on trimethylsulphonium chloride may be different again, and indeed the results of the isotope studies are rather reminiscent of those with *Pseudomonas* AM1. It may be recalled (see p. 154) that extracts of trimethylsulphonium chloride-grown *Pseudomonas* MS are capable of forming methylenetetrahydrofolate from the growth substrate.

3. *Ribose Phosphate Cycle of Formaldehyde Fixation*

It has already been mentioned (see p. 179) that *Mtn. methano-oxidans* uses the serine pathway for incorporation of methane or methanol into cell constituents. This behaviour contrasted sharply with that observed for *Ps. methanica* (Johnson and Quayle, 1965) and it became clear that diversity of carbon assimilation pathway must occur amongst methane-utilizing bacteria. The following sections of this review will describe the assimilation pathway (the ribose phosphate cycle of formaldehyde fixation) first discovered in *Ps. methanica* and will comment on the distribution of the two types of pathway amongst bacteria capable of growth on methane.

a. *Evidence for Operation of the Cycle.* Leadbetter and Foster (1958) observed that the specific radioactivity of cells of various strains of *Ps. methanica* grown under methane plus air plus $^{14}CO_2$ was always much less than that of the exogenous $^{14}CO_2$, showing that the cells could not be synthesized exclusively from exogenous carbon dioxide, or from respiratory carbon dioxide in equilibrium with the exogenous carbon dioxide. They concluded from this that the metabolism of the organism on methane as growth substrate was heterotrophic rather than auto-trophic. This was confirmed by whole-cell experiments and also by the lack of ribulose diphosphate carboxylase in extracts of the organism (Johnson and Quayle, 1965).

i. *Experiments with whole organisms.* Incorporation of labelled one-carbon compounds into methane- and methanol-grown *Ps. methanica* and methane-grown *Mtlc. capsulatus* has been studied by pulse-labelling techniques by Johnson and Quayle (1965), Kemp and Quayle (1967) and Lawrence *et al.* (1970). The experimental combinations which were used are summarized in Table 7. The combined results showed that isotope from [^{14}C]methane, [^{14}C]methanol and [^{14}C]formaldehyde was incorporated mainly into sugar phosphates; [^{14}C]formate was incorpor-ated mainly into serine and malate, and [^{14}C]carbon dioxide appeared mainly in aspartate and malate. There was also a quantitative difference in the amount of ^{14}C fixation which occurred from the different tracers; high fixation occurred between oxidation levels of methane and formalde-hyde, and low fixation between the levels of formate and carbon dioxide. If the assumption is made that the conversions involved in the oxidation

of methane to carbon dioxide are essentially irreversible, and this seems likely for aerobic organisms, then the results point to formaldehyde as the oxidation level at which most of the carbon is assimilated into cell constituents. Phosphates of glucose and fructose together constituted the largest part (60–90%) of the early-labelled phosphorylated compounds. Hence the simplest explanation of these results would be that formaldehyde, derived from methane or methanol by oxidation, is incorporated directly into a hexose phosphate skeleton. Degradation of glucose derived from glucose phosphate after 3-sec. incubation with [^{14}C]methanol of *Ps. methanica* growing on methane showed that over 70% of the radioactivity was located in C-1; as the time of incubation

TABLE 7. Combinations used in Isotopic Experiments with Whole Cells of *Pseudomonas methanica* and *Methylococcus capsulatus*

Organism	Growth substrate	Compounds present during isotope experiment	Reference
Pseudomonas methanica	Methane	$^{14}CH_4$	Johnson and Quayle (1965)
		$CH_4 + {}^{14}CH_3OH$	Johnson and Quayle (1965)
		$CH_4 + H^{14}CHO$	Kemp and Quayle (1967)
		$CH_4 + H^{14}CO_2H$	Kemp and Quayle (1967)
		$CH_4 + {}^{14}CO_2$	Johnson and Quayle (1965)
	Methanol	$^{14}CH_3OH$	Johnson and Quayle (1965)
Methylococcus capsulatus	Methane	$^{14}CH_3OH$	Lawrence *et al.* (1970)

increased the radioactivity spread throughout the molecule (Kemp and Quayle, 1967). These results were clearly consistent with incorporation of formaldehyde into a hexose phosphate skeleton proceeding by a $C_1 + C_5$ condensation.

ii. *Experiments with cell-free extracts.* Following the isotopic work with whole cells, it was found that cell-free extracts of *Ps. methanica* and *Mtlc. capsulatus* catalysed a condensation reaction between formaldehyde and ribose 5-phosphate to give a mixture of sugar phosphates, the principal components being fructose phosphate and an unknown compound (Kemp and Quayle, 1965, 1966; Lawrence *et al.*, 1970). Testing of other pentose phosphates, hexose phosphates and glyceraldehyde 3-phosphate showed that within the range of substrates tested the reaction was specific for ribose 5-phosphate. This suggested that the

METABOLISM OF ONE-CARBON COMPOUNDS BY MICRO-ORGANISMS 191

unknown product might be a 6-phosphate of allose, altrose, allulose (psicose) or hamamelose:

$$
\begin{array}{llll}
\text{CHO} & \text{CHO} & \text{CH}_2\text{OH} & \\
| & | & | & \text{CHO} \\
\text{HC.OH} & \text{HO.CH} & \text{CO} & | \\
| & | & | & \text{HO.CH}_2.\text{C.OH} \\
\text{HC.OH} & \text{HC.OH} & \text{HC.OH} & | \\
| & | & | & \text{HC.OH} \\
\text{HC.OH} & \text{HC.OH} & \text{HC.OH} & | \\
| & | & | & \text{HC.OH} \\
\text{HC.OH} & \text{HC.OH} & \text{HC.OH} & | \\
| & | & | & \text{CH}_2\text{OH} \\
\text{CH}_2\text{OH} & \text{CH}_2\text{OH} & \text{CH}_2\text{OH} & \\
\text{Allose} & \text{Altrose} & \text{Allulose} & \text{Hamamelose}
\end{array}
$$

The unknown product was dephosphorylated enzymically and the resulting free sugar identified as allulose on the basis of co-chromatography, co-crystallization of the derived phenylosazone and dinitrophenylosazone with authentic derivatives of allulose, and behaviour towards oxidation with bromine water (Kemp and Quayle, 1966). Further confirmation of its identity with allulose was provided by electrophoresis in 0·2 M-arsenite buffer, pH 9·6; reduction with borohydride followed by electrophoresis in arsenite buffer; formation of anhydro derivatives by heating the products of borohydride reduction (Kemp, 1966).

The overall condensation reaction (65) represents an acyloin condensation:

$$
\text{H.}\overset{*}{\text{C}}\text{HO} +
\begin{array}{l}
\text{CHO} \\
| \\
\text{HC.OH} \\
| \\
\text{HC.OH} \\
| \\
\text{HC.OH} \\
| \\
\text{CH}_2\text{O.PO}_3\text{H}_2
\end{array}
\longrightarrow
\begin{array}{l}
\overset{*}{\text{C}}\text{H}_2.\text{OH} \\
| \\
\text{CO} \\
| \\
\text{HC.OH} \\
| \\
\text{HC.OH} \\
| \\
\text{HC.OH} \\
| \\
\text{CH}_2\text{O.PO}_3.\text{H}_2
\end{array}
\qquad (65)
$$

This is a familiar type of reaction catalysed by enzymes involving thiamine pyrophosphate (TPP) as coenzyme (Holzer, 1961) and there is evidence that formaldehyde-TPP can condense with glyoxylate to form tartronic acid semi-aldehyde in the presence of carboligase (Kohlhaw et al., 1965; Jaenicke and Koch, 1962):

$$
[\text{TPP.}\overset{*}{\text{C}}\text{HO}] + \text{OHC.COOH} \rightarrow \text{OH}\overset{*}{\text{C}}\text{.CH(OH).COOH} + \text{TPP} \qquad (66)
$$

In these acyloin condensations the aldehyde which is activated by the TPP forms the carbonyl group of the acyloin. Hence if reaction (65) involved [TPP.CHO] as a reactant, it would be expected that an aldohexose, rather than a hexulose, would be formed as the product. From the

192   J. R. QUAYLE

known stereochemistry of the addition of [TPP.CHO.CH$_2$OH] to aldoses, catalysed by transketolase, it may be predicted that the aldohexose formed by condensation of [TPP-CHO] with ribose 5-phosphate would be altrose 6-phosphate and not allulose 6-phosphate. There is no evidence, so far, from work with crude or fractionated bacterial extracts that altrose 6-phosphate is formed (Lawrence, 1971). Hence, if TPP is involved as a coenzyme, it seems more likely that the carbonyl group of ribose 5-phosphate rather than that of formaldehyde would be activated by the coenzyme. This would then be an analogous reaction to the condensation of [TPP.CHO.CH$_2$OH] with formaldehyde which is catalysed by transketolase (Dickens and Williamson, 1958):

$$CH_2OH.CO.COOH + TPP \rightarrow [TPP.CHO.CH_2OH] + CO_2 \qquad (67)$$

$$[TPP.CHO.CH_2OH] + \overset{*}{H}CHO \rightarrow HOH_2\overset{*}{C}.CO.CH_2OH + TPP \qquad (68)$$

Aldol condensation between formaldehyde and acceptor molecules containing a methylene group adjacent to a carbonyl group have frequently been encountered:

$$\overset{*}{H}CHO + CH_3.CO.CO_2H \longrightarrow HOH_2\overset{*}{C}.CH_2.CO.CO_2H \qquad (69)$$

(Hift and Mahler, 1952)

$$\overset{*}{H}CHO + \overset{\displaystyle CH_3}{\underset{\displaystyle CH_3}{\overset{|}{\underset{|}{CH}}}}.CO.CO_2H \longrightarrow HOH_2\overset{*}{C}.\overset{\displaystyle CH_3}{\underset{\displaystyle CH_3}{\overset{|}{\underset{|}{C}}}}.CO.CO_2H \qquad (70)$$

(McIntosh et al., 1957)

Of special interest is the condensation between formaldehyde and dihydroxyacetone phosphate to form L-erythrulose 1-phosphate catalysed by fructose diphosphate aldolase from muscle (Meyerhof et al., 1936; Dische and Landsberg, 1960):

$$\overset{*}{H}CHO + HOH_2C.CO.CH_2OP \rightarrow HOH_2\overset{*}{C}.CH(OH).CO.CH_2OP \qquad (71)$$

Evidence has been obtained for the occurrence of the last reaction in rat liver (Charalampous, 1954), Swiss chard (Mueller et al., 1955) and erythritol-grown Propionibacterium pentosaceum (Wawszkiewiecz and Barker, 1968). Charalampous (1954) purified 40-fold an enzyme which catalysed reaction (71) in rat liver, and concluded that this enzyme was different from fructose diphosphate aldolase. The partially-purified enzyme showed no coenzyme requirement. It seems unlikely that an aldol condensation of this kind could explain the results obtained with crude extracts of Ps. methanica and Mtlc. capsulatus because such hydroxymethylation of ribose 5-phosphate at C-2 would yield hamamelose phosphate whose chemical properties are not consistent with those

METABOLISM OF ONE-CARBON COMPOUNDS BY MICRO-ORGANISMS 193

of the actual reaction product (Kemp and Quayle, 1966; Kemp, 1966). Work with crude extracts cannot exclude the possibility of an aldol condensation between formaldehyde and C-1 of a pentulose phosphate such as ribulose 5-phosphate or xylulose 5-phosphate, formed from ribose 5-phosphate through the action of pentose phosphate isomerase and epimerase. Such reactions would lead to 6-phosphates of rather recondite or unknown 3-ketohexoses. There is no evidence that products of this kind are formed.

The enzymic system catalysing the condensation of formaldehyde with pentose phosphate has been termed "hexose phosphate synthetase" (Lawrence *et al.*, 1970) and attempts are being made to fractionate it (Lawrence, 1971). The system is particle-bound and is unstable in the absence of Mg^{2+} ions and ribose 5-phosphate.

iii. *Formulation of cycle.* The results of the isotopic experiments with whole cells and the work with cell-free extracts led to the proposal (Kemp and Quayle, 1965, 1966, 1967) of a cycle of reactions, termed the ribose phosphate cycle of formaldehyde fixation, effecting the net synthesis of one molecule of triose phosphate from three molecules of formaldehyde:

$$3H.CHO + ATP \rightarrow \text{triose phosphate} + ADP \qquad (72)$$

In this cycle, shown in Fig. 11, allulose 6-phosphate must first be epimerized at C-3 to give fructose 6-phosphate which, after phosphorylation at C-1, is cleaved by fructose diphosphate aldolase to give dihydroxy-acetone 3-phosphate and glyceraldehyde 3-phosphate. Regeneration of the ribose 5-phosphate acceptor from fructose 6-phosphate and glycer-aldehyde 3-phosphate could take place by the scheme of rearrangement reactions which operates in the ribulose diphosphate cycle. One variant of this scheme is shown in Fig. 4 and involves the following enzymes: transketolase, transaldolase, ribulose 5-phosphate 3-epimerase, ribose 5-phosphate ketol-isomerase. Another variant, not shown, involves sedoheptulose 1,7-diphosphate and sedoheptulose diphosphatase in place of transaldolase. The detailed labelling pattern of early-labelled hexose phosphate obtained from *Ps. methanica* incubated with [^{14}C]-methanol is consistent with either of these variants (Kemp and Quayle, 1967). The presence of all the enzymes concerned at activities high enough to account for the rate of growth of the organisms remains to be established. It should be noted that this particular question still poses problems in some autotrophic organisms with respect to some of the same enzymes (Latzko and Gibbs, 1969).

The main difference between the proposed new cycle and the ribulose diphosphate cycle is the by-passing, in formaldehyde fixation, of the reductive step, i.e. phosphoglycerate to glyceraldehyde phosphate, which is necessary to reduce carbon dioxide to the level of formaldehyde.

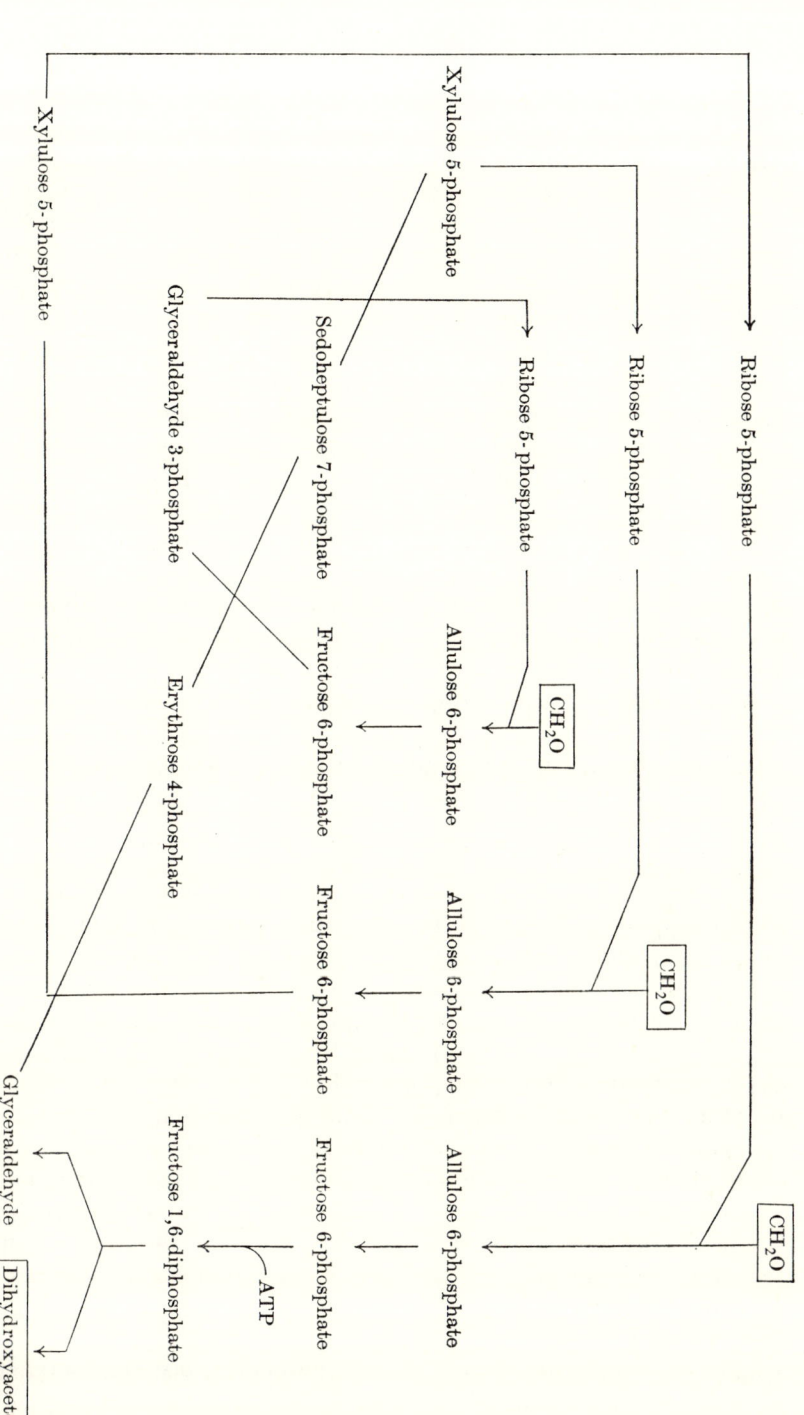

Fig. 11. Ribose phosphate cycle of formaldehyde fixation.

iv. *Reactions of D-allulose in organisms other than methane utilizers.*
Derivatives of D-allulose are not common intermediary metabolites.
The sugar occurs in the antibiotic 6-amino 9-D-psicofuranosylpurine
(Schroeder and Hoeksema, 1959; Sugimori and Suhadolnik, 1965) and
as a component of plants of *Itea* species (Hough and Stacey, 1963).
Allulose 6-phosphate has been implicated as an intermediate in the
conversion of D-allose to fructose 6-phosphate during growth of
Aerobacter aerogenes on D-allose (Gibbins and Simpson, 1963, 1964).
The following scheme of reactions is proposed:

$$\text{D-allose} + \text{ATP} \rightarrow \text{D-allose 6-phosphate} + \text{ADP} \tag{73}$$

$$\text{D-allose 6-phosphate} \rightleftharpoons \text{D-allulose 6-phosphate} \tag{74}$$

$$\text{D-allulose 6-phosphate} \rightleftharpoons \text{D-fructose 6-phosphate} \tag{75}$$

An inducible enzyme, D-allose 6-kinase, catalysing reaction (73), has
been purified 25-fold from extracts of D-allose-grown *Aerobacter
aerogenes*. An enzyme, D-allose 6-phosphate ketol isomerase, catalysing
reaction (74), has also been partially purified from the same source
(Matsushima and Simpson, 1966). The presence of an enzyme which
epimerizes C-3 of D-allulose 6-phosphate to fructose 6-phosphate,
reaction (75), has been surmised but such an enzyme has not been purified.
Gibbins and Simpson (1964) suggest that, in view of the similarity in
configuration between D-allulose 6-phosphate - D-fructose 6-phosphate
and D-ribulose 5-phosphate - D-xylulose 5-phosphate, a single 3-epimerase
might catalyse interconversions of both pairs of sugar phosphates.
b. *Diversity of Assimilation Pathway in Methane-Utilizing Bacteria.*
Two carbon-assimilation pathways have been implicated in growth of
methane-utilizing bacteria, namely the ribose phosphate cycle of
formaldehyde fixation in *Ps. methanica* and *Mtlc. capsulatus* and the
serine pathway in *Mtn. methano-oxidans*. Key enzymes for these two
pathways are, respectively, hexose phosphate synthetase (reaction (65))
and hydroxypyruvate reductase (reaction (53)). Lawrence and Quayle
(1971) screened eight of the new methane-utilizing organisms isolated
by Whittenbury *et al.* (1970b) for the presence or absence of these two
enzymes. The strains were chosen as representatives of each of the sub-
groups designated in Table 2 (p. 132). In every case, the organisms
possessed high levels of activity of either of the two enzymes, but not
both; furthermore, the distribution of the two enzymes precisely
followed the distribution of the two membrane types I and II as shown
in Table 1 (p. 131). It appears that organisms possessing a Type II
membrane structure use the serine pathway whereas those possessing a
Type I membrane structure use the ribose phosphate cycle. This
subdivision is confirmed by yet another criterion, *viz.* the extent of
incorporation of $^{18}O_2$ into cellular material during growth on methane.

It may be predicted that operation of the serine pathway would result in little ^{18}O incorporation as compared with operation of the ribose phosphate cycle. In the former case, prior to assimilation, the oxygen atom of formaldehyde is lost on formation of 5,10-methylenetetrahydrofolate; in the latter case, the oxygen atom of formaldehyde is assimilated as a hydroxyl group at C-1 of hexose. Precisely this behaviour has been found by Bryan-Jones and Wilkinson (personal communication) in the case of "*Methylosinus trichosporium*" (Type II membrane system, low ^{18}O incorporation) as compared to "*Methylomonas albus*" and *Methylomonas agile* (Type I membrane system, high ^{18}O incorporation). Thus the methane-utilizing bacteria may be divided into at least two major groups possessing fundamental differences of both carbon assimilation pathway and internal membrane structure. Such differences suggest that both groups of organisms must have evolved separately and this raises a fascinating question, namely why has their separate evolution resulted in all organisms being obligate methylotrophs?

It will be recalled that, with the possible exception of organism JOB5 (Ooyama and Foster, 1965; Perry, 1968), there is no well-authenticated case of an organism which is capable of growth on methane being able to grow on any other substrate except methanol or dimethyl ether. Methane-utilizing organisms can however carry out incomplete oxidations of other compounds, in some cases only in the presence of methane (co-oxidation) (Leadbetter and Foster, 1958, 1960; Patel *et al.*, 1969). Sugars, amino acids and carboxylic acids can stimulate growth of *Mtlc. capsulatus* at low concentrations but, in general, they are inhibitory to growth (Eroshin *et al.*, 1968). This behaviour is reminiscent of obligate chemolithotrophy in explanation of which several ideas have been put forward, e.g. defects in the ability to oxidize organic compounds or to derive utilizable energy from their oxidation (Smith *et al.*, 1967; for review see Rittenberg, 1969). Patel *et al.* (1969) reported that the incorporation pattern of [^{14}C]acetate by *Mtlc. capsulatus* growing on methane is similar to that of obligate chemolithotrophs and that no α-oxoglutarate dehydrogenase activity could be detected in cell-free extracts.

It is tempting to speculate on possible reactions between obligate methylotrophy and obligate lithotrophy. It is generally considered likely that the primaeval atmosphere contained methane, ammonia and water but no oxygen, the last compound arising by the activity of photosynthetic organisms (see Bernal, 1967). It would only need a small change in the enzymic complement of an autotrophic organism, *viz.* acquisition of an enzyme catalysing reaction (65), to enable the organism to perform net assimilation of formaldehyde. As pointed out by Wilkinson (1971), evolution of a membrane-bound system capable of

oxidizing methane might possibly be connected with the membrane-bound ammonia-oxidizing system of nitrifying bacteria, since the oxidation of methane by methane-utilizing bacteria is competitively inhibited by ammonia and the organisms can also oxidize ammonia. Thus a possible evolutionary development of a methane utilizer from an ammonia-oxidizing lithotroph might be envisaged, in which a common nutritional fastidiousness stemmed from their highly specific enzyme systems capable of deriving utilizable reducing power and ATP from only one kind of substrate.

It has been noted that hexose phosphate synthetase is particle-bound and thus organisms possessing a Type I membrane system may have *in situ* an assimilation trap for the formaldehyde equivalents being generated from methane oxidation. If subsequent work should show that the assimilation trap (tetrahydrofolate?) for formaldehyde equivalents in the serine pathway is cytoplasmic, then it might be envisaged that loss of hexose phosphate synthetase from a Type I membrane system might open the way for a joining of the methane → formaldehyde step with an independently-evolving serine pathway. Perhaps the Type II membrane system is derived from the Type I system by a change of this kind. In a primaeval atmosphere rich in methane there might be no selective advantage for an organism to acquire oxidative systems capable of oxidizing organic compounds less abundant than methane. At a later stage, if the methane content of the atmosphere diminished, there would be a selective advantage in an organism acquiring the ability to oxidize other compounds, leading to the emergence of the group of nutritionally versatile organisms characterized by *Vibrio extorquens*.

Clearly much further work needs to be done on the microbiology and biochemistry of the one-carbon-utilizing organisms before the inter-relationships between obligate methylotrophy and lithotrophy are fully understood. It is nevertheless a field in which rapid progress can be expected in the next few years.

V. Acknowledgements

I am grateful to Dr. C. Wagner and Dr. J. R. Guest for their helpful criticism and advice during the preparation of this review. I am also indebted to the many authors who have made available to me manuscripts and data in advance of publication, in particular Professor P. Hirsch, Dr. D. W. Ribbons, Professor R. Whittenbury, Professor J. F. Wilkinson and their colleagues.

REFERENCES

Anderson, L. and Fuller, R. C. (1967a). *Pl. Physiol.*, Lancaster **42**, 487.
Anderson, L. and Fuller, R. C. (1967b). *Pl. Physiol.*, Lancaster **42**, 491.
Anderson, L. and Fuller, R. C. (1967c). *Pl. Physiol.*, Lancaster **42**, 497.

Andreesen, J. R., Gottschalk, G. and Schlegel, H. G. (1970). *Arch. Mikrobiol.* **72**, 154.

Anthony, C. (1970). *Biochem. J.* **119**, 54P.

Anthony, C. and Zatman, L. J. (1964a). *Biochem. J.* **92**, 609.

Anthony, C. and Zatman, L. J. (1964b). *Biochem. J.* **92**, 614.

Anthony, C. and Zatman, L. J. (1965). *Biochem. J.* **96**, 808.

Anthony, C. and Zatman, L. J. (1967a). *Biochem. J.* **104**, 953.

Anthony, C. and Zatman, L. J. (1967b). *Biochem. J.* **104**, 960.

Archer, M. C. and Scrimgeour, K. G. (1970). *Can. J. Biochem.* **48**, 526.

Bachofen, R., Buchanan, B. B. and Arnon, D. I. (1964). *Proc. natn. Acad. Sci. U.S.A.* **51**, 690.

Barker, H. A. (1940). *Antonie van Leeuwenhoek* **6**, 201.

Barker, H. A. (1956). "Bacterial Fermentations", John Wiley & Sons, Inc., New York and London.

Barker, H. A. (1967). *Biochem. J.* **105**, 1.

Bassalik, K. (1913). *Jb. wiss. Bot.* **53**, 255.

Bassham, J. A., Benson, A. A., Kay, L. D., Harris, A. Z., Wilson, A. T. and Calvin, M. (1954). *J. Am. chem. Soc.* **76**, 1760.

Beijerinck, M. W. and van Delden, A. (1903). *Zentbl. Bakt. ParasitKde (Abt II)* **10**, 33.

Bernal, J. D. (1967). "The Origin of Life", Weidenfeld and Nicolson, London.

Bhat, J. V. and Barker, H. A. (1948). *J. Bact.* **55**, 359.

Blackmore, M. A. and Quayle, J. R. (1970). *Biochem. J.* **118**, 53.

Blackmore, M. A., Quayle, J. R. and Walker, I. O. (1968). *Biochem. J.* **107**, 699.

Blakley R. L. (1969). "The Biochemistry of Folic Acid and Related Pteridines", North-Holland Publishing Co., Amsterdam and London.

Blaylock, B. A. (1968). *Archs Biochem. Biophys.* **124**, 314.

Blaylock, B. A. and Stadtman, T. C. (1963). *Ann. N.Y. Acad. Sci.* **112**, 799.

Blaylock, B. A. and Stadtman, T. C. (1966). *Archs Biochem. Biophys.* **116**, 138.

Breed, R. S., Murray, E. G. D. and Smith, N. R. (1957). "Bergey's Manual of Determinative Bacteriology", The Williams and Wilkins Co., Baltimore.

Brill, W. J., Wolin, E. A. and Wolfe, R. S. (1964). *Science, N.Y.* **144**, 297.

Brot, N. and Weissbach, H. (1965). *J. biol. Chem.* **240**, 3064.

Brown, L. R., Strawinski, R. J. and McCleskey, C. S. (1964). *Can. J. Microbiol.* **10**, 791.

Bryant, M. P., McBride, B. C. and Wolfe, R. S. (1968). *J. Bact.* **95**, 1118.

Bryant, M. P., Wolin, E. A., Wolin, M. J. and Wolfe, R. S. (1967). *Arch. Mikrobiol.* **59**, 20.

Buchanan, B. B. and Evans, M. C. W. (1965). *Proc. natn. Acad. Sci. U.S.A.* **54**, 1212.

Buchanan, B. B., Bachofen, R. and Arnon, D. I. (1964). *Proc. natn. Acad. Sci. U.S.A.* **52**, 839.

Buchanan, B. B., Evans, M. C. W. and Arnon, D. I. (1967). *Arch. Mikrobiol.* **59**, 32.

Budd, J. A. (1969). *Marine Biol.* **4**, 257.

Budd, J. A. and Spencer, C. P. (1968). *Marine Biol.* **2**, 92.

Burton, K. (1961). *In* "Biochemists Handbook", (C. Long, ed.) p. 90, Spon, London.

Cardini, G. and Jurtshuk, P. (1968). *J. biol. Chem.* **243**, 6070.

Chapman, H. M. and Ribbons, D. W. (1968). *J. gen. Microbiol.* **50**, viii.

Chappelle, E. W. (1962). *Biochim. biophys. Acta* **62**, 45.

Charalampous, F. C. (1954). *J. biol. Chem.* **211**, 249.

METABOLISM OF ONE-CARBON COMPOUNDS BY MICRO-ORGANISMS 199

Colby, J. and Zatman, L. J. (1971). *Biochem. J.* **121**, 9P.

Collins, V. G. (1969). *In* "Methods in Microbiology", Norris, J. R. and Ribbons, D. W., eds., Vol. 3B, p. 1, Academic Press, London and New York.

Conti, S. F. and Hirsch, P. (1965). *J. Bact.* **89**, 503.

Davies, S. L. and Whittenbury, R. (1970). *J. gen. Microbiol.* **61**, 227.

Davis, J. B., Coty, V. F. and Stanley, J. P. (1964). *J. Bact.* **88**, 468.

Delwiche, C. C. and Finstein, M. S. (1965). *J. Bact.* **90**, 102.

Dickens, F. and Williamson, D. H. (1958). *Nature, Lond.* **181**, 1790.

Dische, Z. and Landsberg, E. (1960). *Biochim. biophys. Acta* **39**, 144.

den Dooren de Jong, L. E. (1926). "Bijdrage tot de Kennis van het Mineralisatie-proces", Rotterdam: Nijgh and Van Ditmar.

den Dooren de Jong, L. E. (1927). *Zentl. Bakt. ParasitKde* (*Abt II*) **71**, 193.

Dworkin, M. and Foster, J. W. (1956). *J. Bact.* **72**, 646.

Dyer, W. J. (1952). *J. Fish. Res. Bd. Can.* **8**, 314.

Eady, R. R. and Large, P. J. (1968). *Biochem. J.* **106**, 245.

Eady, R. R. and Large, P. J. (1969). *Biochem. J.* **111**, 37P.

Elizarova, T. N. (1963). *Mikrobiologiya* **32**, 1091.

Erebo, L. (1967). *Acta chem. Scand.* **21**, 625.

Eroshin, V. K., Harwood, J. H. and Pirt, S. J. (1968). *J. appl. Bact.* **31**, 560.

Evans, M. C. W. and Buchanan, B. B. (1965). *Proc. natn. Acad. Sci. U.S.A.* **53**, 1420.

Evans, M. C. W. and Whatley, F. R. (1970). *In* "Organization and Control in Prokaryotic and Eukaryotic Cells", (H. P. Charles and B. C. J. G. Knight, eds.), p. 203, *20th Symp. Soc. gen. Microbiol.* Univ. Press, Cambridge.

Evans, M. C. W., Buchanan, B. B. and Arnon, D. I. (1966). *Proc. natn. Acad. Sci. U.S.A.* **55**, 928.

Fitch, W. M. (1963). *Biochemistry, N.Y.* **2**, 1217.

Foster, J. W. and Davis, R. H. (1966). *J. Bact.* **91**, 1924.

Fuller, R. C. and Gibbs, H. (1959). *Pl. Physiol., Lancaster* **34**, 324.

Fuller, R. C., Smillie, R. M., Sisler, E. C. and Kornberg, H. L. (1961). *J. biol. Chem.* **236**, 2140.

Gibbins, L. N. and Simpson, F. J. (1963). *Can. J. Microbiol.* **9**, 770.

Gibbins, L. N. and Simpson, F. J. (1964). *Can. J. Microbiol.* **10**, 829.

Gibbs, M. (1967). *A. Rev. Biochem.* **36**, 757.

Glover, J., Kamen, M. D. and van Genderen, H. (1952). *Archs Biochem. Biophys.* **35**, 384.

Goldstein, D. B. (1959). *J. Bact.* **78**, 695.

Haggstrom, L. (1969). *Biotech. Bioengng.* **11**, 1043.

Harder, W. and Quayle, J. R. (1971a). *Biochem. J.* **121**, 753.

Harder, W. and Quayle, J. R. (1971b). *Biochem. J.* **121**, 763.

Harrington. A. A. and Kallio, R. E. (1960). *Can. J. Microbiol.* **6**, 1.

Harwood, J. H. (1970). Ph.D. thesis: University of London.

Heptinstall, J. and Quayle, J. R. (1969). *J. gen. Microbiol.* **55**, xvi.

Heptinstall, J. and Quayle, J. R. (1970). *Biochem. J.* **117**, 563.

Hift, H. and Mahler, H. R. (1952). *J. biol. Chem.* **198**, 901.

Higgins, I. J. and Quayle, J. R. (1970). *Biochem. J.* **118**, 201.

Hirsch, P. (1968). *Nature, Lond.* **217**, 555.

Hirsch, P. and Conti, S. F. (1964a). *Arch. Mikrobiol.* **48**, 339.

Hirsch, P. and Conti, S. F. (1964b). *Arch. Mikrobiol.* **48**, 358.

Hirsch, P. and Conti, S. F. (1968). *Arch. Mikrobiol.* **62**, 289.

Hirsch, P. and Rheinheimer, G. (1968). *Arch. Mikrobiol.* **62**, 289.

Hoare, D. S. (1963). *Biochem. J.* **87**, 284.
Hollinshead, J. A. (1966). *Biochem. J.* **99**, 389.
Holzer, H. (1961). *Angew. Chem.* **73**, 721.
Höpner, T. and Trautwein, A. (1971). *Arch. Mikrobiol.* **77**, 26.
Hornig, D. and Wagner, C. (1968). *Bact. Proc.* 115.
Hough, L. and Stacey, B. E. (1963). *Phytochem.* **2**, 315.
Huennekens, F. M. (1963). *Biochemistry, N.Y.* **2**, 151.
Ida, S. and Alexander, M. (1965). *J. Bact.* **90**, 151.
Iyer, S. N. and Kallio, R. E. (1958). *Archs Biochem. Biophys.* **76**, 295.
Jaenicke, L. and Koch, J. (1962). *Biochem. Z.* **336**, 432.
Janota, L. (1950). *Med. Doswiadezalna Mikrobiol.* **2**, 131.
Jarman, T. R., Eady, R. R. and Large, P. J. (1970). *Biochem. J.* **119**, 55P.
Johnson, M. J. (1967). *Science, N.Y.* **155**, 1515.
Johnson, P. A. and Quayle, J. R. (1964). *Biochem. J.* **93**, 281.
Johnson, P. A. and Quayle, J. R. (1965). *Biochem. J.* **95**, 859.
Johnson, P. A., Jones-Mortimer, M. C. and Quayle, J. R. (1964). *Biochim. biophys. Acta* **89**, 351.
Jones, K. M. and Bridgeland, E. S. (1966). *Biochem. J.* **99**, 25P.
Kaneda, T. and Roxburgh, J. M. (1959a). *Can. J. Microbiol.* **5**, 87.
Kaneda, T. and Roxburgh, J. M. (1959b). *Biochim. biophys. Acta* **33**, 106.
Kaneda, T. and Roxburgh, J. M. (1959c). *Can. J. Microbiol.* **5**, 187.
Karlsson, J. L., Volcani, B. E. and Barker, H. A. (1948). *J. Bact.* **56**, 781.
Kaserer, H. (1906). *Zentbl. Bakt. ParasitKde* (Abt. II) **16**, 681.
Katagiri, M., Ganguli, B. N. and Gunsalus, I. C. (1968). *J. biol. Chem.* **243**, 3543.
Kaufman, S. (1969). *J. biol. Chem.* **234**, 2677.
Kaufman, S. and Levenberg, B. (1969). *J. biol. Chem.* **234**, 2683.
Kelly, D. P. (1967). *Sci. Prog., Oxf.* **55**, 35.
Kemp, M. B. (1966). Ph.D. thesis: University of Sheffield.
Kemp, M. B. and Quayle, J. R. (1965). *Biochim. biophys. Acta* **107**, 174.
Kemp, M. B. and Quayle, J. R. (1966). *Biochem. J.* **99**, 41.
Kemp, M. B. and Quayle, J. R. (1967). *Biochem. J.* **102**, 94.
Kersten, D. K. (1964). *Mikrobiologiya* **33**, 31.
Khambata, S. R. and Bhat, J. V. (1953). *J. Bact.* **66**, 505.
Kiesow, L. A. (1967). *Curr. Topics Bioenergetics* **2**, 195.
Kistner, A. (1953). *Proc. K. ned. Akad. Wet. Ser. C.* **56**, 443.
Kistner, A. (1954). *Proc. K. ned. Akad. Wet. Ser. C.* **57**, 186.
Klein, S. M. and Sagers, R. D. (1966). *J. biol. Chem.* **241**, 197.
Kluyver, A. J. and Schnellen, C. G. T. P. (1947). *Archs Biochem. Biophys.* **14**, 57.
Kochi, H. and Kikuchi, G. (1969). *Archs Biochem. Biophys.* **132**, 359.
Kohlhaw, G., Deus, B. and Holzer, H. (1965). *J. biol. Chem.* **240**, 2135.
Kondrat'eva, E. N. and Trotsenko, Yu.A. (1969). *Dokl. Akad. Nauk SSSR* **185**, 202.
Kornberg, H. L. and Gotto, A. M. (1961). *Nature, Lond.* **183**, 1791.
Kornberg, H. L. and Krebs, H. A. (1957). *Nature, Lond.* **179**, 988.
Kortstee, G. J. J. (1970). *Arch. Mikrobiol.* **71**, 235.
Kozlova, E. I., Vorob'eva, L. I., Arkad'eva, Z. A. and Rozanova, L. I. (1969). *Mikrobiologiya* **38**, 251.
Kung, H. (1969). Ph.D. thesis: Vanderbilt University, U.S.A.
Kung, H. and Wagner, C. (1969). *J. biol. Chem.* **244**, 4136.
Kung, H. and Wagner, C. (1970a). *Biochim. biophys. Acta* **201**, 513.
Kung, H. and Wagner, C. (1970b). *Biochem. J.* **116**, 357.

METABOLISM OF ONE-CARBON COMPOUNDS BY MICRO-ORGANISMS 201

Ladner, A. and Zatman, L. J. (1969). *J. gen. Microbiol.* **55**, xvi.

Langenberg, K. F., Bryant, M. P. and Wolfe, R. S. (1968). *J. Bact.* **95**, 1124.

Large, P. J. and Quayle, J. R. (1963). *Biochem. J.* **87**, 386.

Large, P. J., Eady, R. R. and Murden, D. J. (1969). *Analyt. Biochem.* **32**, 402.

Large, P. J., Peel, D. and Quayle, J. R. (1961). *Biochem. J.* **81**, 470.

Large, P. J., Peel, D. and Quayle, J. R. (1962a). *Biochem. J.* **82**, 483.

Large, P. J., Peel, D. and Quayle, J. R. (1962b). *Biochem. J.* **85**, 243.

Latzko, E. and Gibbs, M. (1969). *Pl. Physiol., Lancaster* **44**, 295.

Lawrence, A. J. (1971). Ph.D. thesis: University of Sheffield.

Lawrence, A. J. and Quayle, J. R. (1970). *J. gen. Microbiol.* **63**, 371.

Lawrence, A. J., Kemp, M. B. and Quayle, J. R. (1970). *Biochem. J.* **116**, 631.

Leadbetter, E. R. and Foster, J. W. (1958). *Arch. Mikrobiol.* **30**, 91.

Leadbetter, E. R. and Foster, J. W. (1959). *Nature, Lond.* **184**, 1428.

Leadbetter, E. R. and Foster, J. W. (1960). *Arch. Mikrobiol.* **35**, 92.

Leadbetter, E. R. and Gottlieb, J. A. (1967). *Arch. Mikrobiol.* **59**, 211.

Levintow, L. and Meister, A. (1954). *J. biol. Chem.* **209**, 265.

Linke, H. A. B. (1969). *Arch. Mikrobiol.* **64**, 203.

Ljungdahl, L. G. and Wood, H. G. (1965). *J. Bact.* **89**, 1055.

Ljungdahl, L. G. and Wood, H. G. (1969). *A. Rev. Microbiol.* **23**, 515.

Ljungdahl, L. G., Irion, E. and Wood, H. G. (1965). *Biochemistry, N.Y.* **4**, 2771.

Malavolta, E., Delwiche, C. C. and Burge, W. D. (1962). *Biochim. biophys. Acta* **57**, 347.

Matsushima, K. and Simpson, F. J. (1966). *Can. J. Microbiol.* **12**, 313.

McIntosh, E. N., Purko, M. and Wood, W. A. (1957). *J. biol. Chem.* **228**, 499.

Mevius, W. (1953). *Arch. Mikrobiol.* **19**, 1.

Meyerhof, K., Lohmann, K. and Schuster, P. (1936). *Biochem. Z.* **286**, 301.

Mueller, G. C., Quinn, E. M. and Rueckert, R. R. (1955). *Archs Biochem. Biophys.* **55**, 408.

Myers, P. A. and Zatman, L. J. (1971). *Biochem. J.* **121**, 10P.

O'Kelley, J. C. and Nason, A. (1970). *Biochim. biophys. Acta* **205**, 426.

O'Kelley, J. C., Becker, G. E. and Nason, A. (1970). *Biochim. biophys. Acta* **205**, 409.

Ooyama, J. and Foster, J. W. (1965). *Antonie van Leeuwenhoek* **31**, 45.

Orla-Jensen, S. (1909). *Zentbl. Bakt. ParasitKde (Abt II)* **22**, 311.

Patel, R., Hoare, D. S. and Taylor, B. F. (1969). *Bact. Proc.* 128.

Paynter, M. J. B. and Hungate, R. E. (1968). *J. Bact.* **95**, 1943.

Peck, H. D. (1968). *A. Rev. Microbiol.* **22**, 489.

Peel, D. and Quayle, J. R. (1961). *Biochem. J.* **81**, 465.

Perry, J. J. (1968). *Antonie van Leeuwenhoek* **34**, 27.

Peterkofsky, A. and Racker, E. R. (1961). *Pl. Physiol., Lancaster* **36**, 409.

Peterson, J. A., Kusunose, M., Kusunose, E. and Coon, M. J. (1967). *J. biol. Chem.* **242**, 4334.

Pfennig, N. (1967). *A. Rev. Microbiol.* **21**, 285.

Pizer, L. I. (1963). *J. biol. Chem.* **238**, 3934.

Pizer, L. I. (1965). *J. Bact.* **89**, 1145.

Pizer, L. I. and Potochny, M. L. (1964). *J. Bact.* **88**, 611.

Pizer, L. I., Ponce-de-Leon, M. and Michalka, J. (1969). *J. Bact.* **97**, 1357.

Poston, M. J., Kuratomi, K. and Stadtman, E. R. (1964). *Proc. natn. Acad. Sci. U.S.A.* **112**, 804.

Proctor, H. M., Norris, J. R. and Ribbons, D. W. (1969). *J. appl. Bact.* **32**, 118.

Qadri, S. M. H. and Hoare, D. S. (1968). *J. Bact.* **95**, 2344.

Quayle, J. R. (1961). *A. Rev. Microbiol.* **15**, 119.
Quayle, J. R. (1971). *In* "Methods in Microbiology", (J. R. Norris and D. W. Ribbons, eds.), Vol. 6, in press, Academic Press, London and New York.
Quayle, J. R. and Keech, D. B. (1959a). *Biochem. J.* **72**, 623.
Quayle, J. R. and Keech, D. B. (1959b). *Biochem. J.* **72**, 631.
Ribbons, D. W. (1968). *J. Inst. Petroleum* (Microbiology) 47.
Ribbons, D. W. (1970). *FEBS Letters* **8**, 101.
Ribbons, D. W. and Michalover, J. L. (1970). *FEBS Letters* **11**, 41.
Ribbons, D. W., Harrison, J. E. and Wadzinski, A. M. (1970). *A. Rev. Microbiol.* **24**, 135.
Rittenberg, S. C. (1969). *Adv. microbial Physiol.* **3**, 159.
Roberton, A. M. and Wolfe, R. S. (1969). *Biochem. biophys. Acta* **192**, 420.
Roberton, A. M. and Wolfe, R. S. (1970). *J. Bact.* **102**, 43.
Rolls, J. P. and Lindstrom, E. S. (1967). *J. Bact.* **94**, 860.
Rose, Z. B. and Racker, E. (1962). *J. biol. Chem.* **237**, 3279.
Ruffo, A., Testa, E., Adinolfi, A. and Pelizza, G. (1962). *Biochem. J.* **85**, 588.
Schnellen, C. G. T. P. (1947). Doctoral thesis: Technical University at Delft, The Netherlands.
Schön, G. (1965). *Arch. Mikrobiol.* **50**, 111.
Schrauzer, G. N. (1968). *Accounts of Chemical Research* **1**, 97.
Schroeder, W. and Hoeksema, H. (1959). *J. Am. chem. Soc.* **81**, 1767.
Shaposhnikov, V. N. and Loriya, Zh. K. (1964). *Dokl. Akad. Nauk SSSR* **156**,1201.
Shaposhnikov, V. N. and Loriya, Zh. K. (1967). *Dokl. Akad. Nauk SSSR* **175**, 721.
Shaw, W. V., Tsai, L. and Stadtman, E. R. (1966). *J. biol. Chem.* **241**, 935.
Shewan, J. M. (1951). *Biochem. Soc. Symp.* **6**, 28.
Shieh, H. S. (1964). *Can. J. Microbiol.* **10**, 837.
Shieh, H. S. (1965). *Can. J. Microbiol.* **11**, 375.
Shieh, H. S. (1966a). *Nature, Lond.* **212**, 1608.
Shieh, H. S. (1966b). *Can. J. Microbiol.* **12**, 299.
Shieh, H. S. (1968). *Can. J. Biochem.* **46**, 21.
Silver, W. S. (1960). *Nature, Lond.* **185**, 555.
Sirevåg, R. and Ormerod, J. G. (1970). *Biochem. J.* **120**, 399.
Smillie, R. M., Rigopoulos, N. and Kelly, H. (1962). *Biochim. biophys. Acta* **56**, 612.
Smith, A. J., London, J. and Stanier, R. Y. (1967). *J. Bact.* **94**, 972.
Smith, P. H. and Hungate, R. E. (1958). *J. Bact.* **75**, 713.
Smith, U. and Ribbons, D. W. (1970). *Arch. Mikrobiol.* **74**, 116.
Smith, U., Ribbons, D. W. and Smith, D. S. (1970). *Tissue and Cell* **2**, 513.
Söhngen, N. L. (1906). *Zentbl. Bakt. ParasitKde. (Abt. II)* **15**, 513.
Sorokin, Y. I. (1961). *Mikrobiologia* **30**, 337.
Speck, J. F. (1949). *J. biol. Chem.* **179**, 1405.
Stadtman, T. C. (1967). *A. Rev. Microbiol.* **21**, 121.
Stadtman, T. C. and Barker, H. A. (1951). *J. Bact.* **62**, 269.
Stanier, R. Y., Palleroni, N. J. and Doudoroff, M. (1966). *J. gen. Microbiol.* **43**, 159.
Stocks, P. K. and McCleskey, C. S. (1964a). *J. Bact.* **88**, 1065.
Stocks, P. K. and McCleskey, C. S. (1964b). *J. Bact.* **88**, 1071.
Stokes, J. E. and Hoare, D. S. (1969). *J. Bact.* **100**, 890.
Stokstad, E. L. R. and Koch, J. (1967). *Physiol. Rev.* **47**, 83.
Stouthamer, A. H. (1969). *In* "Methods in Microbiology", (J. R. Norris and D. W. Ribbons, eds.). Vol. 1, p. 629, Academic Press, London.
Strittmatter, P. and Ball, E. G. (1955). *J. biol. Chem.* **213**, 445.
Sugimori, T. and Suhadolnik, R. J. (1965). *J. Am. chem. Soc.* **87**, 1136.

METABOLISM OF ONE-CARBON COMPOUNDS BY MICRO-ORGANISMS

Sun, A. Y., Ljungdahl, L. G. and Wood, H. G. (1969). *J. Bact.* **98**, 842.
Umbarger, H. E. and Umbarger, M. A. (1962). *Biochim. biophys. Acta* **62**, 193.
Umbarger, H. E., Umbarger, M. A. and Siu, P. M. L. (1963). *J. Bact.* **85**, 1431.
Urushibara, T. and Forrest, H. S. (1970). *Boichem. biophys. Res. Commun.* **40**, 1189.
van Gool, A. and Laudelot, H. (1966). *Biochim. biophys. Acta* **127**, 295.
van Niel, C. B. (1944). *Bact. Rev.* **8**, 1.
Vary, P. S. and Johnson, M. J. (1967). *J. Bact.* **15**, 1473.
Wagner, C. (1964). *Bact. Proc.* 103.
Wagner, C., Lusty, S. M., Kung, H. and Rogers, N. L. (1966). *J. biol. Chem.* **241**, 1923.
Wagner, C., Lusty, S. M., Kung, H. and Rogers, N. L. (1967). *J. biol. Chem.* **242**, 1287.
Walker, D. A. and Crofts, A. R. (1970). *A. Rev. Biochem.* **39**, 389.
Ware, G. C. and Painter, H. A. (1955). *Nature, Lond.* **175**, 900.
Wawszkiewicz, E. J. and Barker, H. A. (1968). *J. biol. Chem.* **243**, 1948.
Weissbach, H. and Taylor, R. T. (1968). *Vitamins and Hormones* **26**, 395.
Wertlieb, D. and Vishniac, W. (1967). *J. Bact.* **93**, 1722.
Whittenbury, R. (1969). *Process Biochem.* **4** (no. 1), 51.
Whittenbury, R., Davies, S. L. and Davey, J. F. (1970). *J. gen. Microbiol.* **61**, 219.
Whittenbury, R., Phillips, K. C. and Wilkinson, J. F. (1970). *J. gen. Microbiol.* **61**, 205.
Wieringa, K. T. (1940). *Antonie van Leeuwenhoek* **6**, 251.
Wilkinson, J. F. (1971). *Symp. Soc. gen. Microbiol.* **21**, 15.
Wolin, E. A., Wolin, M. J. and Wolfe, R. S. (1963a). *J. biol. Chem.* **238**, 2882.
Wolin, M. J., Wolin, E. A. and Wolfe, R. S. (1963b). *Biochem. biophys. Res. Commun.* **12**, 464.
Wood, H. G. (1952). *J. biol. Chem.* **194**, 905.
Wood, J. M. and Wolfe, R. S. (1965). *Biochem. biophys. Res. Commun.* **19**, 306.
Wood, J. M. and Wolfe, R. S. (1966a). *Biochem. biophys. Res. Commun.* **22**, 119.
Wood, J. M. and Wolfe, R. S. (1966b). *Biochemistry, N.Y.* **5**, 3598.
Wood, J. M. and Wolfe, R. S. (1966c). *J. Bact.* **92**, 696.
Wood, J. M., Allam, A. M., Brill, W. J. and Wolfe, R. S. (1965). *J. biol. Chem.* **240**, 4564.
Woolfolk, C. A., Shapiro, B. and Stadtman, E. R. (1966). *Archs Biochem. Biophys.* **116**, 177.
Yagi, T. (1958). *Biochim. biophys. Acta* **30**, 194.
Yagi T. (1959). *J. Biochem., Tokyo* **46**, 949.
Yagi, T. and Tamiya, N. (1962). *Biochim. biophys. Acta* **65**, 508.
Yoch, D. C. and Lindstrom, E. S. (1967). *Biochem. biophys. Res. Commun.* **28**, 65.
Yoch, D. C. and Lindstrom, E. S. (1969). *Arch. Mikrobiol.* **67**, 182.
Zajic, J. E., Volesky, B. and Wellman, A. (1969). *Can. J. Microbiol.* **15**, 1231.

Note Added in Proof

Recent work by Höpner and Trautwein (1971) suggests that growth of *Pseudomonas oxalaticus* on formate is dependent on the presence of small quantities of a cosubstrate such as pyruvate, lactate or acetate (see pp. 135, 159 and 170).

Regulatory Phenomena in the Metabolism of Knallgasbacteria

H. G. SCHLEGEL and U. EBERHARDT

Institute for Microbiology of the Gesellschaft für Strahlenforschung mbH, München, in Göttingen and Institute for Microbiology, University of Göttingen, West Germany

I. Introduction 205
II. Chemolithotrophic Metabolism 206
    A. General Survey 206
    B. Hydrogen Oxidation 207
    C. Hydrogenase Synthesis 210
    D. Carbon Dioxide Fixation 215
    E. Ribulose Diphosphate Carboxylase and Phosphoribulokinase
    Synthesis 216
III. Chemo-Organotrophic and Mixotrophic Metabolism. . . . 219
    A. Enzymes Involved in Hexose and Gluconate Utilization . . 219
    B. Effect of Hydrogen on Hexose Degradation 222
    C. Formation and Function of the Tricarboxylic Acid Cycle Enzymes 229
    D. Utilization of Nitrogenous Compounds 230
IV. Biosynthesis of Amino Acids. 234
V. Final Considerations 238
VI. Acknowledgements 239
    References 240

I. Introduction

The knallgasbacteria have been known to be facultative chemolithotrophic bacteria since Kaserer (1906) isolated the first strain, *Bacillus pantotrophus*. This name clearly expresses the capability of the organisms to grow either with hydrogen, oxygen and carbon dioxide, or to use organic compounds as sources of carbon and energy. All of the hydrogen-oxidizing bacteria that have been studied so far exhibit this diversity in nutrition and many investigations of them have been initiated with the intent of clarifying the relationships between autotrophic and heterotrophic metabolism. Some interactions had already been observed and recorded before metabolic regulations in bacteria became a fashionable field of research. Therefore, it seemed to be attractive to review and discuss the currently known regulatory interactions of heterotrophic

and autotrophic metabolism in these organisms. With respect to the general physiology and biochemistry of knallgasbacteria, the reader is referred to some recent reviews published by Repaske (1966), Schlegel (1966), Peck (1968), and Rittenberg (1969).

The group of hydrogen bacteria, or knallgasbacteria, is physiologically defined and, with regard to taxonomic relations, is very heterogenous (cf. Schlegel, 1966). Even the genus *Hydrogenomonas*, earlier considered to be typical of the whole group, includes strains of two different genera (Davis *et al.*, 1970). *Hydrogenomonas facilis* and *H. ruhlandii* with *Pseudomonas saccharophila* represent the group of real pseudomonads. In contrast, *Hydrogenomonas eutropha* and some other strains including *Hydrogenomonas* strains H 16, H 1 and H 20 are thought to belong to the genus *Alcaligenes*. The distinction between *Pseudomonas* and *Alcaligenes* strains is based, among a few other properties, on differences in flagellation, substrate utilization and organotrophic denitrification. With regard to the chemolithotrophic metabolism, both groups appear to be very similar. Nevertheless the taxonomic diversity should be kept in mind when one attempts to generalize from results obtained with one strain. In this review the name *Hydrogenomonas* still is used for convenience. *Hydrogenomonas eutropha* strain H 16, which has been used in the authors' laboratory, is referred to as *Hydrogenomonas* H 16 in order to differentiate this strain from the type species strain.

II. Chemolithotrophic Metabolism

A. General Survey

The main path of carbon dioxide assimilation in the strains of hydrogen bacteria tested so far is the reductive pentosephosphate cycle. This process has been fairly well elucidated. With respect to the electron-transport system and the activation of hydrogen the mechanism is not as clear. Two types of hydrogenases have been found. Apparently the first type has been studied most intensively with *H. eutropha* (H 16) (Eberhardt, 1966a; Pfitzner, 1969). This type of hydrogenase is bound to membranes which oxidize hydrogen by a complete respiratory chain; oxygen or nitrate can serve as the terminal acceptors. The primary hydrogen acceptor is not known and the activity of the hydrogenase is usually tested with methylene blue (Pfitzner, 1969). There is no doubt that this hydrogenase provides hydrogen for respiration and coupled ATP synthesis. Similar enzymes were found in *Ps. saccharophila* and *H. ruhlandii* (Vishniac and Trudinger, 1962), *H. facilis* (Atkinson and McFadden, 1954), *H. eutropha* (Wittenberger and Repaske, 1961; Mower, personal communication, 1969) and finally with the unidentified strains 12/60/x and 11/x (Eberhardt, 1969).

In addition to the membrane-bound hydrogenase, *Ps. saccharophila* and *H. ruhlandii* (Bone *et al.*, 1963) and *H. eutropha* (H 16) (Pfitzner *et al.*, 1970) contain a second type of hydrogenase. This enzyme is localized in the soluble fraction and catalyses reversibly, in one step, the reduction of NAD by hydrogen. No cofactor participating in NAD reduction could be found. Consequently, Bernstein and Vishniac proposed the designation hydrogen dehydrogenase for this enzyme. This name is used in the following, to differentiate between the two hydrogenases.

Surely the function of hydrogen dehydrogenase is to provide the reductive pentosephosphate cycle with $NADH_2$. Some $NADH_2$ is possibly oxidized by the respiratory chain, since Pfitzner (1969) found $NADH_2$-oxidase activity with autotrophically grown cells of *Hydrogenomonas* H 16.

Hydrogenomonas eutropha also contains a soluble, NAD-reducing hydrogenase (system). According to Wittenberger and Repaske (1961) and Repaske and Dans (1968), however, the reduction of NAD is a multi-step reaction with participation of coupling factors. Therefore it is doubtful whether the mechanism of NAD reduction in *H. eutropha* is the same as in *Hydrogenomonas* H 16, *H. ruhlandii* and *Ps. saccharophila*.

No NAD-reducing hydrogenase was found in cell-free extracts from *H. facilis* (B. A. McFadden, personal communication, 1968) or from strains 12/60/x and 11/x (Eberhardt, 1969). It is not clear whether such an enzyme is present in the intact cells, but is inactivated during extraction, or whether the membrane-bound hydrogenase is able to fulfil both functions (i.e. generation of $NADH_2$ for the reductive pentosephosphate cycle and generation of electrons for the production of ATP via respiration).

B. Hydrogen Oxidation

In early studies on hydrogen bacteria it was observed that oxidation of hydrogen also occurred in the absence of carbon dioxide (Lebedeff, 1910; Ruhland, 1924). As a matter of fact, there are strains which oxidize hydrogen with equal rate whether in the presence or absence of carbon dioxide (Eberhardt, 1969). On the other hand, as has been found by Bartha (1962), the *Hydrogenomonas eutropha* strains H 1, H 20 and H 16 respond to the withdrawal of carbon dioxide by a drastic reduction of hydrogen and oxygen uptake. *Hydrogenomonas ruhlandii* exhibits the same phenomenon. Under certain conditions, the gas consumption (hydrogen, oxygen, and carbon dioxide or hydrogen and oxygen, respectively) was decreased by 75–80%. This was true of resting as well as

growing cells. Gas uptake in the absence of carbon dioxide is referred to as "Leerlauf oxidation".

Bartha thought the inhibition of gas uptake to be a feedback effect caused by the ADP level of the cell. Assuming that electron transport and oxidative phosphorylation to be tightly coupled, the rate of hydrogen oxidation would depend on the availability of ADP. In the presence of carbon dioxide, ADP is constantly regenerated by energy-dependent growth or, in resting cells, by the formation of poly-β-hydroxybutyric acid, so that hydrogen can be oxidized at a maximal rate. In the absence of carbon dioxide, hydrogen oxidation is limited by other energy-consuming reactions (e.g. maintenance metabolism). Bartha's hypothesis was confirmed to some extent by ATP and $NADH_2$ measurements (Ahrens, 1966). After the withdrawal of carbon dioxide, the intracellular ATP concentration increased by 30% and the ratio $NADH_2/NAD$ rose from 0·5 to 1·4. The stimulation of hydrogen oxidation is surely not as high as it may appear from the ratio of gas uptake in the presence of carbon dioxide to gas uptake in the absence of carbon dioxide (4·8), for the total gas uptake includes not only the hydrogen consumed in the reduction of oxygen, but also that used for the reduction of carbon dioxide and the carbon dioxide itself. Based on a ratio 6:2:1 for hydrogen, oxygen and carbon dioxide, we may assume that only about 45% of the total gas uptake corresponds to oxidized hydrogen; in the absence of carbon dioxide, 66% of the gas taken up is hydrogen. Finally it is apparent that the rate of H_2 oxidation is three times higher in the presence of carbon dioxide than in its absence. A possible contributing factor is that, in the absence of carbon dioxide, the oxidation of $NADH_2$ should not occur, thereby failing to generate the acceptor for hydrogen dehydrogenase and reducing hydrogen consumption. If Bartha's hypothesis is true, two experimental tests are suggested: (1) the "Leerlauf oxidation" should be stimulated if the electron transport is artificially uncoupled; and (2) the gas uptake in the presence of carbon dioxide should be lowered drastically when carbon dioxide fixation is inhibited. However, Hippe (1967) found that 2,4-dinitrophenol had no influence on the "Leerlauf oxidation" but that carbonylcyanide-m-chlorophenyl-hydrazone stimulated gas consumption 2·5 fold. The inhibition of carbon dioxide fixation was tested in two ways. Bartha obtained an inhibition of carbon dioxide fixation of 99% by adding monoiodoacetic acid. Under these conditions the gas uptake in the presence of carbon dioxide corresponded exactly to the "Leerlauf oxidation". Results from experiments with mutants, however, seemingly do not support Bartha's hypothesis (Schlegel et al., 1970). Mutants of strain *Hydrogenomonas* H 16 were isolated which were able to grow autotrophically but unable to form poly-β-hydroxybutyric acid. Resting cells of this mutant

were unable to assimilate carbon dioxide; nevertheless these mutants exhibited stimulated gas uptake in the presence of carbon dioxide, although not to the same extent as the wild-type cells. Since the mutants did not excrete ^{14}C-compounds into the medium, the authors drew the conclusion that the carbon dioxide effect is more direct. However, the fate of the ATP or hydrogen respectively obtained by the mutants during the carbon dioxide-stimulated gas uptake was not determined, and is still unclear.

The possible direct influence of carbon dioxide or of a product of the carbon dioxide assimilation, on the hydrogenase, or the respiratory chain, has not been studied. The hydrogen dehydrogenase is not influenced by bicarbonate and slightly stimulated by ATP and $NADH_2$ (Pfitzner et al., 1970).

A problem that may be important also for the explanation of carbon dioxide-stimulated gas uptake is that of the "Leerlauf oxidation". Lebedeff (1910) and Ruhland (1924) studied the stoichiometry and found the consumption of hydrogen and oxygen to be in the proportions 2:1. Surely this is true of other knallgasbacteria. There are two possibilities to explain the hydrogen oxidation: (1) hydrogen is oxidized by electron transport not coupled to phosphorylation; or (2) ATP is hydrolysed by ATPases or consumed in energy-dependent reactions not related to synthesis of cell material. In this connection the formation of polyphosphates and maintenance metabolism should come into consideration. The formation of polyphosphates by *Hydrogenomonas* H 20 has been demonstrated (Kaltwasser, 1962). Polyphosphates, however, arc not formed in the absence of carbon dioxide (Schlegel, 1955). If the maintenance metabolism consumes the ATP, we should expect that in the presence and absence of carbon dioxide the same amount of ATP (or hydrogen/oxygen), proportional to the cell mass, is spent for it. This is not so with *H. eutropha* (Bongers, 1970). Bongers determined the uptake of hydrogen, oxygen and carbon dioxide, limiting the growth rate by either one of these components. He showed that, during limitation by hydrogen or oxygen, the total uptake of all three gases is less, compared to that of "unlimited" cells, but the proportions of hydrogen:oxygen:carbon dioxide were the same. During limitation by carbon dioxide, the consumption of hydrogen, oxygen and carbon dioxide was lowered, but, based on the carbon dioxide fixed, more hydrogen and oxygen was consumed. If *Hydrogenomonas* H 16 is really unable to form polyphosphates in the absence of carbon dioxide, then there remains the simpler explanation of an electron-transport chain not coupled to oxidative phosphorylation. In the absence of carbon dioxide, a good portion of the coupled chain may be by-passed by a reaction of flavine, or cytochrome *b*, directly with oxygen. The

uncoupled segment may not work at a maximal rate in the presence of carbon dioxide as a result of competition by the ATP-generating respiratory chain for electrons. In the absence of carbon dioxide, when the coupled electron transport may be inhibited (e.g. by unavailability of ADP), the by-pass may act as a pressure value and thereby permit "Leerlauf oxidation".

Whether these considerations are applicable to those strains whose gas uptake is independent of the presence or absence of carbon dioxide remains unclear. The strain 12/60/x, for example, contains very large electron-dense inclusions that usually are taken for polyphosphate (U. Eberhardt, unpublished data). Therefore the synthesis of considerable amounts of polyphosphates in the absence of carbon dioxide might possibly occur. Thus, these strains may have the same regulation mechanism as *H. eutropha* but it may be masked by other processes.

Results analogous to the findings of Bartha have been obtained with *Nitrobacter winogradskyi* (Bock and Engel, 1966). These authors, too, assume a tight coupling between electron transport and phosphorylating reactions.

C. HYDROGENASE SYNTHESIS

The regulation of hydrogenase formation has been discussed repeatedly (Schlegel, 1966; Peck, 1968; Rittenberg, 1969). This review therefore deals only with certain aspects that have received less attention.

Hydrogenase activity has often been determined by measuring the rate of methylene blue reduction by intact cells. In experiments aiming at studying the adaptive behaviour of these organisms, this method can only be applicable, if at all, to those strains which, like *H. facilis*, apparently contain only the membrane-bound hydrogenase. With *H. eutropha* the existence of two different hydrogenases necessitates a separate determination of hydrogenase and hydrogen dehydrogenase in cell-free extracts. For it seems to be doubtful whether or not the hydrogen dehydrogenase in intact cells is able to reduce methylene blue. The purified hydrogen dehydrogenase is unable to reduce methylene blue unless catalytic amounts of NAD are present (Vishniac and Trudinger, 1962; Pfitzner *et al.*, 1970). In any case, it is not possible to distinguish the activities of the two enzymes using intact cells. Similarly, the measurement of total gas uptake does not provide for a distinction to be made between the two hydrogenases. Thus, many of the published data are of limited value for the discussion of regulation mechanisms. In *H. ruhlandii*, *Hydrogenomonas* H 16 and *Ps. saccharophila*, the levels of both types of hydrogenases increase and decrease adaptively when the cells are transferred from autotrophic to heterotrophic condi-

REGULATORY PHENOMENA IN METABOLISM OF KNALLGASBACTERIA 211

tions or *vice versa* (Vishniac and Trudinger, 1962; Eberhardt, 1966b). Thus there seems to be little doubt that synthesis of hydrogenases is subject to some sort of regulation. Unfortunately, the data collected so far are not sufficient to determine conclusively the principles of regulation. Proceeding from the observation that the hydrogenase activity is always less in heterotrophically grown cells, several hypotheses are suggested: (a) the hydrogenases are induced by hydrogen, or hydrogen and carbon dioxide, and are not subject to catabolite repression by organic compounds; (b) the formation of hydrogenases is independent of the presence of hydrogen, but is repressed by organic compounds; (c) the hydrogenases are induced by hydrogen, the induction being repressed by organic compounds.

It is well established, with *Hydrogenomonas* H 16 that both hydrogenases are synthesized during heterotrophic growth with certain organic substrates (Eberhardt, 1966b; Frings, 1969) though at a diminished rate. Thus, if "endogenous hydrogen" (Rittenberg, 1969) is excluded as a possible inducer of hydrogenases, hypotheses (a) and (c) are irrelevant since in these cases hydrogenase synthesis should depend on the presence of hydrogen. It is inferred that the formation of the hydrogenases is independent of the presence of hydrogen. Three reasons must be taken into consideration: (aa) the hydrogenases are not subject to induction-repression regulation and are formed as long as no compound repressing by catabolite repression is present; (bb) the hydrogenases are semiconstitutive (Pardee and Beckwith, 1963), i.e. they are inducible enzymes formed with a high minimum rate of synthesis in the absence of the inducer hydrogen; (cc) the synthesis of hydrogenases is controlled by endogenous factors that are mediately related to the presence of hydrogen or organic compounds, e.g. the size of the cellular pool of ATP or NADH$_2$. In the latter case variations of hydrogenase activity should also be expected during autotrophic growth. The supposition that hydrogen has no direct influence on the regulation of hydrogenase formation was confirmed, to some extent, by the observation that hydrogen did not stimulate hydrogenase formation in cultures growing with lactate (Rittenberg and Goodman, 1969). *Hydrogenomonas eutropha* also forms considerable amounts of hydrogenase under heterotrophic growth conditions (Bovell, 1957; Rittenberg, 1969; Stukus and DeCicco, 1970). To confirm the conclusion that the formation of hydrogenases is controlled by factors other than hydrogen, a number of experiments should be performed. In particular, the activities of hydrogenases should be determined with cells grown with a variety of organic substrates in the presence and absence of hydrogen. Moreover, it should be determined whether the presence of hydrogen (or autotrophic conditions in general) is the

indispensable prerequisite for the formation of hydrogenases after their complete loss during continued heterotrophic growth.

With *Micrococcus denitrificans* the stimulation of hydrogenase formation by hydrogen in the presence of glucose was reported (Fewson and Nicholas, 1961). This result indicates semiconstitutivity or inducibility of hydrogenase.

A very important type of experiment was performed by Lascelles and Rittenberg (cf. Rittenberg, 1969). *Hydrogenomonas eutropha* was grown in a lactate-containing medium and was then transferred to media containing other organic substrates. With some substrates the hydrogenase activity was higher than with lactate. For instance lactate-grown cells had hydrogenase activity corresponding to 31 μl H_2/hr./mg. protein. Changing to glutamate (or fructose) resulted in an increase of hydrogenase activity to 191 μl H_2/hr./mg. protein (356 μl H_2/hr./mg. protein in the case of fructose) without exposure to autotrophic conditions. In the stationary phase, the hydrogenase activity increased with all substrates used to about 600 μl H_2/hr./mg. protein. These results suggest a quantitatively different repression of hydrogenase formation by organic compounds. Unfortunately, in the experiments of Lascelles and Rittenberg, hydrogenase activity was determined using intact cells and methylene blue.

Independent of the repression by organic substrates, high contents of oxygen in the atmosphere may repress the hydrogenase. Just as the repression by organic compounds may quantitatively vary, so repression by oxygen varies, depending on the pressure. This is suggested by the observation (Wilson *et al.*, 1953) that *H. facilis* synthesizes hydrogenase during heterotrophic growth only when the oxygen content of the atmosphere is decreased to 5%. In conformity with this result Linday and Syrett (1958) found, with the same strain, a complete loss of hydrogenase during heterotrophic growth under air. The reformation of hydrogenase in the absence of the organic substrate was dependent on the presence of hydrogen. Thus the hydrogenase behaved as a truly inducible enzyme although it was formed in the absence of hydrogen in the experiments of Wilson *et al.* (1953). It is, however, questionable whether the presence of hydrogen was a prerequisite for induction of hydrogenase *per se* or for generation of energy for further enzyme formation.

Besides organic substrates and oxygen pressure, the presence and the type of nitrogen source may influence hydrogenase formation. Though lactate was able to inhibit the synthesis of hydrogenase in *H. facilis*, after heterotrophic growth the inhibition was relieved by the addition of ammonium sulphate (Linday and Syrett, 1958). An interesting effect of the type of the nitrogen source was found with *Hydrogenomonas*

H 16 (Pfitzner, 1969). In these experiments the strain lost all of the membrane-bound hydrogenase when grown with fructose and ammonium chloride. In contrast, the hydrogenase was fully present in cells grown with fructose and potassium nitrate. Moreover, the membrane fragments from nitrate-grown cells contained a nitrate reductase so that the membrane particles were able to reduce nitrate with hydrogen. This observation is reminiscent of the finding that *Azotobacter* produces high levels of hydrogenase when grown with atmospheric nitrogen, but only low levels when grown with ammonium salts (Lee and Wilson, 1943). Thus it may be that, if nitrogen is provided in an oxidized form (nitrogen in the case of *Azotobacter*, nitrate in the case of *Hydrogenomonas*), the need for reducing power "induces" the hydrogenase even if hydrogen is not present. It is, however, not clear how nitrate, in the case of *Hydrogenomonas*, overcomes the repression of hydrogenase by fructose. Moreover, it must be clarified whether the "nitrate-induced" hydrogenase is different from the hydrogenase formed in cells growing with ammonium salts. Hydrogenase synthesis is not repressed to the full extent immediately after a change from autotrophic to heterotrophic growth. Rather, the rate of hydrogenase synthesis decreases progessively in the course of many generations (Eberhardt, 1966b). A similar phenomenon was observed with the ribulosediphosphate carboxylase and, so far, neither of these observations can be explained.

As mentioned above, the activities of both hydrogenases vary in a parallel manner, depending on growth conditions. This does not necessarily mean that the synthesis of each enzyme is coupled under all conditions. To be sure, no growth conditions were found that caused the loss of one of the hydrogenases without affecting the other. For instance, during growth with an organic substrate in the presence of hydrogen, the formation of hydrogen dehydrogenase would be redundant, since hydrogen would be used exclusively as an energy source (Gottschalk, 1964). However, we found less than 2% of either hydrogenase in cells cultured with succinate-air or succinate-knallgas. Moreover, it was not possible to find mutants that had lost one of the hydrogenases and hence were unable to grow autotrophically. In such experiments, however, is is difficult to find conditions that allow for growth of mutants heterotrophically or mixotrophically (in the sense of Rittenberg, 1969) without repressing the remaining hydrogenase. On the other hand, a number of mutants of *H. eutropha* (H 16) were isolated that formed the hydrogenases in widely varying proportions (Table 1; Frings, 1969). Variations in the proportions of hydrogenase and hydrogen dehydrogenase activity were also found with autotrophically-grown cells that had been cultured in a chemostat, at different growth

TABLE 1. Formation of Hydrogenase and Hydrogen Dehydrogenase by Mutants of *Hydrogenomonas* H 16 (Frings, 1969)

Mutant	Enzyme Activity (μl H_2/hr./mg. protein)	
	Hydrogenase	Hydrogen dehydrogenase
F 7	650	320
N 4	1,320	1,415
L 1	543	3,320
B 7	1,450	3,940
C 3	1,100	520
G 1	1,210	5,700
I 5	690	320
wild strain (Eberhardt, 1966a)	1,100	1,800

rates, with hydrogen as the growth-limiting factor (E. Schuster and U. Eberhardt, unpublished observation; Table 2). We think these results to be an inducement to study further the co-ordination of hydrogenase synthesis.

TABLE 2. Activities of Hydrogenases in *Hydrogenomonas* H 16 Grown at Different Growth Rates (Hydrogen-Limited Growth)

Growth rate ($\mu = D$)	Enzyme activity (μl H_2/hr./mg. protein)	
	Hydrogenase	Hydrogen dehydrogenase
0·04	5,200	1,480
0·16	3,700	2,340
batch culture	1,100	1,800

As shown in Table 2, cells grown in hydrogen-limitated chemostat culture have much higher hydrogenase activity than cells grown in batch culture. The hydrogenase activity (determined as rate of reduction of methylene blue by washed cells) varied, under the conditions described above, with the growth rate in a complicated manner (Schuster and Schlegel, 1967). With decreasing growth rates ($\mu = D$) from 0·169

to 0.06 hr.$^{-1}$ the hydrogenase paralleled the decreasing hydrogen supply in the cultures (hydrogenase activity close to 2,500 μl H_2/hr./mg. protein). From $D = 0.06$ to 0.04 hr.$^{-1}$ the activity increased very rapidly and was as high as 4,000 μl H_2/hr./mg. protein. This suggests a varying extent of derepression of hydrogenase in autotrophic culture. According to Schuster and Schlegel (1967) the derepression is caused either by shortage of energy or by shortage of reducing power. Then *Hydrogenomonas* H 16 was grown at the same growth rate (0.08 hr.$^{-1}$) under hydrogen limitation and under oxygen limitation. It turned out that cells from the hydrogen-limited culture had an hydrogenase activity corresponding to 2,760 μl H_2/hr./mg. protein as compared with 470 μl H_2/hr./mg. protein for cells grown under oxygen limitation. Since the latter condition limits the generation of energy without derepressing the hydrogenase, it was concluded that the derepressing factor in cells growing under hydrogen limitation is the shortage of reducing power (Schuster and Schlegel, 1967). However, in this case one should expect a large increase in hydrogen dehydrogenase activity rather than in hydrogenase activity (Table 2). Thus the problem of regulation of hydrogenase formation in autotrophic as well as in heterotrophic culture remains unsolved.

D. CARBON DIOXIDE FIXATION

The main path of carbon dioxide assimilation in knallgasbacteria is the reductive pentosephosphate cycle. The enzymes unique to this pathway are ribulose 1,5-diphosphate carboxylase (RDPC) and phosphoribulokinase (PRK). If the functioning of the Calvin cycle is subject to regulation one or both of these enzymes should be the site of control. Regulation of carbon dioxide fixation seems to be advantageous when limiting amounts of hydrogen or oxygen are available to the cell.

McFadden and Tu (1965), studying extracts from *H. facilis*, found carbon dioxide fixation to be independent of exogenous ribulose diphosphate, but dependent on the addition of ATP and $NADH_2$. This carbon dioxide fixation in cell-free extracts was shown to occur *via* the reductive pentose phosphate cycle; it was sensitive to AMP. According to MacElroy *et al.* (1969), $NADH_2$ does not enhance the carbon dioxide fixation since it is the reductant of 3-phosphoglycerate, but rather it specifically stimulates PRK. In addition, PRK is affected allosterically by ATP and is inhibited by AMP. With ribose 5-phosphate and ATP as substrates the reaction sequence

ribose 5-phosphate $\rightarrow$ ribulose 5-phosphate $\rightarrow$

ribulose 1,5-diphosphate $\rightarrow$ 3-phosphoglyceric acid

was stimulated 7–20 fold by the addition of $NADH_2$ (1 mM). Adenosine monophosphate (5 mM) abolished the $NADH_2$ effect. In contrast, neither the carboxylation of ribulose diphosphate nor the isomerization of ribose 5-phosphate was influenced by the addition of $NADH_2$ and/or AMP.

These experiments were done using cell-free extracts from fructose-grown cells. There is, however, no reason to believe that PRK from autotrophically or mixotrophically-grown cells does not react similarly, or identically. As to the characteristics of their regulatory control, the PRKs of *H. facilis* and of other chemolithotrophic bacteria are similar (MacElroy *et al.*, 1968).

McFadden and Tu (1965) assumed that, in heterotrophically-grown cells, carbon dioxide fixation is limited by RDPC activity. Regardless of the regulatory control of PRK, this may be the case if, with certain substrates, the RDPC activity drops to very low levels while PRK activity remains comparatively high (see Section E, p. 218).

E. Ribulose Diphosphate Carboxylase and Phosphoribulokinase Synthesis

The synthesis of RDPC in the absence of hydrogen and carbon dioxide has been clearly established. Accordingly, autotrophic conditions are not necessary to permit RDPC formation. Thus, it is obvious to suppose that RDPC is a constitutive enzyme, although the possibility that some factor other than hydrogen or carbon dioxide is an inducer cannot be excluded. Clearly the rate of RDPC formation in hetero-trophic culture is specifically determined by the substrate used and, at least in *H. facilis*, by the oxygen pressure. The same statements seem to be true of PRK, although experimental evidence is scant.

McFadden and Tu (1967), Kuehn and McFadden (1968) and Stukus and DeCicco (1970) have determined the RDPC activity of both *H. facilis* and *H. eutropha*. It is very striking that both strains, if grown with fructose, have RDPC activity that is only a little lower than that of autotrophically grown cells. In the case of *H. facilis*, this is also true of cells grown with glucose. As a matter of fact, the RDPC of both organisms was purified from fructose-grown cells (Kuehn and McFadden, 1969). With *Hydrogenomonas* H 16 the residual RDPC activity after growth with fructose was about 8–12% of the autotrophic activity (Eberhardt, 1966b). This relatively large decrease was possibly caused by strong aeration.

Other substrates allowing high RDPC activity in *H. eutropha* were glutamate and alanine (36 and 27%, respectively, of the autotrophic value); in contrast acetate and pyruvate repressed enzyme formation

REGULATORY PHENOMENA IN METABOLISM OF KNALLGASBACTERIA   217

completely (Stukus and DeCicco, 1970). *Hydrogenomonas facilis* grown
with glutamate or acetate had about 4% of the autotrophic RDPC
activity as compared with 25% in ribose-grown cells (McFadden and
Tu, 1967). An additional indication that the residual RDPC activity is
specifically determined by the organic substrate used is found in the
article of Rittenberg (1969). According to Lascelles and Rittenberg
H. eutropha, when grown with lactate, had an RDPC activity corresponding to less than 63 mμmoles PGA/hr./mg. protein. After transfer of
lactate-grown cells to fructose, in the absence of hydrogen plus oxygen,
the activity increased to 184 mμmoles PGA/hr./mg. protein. During the
stationary phase there was a further increase to 338 mμmoles
PGA/hr./mg. protein. Even if autotrophic conditions are not a prerequisite for the synthesis of RDPC, incubation of cells with alanine or
glutamate, under a hydrogen-oxygen atmosphere, resulted in a largely
increased RDPC activity compared to cells grown with the same substrates in air (Stukus and DeCicco, 1970). This, of course, does not
describe exactly a constitutive enzyme, but rather a semiconstitutive
one (see p. 211), autotrophic condition being the inducer.

Kuehn and McFadden (1968) studied the conditions that allow for
high RDPC activity in *H. eutropha* and *H. facilis* during growth with
fructose. The required conditions are: (a) the cultures must not be
strongly aerated; (b) they must contain ferric chloride; and (c) the
cells must be transferred in the early exponential phase. Following these
precautions it was possible to hold the RDPC activity at the autotrophic level during six sub-cultures. Then, cells from the middle- or
late-log phase were transferred which resulted in a continuing decrease
of RDPC activity through four sub-cultures, to 40 % of the initial activity.
With *H. facilis* the same observations were made. The iron requirement
for RDPC synthesis was discovered in a study of the formation of a
c-type cytochrome. Ribulose diphosphate carboxylase and the cytochrome are synthesized in proportional amounts as a function of the
growth rate: the slower the growth, the more RDPC and cytochrome
are formed. The reason for the co-ordinated synthesis is not known;
one is, however, reminded of the hypothesis of Lascelles (1960) regarding
the relationship between the electron-transport chain and regulation
of RDPC synthesis. Strong aeration stopped RDPC synthesis and, as
in the experiments of Eberhardt (1966b), the enzyme was diluted out
with growth. After exhaustion of the fructose, RDPC was completely
inactivated within 22 hr. in spite of low aeration; a heat-labile factor
participated in this process. Inactivation was not observed with autotrophic cultures.

At present we do not see any clue to the explanation of the varying
rate of RDPC synthesis during heterotrophic growth, and of the influence

of the substrate used by the cells. Possibly, RDPC formation depends on the growth rate with the substrate in question: the faster the growth, the slower the rate of RDPC synthesis. Such a relationship has been found with fructose (Kuehn and McFadden, 1968). *Hydrogenomonas* H 16 grows much faster with acetate and pyruvate than with fructose (Frings, 1969). The generation times observed were 1·0 hr. (acetate), 1·3 hr. (pyruvate), 1·6 hr. (glutamate) and 2·8 hr. (fructose). A similar gradation was found with regard to the RDPC activity in *H. eutropha*: no activity in pyruvate- or acetate-grown cells, little activity in cells grown with glutamate, and high activity in fructose-grown cells. Furthermore there might be a relationship between the extent of RDPC repression and the suitability of the substrate in question to regenerate ribulose diphosphate and thereby to permit functioning of the Calvin cycle. The formation of ribulose diphosphate from pyruvate or acetate needs more steps than from fructose; accordingly RDPC synthesis (in *H. eutropha*) is completely repressed. In contrast, fructose, glucose, and ribose induce high activities of RDPC. These sugars are easily converted to ribulose diphosphate. They support the assimilation of carbon dioxide in *Hydrogenomones* H 16 and *H. facilis* as shown by Gottschalk *et al.* (1964), Schlegel and Gottschalk (1965) and McFadden *et al.* (1967). In fact, on certain premises, carbon dioxide assimilation *via* the Calvin cycle might considerably contribute to the synthesis of cell material during fructose utilization. The growth of *Hydrogenomonas* H 16 was largely stimulated by 1% (v/v) carbon dioxide; the generation time shortened from 2·8 to 1·6 hr. (Frings, 1969). However, with acetate or pyruvate as substrates, there was no, or only a very small, shortening of generation time by carbon dioxide. Unfortunately the relatively high RDPC activity after growth with glutamate or alanine is difficult to explain in terms of this hypothesis.

Seemingly, the presence of RDPC activity has often been taken as evidence for the functioning of the Calvin cycle since studies on the synthesis of PRK, and of other enzymes, are very rare. To our knowledge, data on PRK activities were only published by McFadden and Tu (1967). *Hydrogenomonas facilis* contains, after growth with fructose or glucose, just as much PRK activity as after autotrophic growth; after growth with ribose the activity is decreased about 50%. This corresponds to the values found for RDPC. However, in the presence of glutamate and succinate, RDPC activity drops to about 3% of the autotrophic level whereas PRK decreases only by 50% (the lowest PRK activity was found in cells grown with acetate; 11% of the autotrophic value). According to the authors these differences between RDPC and PRK levels may indicate unco-ordinate synthesis of the two enzymes. Surely further studies are needed to confirm this conclusion.

Probably the activity of other enzymes of the Calvin cycle may vary with the growth condition (c.f. triosephosphate dehydrogenase; Gottschalk *et al.*, 1964). Systematic studies, however, were not published.

It would be interesting to know whether RDPC (and PRK) and the hydrogenases (in particular the hydrogen dehydrogenases) are synthesized co-ordinately. Some observations suggest that this is not the case. With *Hydrogenomonas* H 16 it was found that, after change from autotrophic to heterotrophic conditions (fructose), both the hydrogenases were continuously formed although at a diminished rate, whereas RDPC synthesis stopped completely, probably as a result of strong aeration, and was diluted out to the basal level (Eberhardt, 1966b). Conversely, changing from heterotrophic to autotrophic conditions caused the specific RDPC activity to increase from 15 to 190 mU./mg. protein within 10 hr., while the activity of both hydrogenases remained almost unchanged at about 40% of the autotrophic level (Eberhardt, 1965). Furthermore, DeCicco and Stukus (1968) and Stukus and De-Cicco (1970) observed the hydrogenase to be active and functioning during incubation of *H. eutropha* with pyruvate (and acetate) and knallgas. In contrast to hydrogenase, the cells had no RDPC activity during this growth phase and had to synthesize the enzyme after exhaustion of the organic substrate. However (and also this might be an important feature in the regulation of RDPC synthesis) to our knowledge on no occasion were cells found to contain RDPC but no hydrogenase. *Mutatis mutandis*, this seems to be true of other chemolithotrophic bacteria, too.

III. Chemo-Organotrophic and Mixotrophic Metabolism

A. Enzymes Involved in Hexose and Gluconate Utilization

Although *Hydrogenomonas* strain H 16 is not able to use glucose, mannose, ribose or xylose as a growth substrate (Wilde, 1962) it grows very well on fructose (Gottschalk *et al.*, 1964). *Hydrogenomonas eutropha* has been found to grow on fructose, which is always present in autoclaved glucose (Cook *et al.*, 1967). No other sugars are used by the wild-type strains of the *Hydrogenomonas-Alcaligenes* group (Davis *et al.*, 1969). Fructose is degraded *via* the Entner-Doudoroff pathway. Fructose-grown cells of *Hydroglucomonas* H 16 contain a high activity of phosphoglucose isomerase, glucose 6-phosphate dehydrogenase and of the enzymes of the Entner-Doudoroff pathway (6-phosphogluconate dehydratase and 2-keto-3-deoxy-6-phosphogluconate dehydrogenase) while in autotrophically grown cells the activity of the Entner-Doudoroff enzymes is negligible. When the latter cells are transferred to a fructose-containing medium, some of the enzymes undergo characteristic changes.

The initial production of phosphoglucose isomerase and glucose 6-phosphate dehydrogenase is followed by formation of the enzymes of the Entner-Doudoroff system; their synthesis is accompanied by a decrease of ribulose diphosphate carboxylase and fructose diphosphate aldolase (Gottschalk et al., 1964).

Fructose-grown cells of the wild-type *Hydrogenomonas* strain H 16 contain all the enzymes necessary for fructose utilization. A crude extract phosphorylates glucose, fructose, and mannose at rates characteristic for other microbial and animal hexokinases (100:80:46, respectively); consequently the phosphorylation of fructose is not bound to any specific fructokinase. However, glucose is not utilized. Oxidation of glucose could not even be effected by gradually increasing the glucose concentration in the medium from 0·5 to 10% (w/v). Glucose is not transported into the cell as was shown by "free space" experiments. The cells are cryptic for glucose (Gottschalk, 1964). Mutants, which can utilize glucose, have been produced from *Hydrogenomonas* strains H 16 and H 20 by ultraviolet irradiation and by treatment with nitrite (Schlegel and Gottschalk, 1965). Further mutants arose spontaneously when cells of *Hydrogenomonas* H 16 were grown in the chemostat with fructose as the limiting factor (König et al., 1969). From other wild-type strains similar to *H. eutropha* additional glucose-positive mutants have been isolated after nitrite treatment.

In all of these mutants, uptake of glucose and fructose into the cells depends on specific transport factors. The substrate saturation curve of the respiration of glucose is unlike the fructose curve. While the respiratory rate with fructose as a substrate is already maximal at 1·66 mM, the respiratory rate for glucose as a substrate (at 1·66 mM glucose) amounts only to 15–55% of the maximal rate which is reached at a glucose concentration of 10–30 mM. Furthermore, glucose-grown mutants do not incorporate fructose and *vice versa*, as has been shown using radioactively labelled sugars and chloramphenicol as an inhibitor of enzyme synthesis. The experiments performed with the glucose-positive mutants of *Hydrogenomonas* strain H 16 can be summarized as follows. The sugars induce the enzymes necessary for their utilization; cells grown chemolithotrophically do not oxidize them. The induction is specific; other organic substrates tested do not have an inductive effect. Protein synthesis is involved in each enzyme induction; in the presence of chloramphenicol, there is no induction of either glucose or fructose utilization. Consequently, there are specific permeases responsible for the uptake of fructose and glucose into cells of *Hydrogenomonas* strain H 16g[+].

As can be concluded from the substrate saturation curves for fructose and glucose (König et al., 1969), the transport mechanisms are different.

REGULATORY PHENOMENA IN METABOLISM OF KNALLGASBACTERIA 221

While the uptake of fructose can be characterized as an active transport, that of glucose may be designated as a facilitated diffusion. It is worth mentioning that all of the glucose-positive mutants are constitutively derepressed for the formation of glucose 6-phosphate dehydrogenase; after chemolithotrophic growth, the activity of this enzyme amounts to 400–700 units/g. protein, while the wild type only contains 10–20 units/g. protein. The relationship between glucose utilization and constitutive glucose 6-phosphate dehydrogenase remains unknown.

Gluconate also serves as a substrate and enables the cells to grow and accumulate storage material (PHB) even faster than with fructose (Schlegel and Blackkolb, 1967). For growth of *Hydrogenomonas* H 16 on gluconate, a gluconokinase is induced and 6-phosphogluconate is metabolized *via* the Entner-Doudoroff pathway. The following results indicate that gluconate is degraded exclusively *via* the Entner-Doudoroff pathway: (a) the specific activity of the Entner-Doudoroff enzymes in the crude extract of gluconate-grown cells is as high as in that of fructose-grown cells; (b) single-step mutants which are lacking 2-keto-3-deoxy-6-phosphogluconate aldolase grow neither on fructose nor on gluconate; and (c) 6-phosphogluconate dehydrogenase is present neither in fructose-nor in gluconate-grown cells. From (b) and (c) it can be deduced that gluconate cannot be degraded via the oxidative pentosephosphate pathway (Blackkolb and Schlegel, 1968a). Phosphoglucose isomerase and glucose 6-phosphate dehydrogenase are not induced during growth on gluconate, their specific activity being only slightly higher than that of chemolithotrophically grown cells. In this respect the enzyme induction pattern of *Hydrogenomonas* H 16 differs markedly from that of *Hydrogenomonas facilis*. In this species, during growth on gluconate the enzymes hexokinase, phosphoglucose isomerase and glucose 6-phosphate dehydrogenase (NAD- and NADP-dependent) are induced in addition to gluconokinase and the enzymes of the Entner-Doudoroff pathway (Bowien, 1970). Contrary to *Hydrogenomonas* H 16, *H. facilis* is able to use glucose (Crouch and Ramsey, 1962) and, in addition to the enzymes of the Entner-Doudoroff pathway, contains an NAD-dependent 6-phosphogluconate dehydrogenase (Ramsey, 1968) which is induced during growth on either the hexoses or gluconate (Bowien, 1970).

In addition, *Hydrogenomonas* H 16 as well as other strains of *H. eutropha* are able to grow on 2-ketogluconate and saccharate (Davis *et al.*, 1970); however, neither the degradative pathways nor the regulation pattern for these sugars have been investigated so far.

Acetate is a good growth substrate for *Hydrogenomonas*. However *Hydrogenomonas* strain H 16 oxidized acetate at the maximum rate only at concentrations of up to 0·5% (w/v) (i.e. 0·0367 M-acetate). Above this concentration, self inhibition occurs. Acetate is activated by means

of acetyl-CoA kinase (acetate thiokinase) (Schindler, 1964). With *Hydrogenomonas* strain H 16g$^+$ the possibility of the activation of acetate by a combined functioning of acetokinase and phosphotransacetylase was excluded (Schlegel and Trüper, 1966).

When cells grown either chemolithotrophically or with fructose as substrate are transferred into an acetate-containing medium, induction of the acetate-metabolizing enzymes occurs very rapidly (within 20–30 min.). The specific activities of three representative enzymes of acetate metabolism (acetyl-CoA kinase and the key enzymes of the glyoxylate cycle, isocitrate lyase and malate synthase) increase up to their half maximal level within 20 min. and reach their upper level after 90 min. The results obtained for *Hydrogenomonas* H 16g$^+$ are in accordance with those found for *Hydrogenomonas facilis* (McFadden and Howes, 1962).

B. Effect of Hydrogen on Hexose Degradation

The aerobic heterotrophic mode of life is based on the utilization and oxidation of organic carbon compounds. Organic substrates have two functions; they serve as the sources of carbon skeletons for the synthesis of cellular material as well as energy sources. From the viewpoints of energetics, organic substrates have to be regarded as sources of hydrogen or reducing power which, by slow combustion, give rise to the generation of metabolic energy (more specifically, to the production of ATP). As is well known, the degradation of hexose *via* the Embden-Meyerhof pathway is regulated by the energy charge of the cell, ATP, ADP and AMP being the indicators and effectors. In this way, adenylates determine the functional activity of enzymes and the rate of metabolic flow.

Facultative litho-autotrophic bacteria are characterized by their ability to utilize either organic or inorganic substrates. In the latter case, the carbon and the hydrogen sources are separate compounds. While carbon skeletons are synthesized *via* autotrophic carbon dioxide fixation, metabolic energy is produced by oxidation of the hydrogen source. Therefore, lithotrophic organisms can be supplied with energy in the absence of a carbon source and hence, for theoretical reasons, lend themselves for studies on the influence of energy charge on the regulation of metabolic processes such as substrate degradation, intermediary metabolism, biosynthesis, and accumulation of storage materials. Such studies have been done using *Hydrogenomonas* strain H 16 and other closely related strains.

1. *Effect on Enzyme Formation*

An effect of hydrogen on the utilization of organic substrates was first observed for the synthesis of poly-β-hydroxybutyrate (PHB). Chemolithotrophically grown cells incorporate β-hydroxybutyrate (Schlegel *et al.*, 1961) and acetate (Wilde, 1962) into PHB more slowly in a hydrogen-oxygen (knallgas) atmosphere than in air. The suppressing effect of hydrogen on growth and on PHB synthesis is very pronounced with fructose as the organic substrate (Gottschalk, 1965; Schlegel and Trüper, 1966). While the enzymes involved in fructose degradation are rapidly synthesized when chemolithotrophically grown cells are shaken or stirred in a fructose nutrient medium in air, formation of these enzymes is completely suppressed in a hydrogen-oxygen atmosphere (80:20%, respectively). Even after 20 hours incubation in knallgas, significant activity of the enzymes of the Entner-Doudoroff system could not be found. Growth experiments have been performed lasting several days. Cells were grown chemolithotrophically in a mineral medium, in a hydrogen-oxygen atmosphere containing 10% carbon dioxide, up to a density of about 0·1 mg. dry weight organisms/ml. After the gaseous carbon dioxide had been removed, and the dissolved carbon dioxide had been used up, fructose was added. Although an ample supply of an appropriate substrate (fructose) and the usual hydrogen acceptor (oxygen) were available, no increase of turbidity occurred. Only when, after three days, the hydrogen-oxygen mixture had been replaced by air did the cells start to grow. The suppression of growth by hydrogen can continue for days. The "hydrogen effect" is also observed on fructose-agar plates; while the cells form colonies during incubation in air, growth is completely suppressed in a knallgas (carbon dioxide-free) atmosphere.

A similar response to hydrogen has been observed in all of the strains of knallgasbacteria (58) tested so far; these included the Repaske strain of *H. eutropha*, *H. ruhlandii* (Packer and Vishniac, 1955), strains 11/x, 12/x and 12/60/x (Eberhardt, 1969) and many strains closely related to, or identical with, *H. eutropha*. The repressing effect of hydrogen on the formation of fructose-utilizing enzymes appears to be a general phenomenon. Only chemolithotrophically grown cells of *Micrococcus denitrificans* are not affected by the hydrogen-containing gas mixture and grow on fructose at equal growth rates, either in air or in knallgas (Blackkolb and Schlegel, 1968a).

The suppressing action of hydrogen becomes effective not only when chemolithotrophically-grown cells are transferred to a fructose medium; synthesis of the enzymes necessary for fructose degradation is even

repressed by hydrogen when fructose-grown cells are exposed to hydrogen. In a fructose-containing medium under knallgas, these cells grow arithmetically rather than exponentially, and the specific activity of the Entner-Doudoroff enzymes in the cells decreases as a result of dilution.

Hydrogen does not prevent the adaptation to all substrates. If chemolithotropically-grown cells are transferred to organic media containing lactate, succinate, pyruvate or crotonate growth is equally well initiated in knallgas or in air. However, initiation of growth is suppressed when fructose, acetate, glutamate or aspartate serves as substrates (Gottschalk, 1965). Additional substrates have been tested and found to give a similar response to fructose; these are gluconate, fumarate, citrate, L-leucine, DL-isoleucine, isovalerate, asparagine, L-histidine, L-tyrosine, L-phenylalanine, and L-proline. Among the additional substrates tested so far, only uric acid and DL-tryptophan supported growth in the presence of hydrogen (Blackkolb and Schlegel, 1968a). Also enzyme synthesis induced by uric acid was not subject to fructose repression (Kaltwasser, 1968b).

Using *Hydrogenomonas* strain H 16g^{+} it was found that addition of carbon dioxide, peptone, or lactate to a chemolithotrophically-grown suspension of cells in a fructose-containing medium does not relieve the repressing effect of hydrogen with respect to the formation of the enzymes of the Entner-Doudoroff pathway. The cells grow, however, at the expense of the additional substrates added. Lactate, acetate, glutamate, and pyruvate were found not to repress enzymes of the Entner-Doudoroff pathway under these conditions (Schlegel and Trüper, 1966).

Mutants of *Hydrogenomonas* strain H 16, which lack the hydrogen-activating enzyme hydrogenase and, therefore, do not grow chemolithotrophically, did not respond to the hydrogen-oxygen gas mixture and grew equally well on fructose, either in air or in knallgas. Consequently, neither gaseous contaminations nor the hydrogen itself exert toxic or regulatory effects; the adaptation to, and utilization of, fructose is suppressed only in cells which are able to activate hydrogen.

As outlined above, acetate belongs to the "repressible substrates". When chemolithotrophically-grown cells of *Hydrogenomonas* H 16 are transferred to an acetate-containing medium under knallgas, growth is not initiated. Using *Hydrogenomonas* H 16g^{+}, it has been shown that, under these conditions, the activities of representative enzymes (acetyl-CoA kinase, isocitrate lyase, and malate synthase) remained constant for at least 24 hr., and instantaneously increased when knallgas was replaced by air (Schlegel and Trüper, 1966). With acetate as substrate, the repressing effect of hydrogen is similarly pronounced as in the case of fructose. Suppression of growth can even be demonstrated on solid

acetate-containing media, which, after inoculation, are incubated in either knallgas or air.

The "hydrogen-effect" is similar to the "glucose effect" (Gale, 1943) and has been interpreted as a "catabolite repression" (Magasanik, 1961). In many organotrophic bacteria, yeasts and green algae, glucose prevents the adaptation to other organic substrates (e.g. the formation of the enzymes necessary for acetate utilization).

The hydrogen effect as a regulatory mechanism appeared to be completely analogous to the glucose effect. In order to test whether this comparison is justified, possible existing repressing effects exhibited by organic substrates have been investigated using *Hydrogenomonas* strain H 16g$^+$ (Schlegel and Trüper, 1966). Fructose-grown cells of this mutant do not utilize glucose, since the specific glucose "permease" has not been induced. In the presence of fructose and glucose, fructose is utilized first. The growth curve (increase of protein) shows the characteristic shape of biphasic growth, thus demonstrating that fructose represses the synthesis of glucose permease. Further manometric experiments lead to the conclusion that fructose suppresses the adaptation to glucose; however, glucose does not suppress the induction of the fructose permease. Under the conditions applied, fructose is the preferred substrate.

The formation of the acetate-metabolizing enzymes is equally suppressed by fructose and by hydrogen; in the presence of acetate and fructose under air, their activities start to increase after the fructose has been consumed. From further experiments of this kind it can be deduced that both glutamate and aspartate utilization are suppressed by the presence of fructose. In contrast, however, lactate utilization is not suppressed by fructose.

For the substrates which are subject to the hydrogen effect, an order of preferential utilization may be established. Hydrogen is the most repressive energy source, followed by substrates in the order fructose, glucose and acetate. The results reported here support the assumption that, not only are the relative concentrations of intermediates in certain pools responsible for enzyme regulation, but that there is a strong substrate specificity. When various substrates are offered concomitantly, the order in which degrading enzymes will be formed is predetermined genetically. Whether the order is determined at the transcriptional level, or by permeability differences at the cytoplasmic membrane, remains to be elucidated.

Many investigations aim at elucidation of the mechanisms which cause the catabolite repression. It has been postulated (Magasanik, 1961) that the extent of the repression of an operon is dependent on the concentration of one or several metabolites formed during substrate

degradation in the cell. The results obtained using hydrogen bacteria, especially *Hydrogenomonas* strain H 16, enable us to delineate the group of substances possibly acting as "repressors" within the cell. In the presence of hydrogen, the cells are able to reduce NAD and to regenerate ATP. When the cells are exposed to a carbon dioxide-free knallgas mixture, the adenylates are mainly present in the form of their triphosphates and the nicotinamide nucleotides in their reduced state (Ahrens, 1966). These facts support the hypothesis that, in *H. eutropha* strain H 16, ATP or $NADH_2$, or other metabolites which are in equilibrium with these metabolites, are acting as signals and low molecular weight co-repressors.

The localization and causal analysis of the "hydrogen-effect" has so far resisted all experimental efforts of clarification. Attempts to isolate mutants which are resistant to this type of catabolite repression proved unsuccessful. The difficulties are partially due to a second effect of hydrogen in that hydrogen not only represses the formation of the fructose-degrading enzymes but inhibits the function of the enzyme(s) already present.

2. *Effect on Enzyme Function*

In *Hydrogenomonas* strain H 16, hydrogen not only represses the formation of the enzymes of the Entner-Doudoroff degradative pathway but also inhibits the utilization of fructose in fully induced cells grown on fructose (Schlegel and Blackkolb, 1967). This inhibition results in a decrease in the rates of growth or of formation of PHB by 80%. This observation was indicative of an inhibitory effect of a metabolite affecting one of the enzymes involved in fructose degradation. Since the utilization of gluconate by gluconate-grown cells is not impaired by hydrogen, and since gluconate in *Hydrogenomonas* H 16 is introduced into the Entner-Doudoroff pathway *via* 6-phosphogluconate, the site of inhibition affected by hydrogen had to be located prior to 6-phosphogluconate. A further deduction concerning the effector could be made; inhibition of fructose degradation depends only on hydrogen; carbon dioxide, the concomitant carbon source, is dispensible. Therefore, it appeared more probable that the inhibition of the initiating conversions of fructose to 6-phosphogluconate is attributable to a product of hydrogen oxidation, ATP and/or $NADH_2$, rather than to intermediatry products of the carbon metabolism. Preliminary experiments showed neither hexokinase nor phosphoglucose isomerase being significantly inhibited. Finally, glucose 6-phosphate dehydrogenase proved to be sensitive to ATP and $NADH_2$ (Blackkolb and Schlegel, 1968b).

REGULATORY PHENOMENA IN METABOLISM OF KNALLGASBACTERIA 227

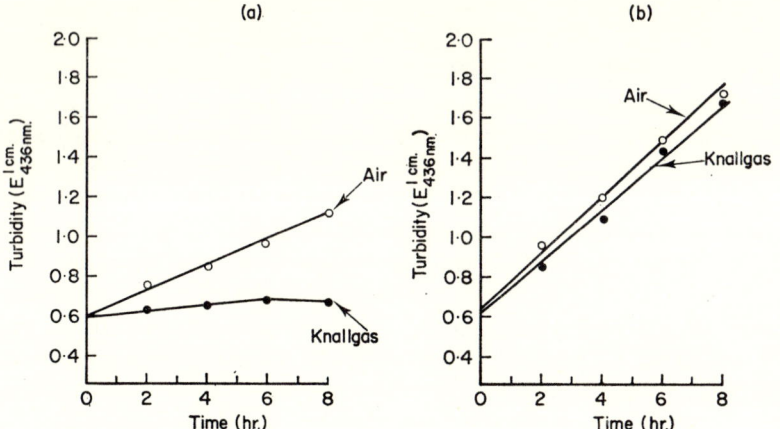

FIG. 1. The effect of hydrogen on the utilization of (a) fructose or (b) gluconate by fully adapted cells of *Hydrogenomonas eutropha* strain H 16. The cells were grown on the corresponding substrates, washed, and incubated in a nitrogen-free mineral solution under air or knallgas (80% hydrogen + 20% oxygen), respectively. The increase of turbidity caused by the accumulation of poly-β-hydroxy-butyric acid within the cells was determined. While the utilization of fructose is suppressed by hydrogen, the utilization of gluconate is not impaired.

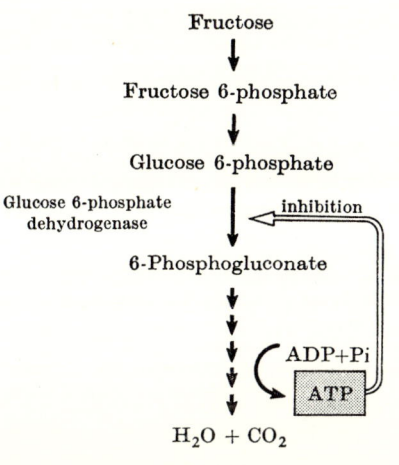

FIG. 2. Regulation of fructose degradation *via* the Entner-Doudoroff pathway by allosteric inhibition by adenosine triphosphate of the enzyme glucose 6-phosphate dehydrogenase.

Glucose 6-phosphate dehydrogenase has been purified about 100-fold from extracts (French press) of fructose-grown *Hydrogenomonas* H 16 cells. The enzyme differs from the yeast enzyme by a lower substrate affinity, a sigmoid substrate-saturation curve, and by a specificity for

both NADP and NAD. In the presence of Mg^{2+}, half maximal activity is reached at $K_{0.5} = 1 \times 10^{-3}$ M glucose 6-phosphate. The enzyme is inhibited by ATP and $NADH_2$. The sigmoid shape of the substrate-saturation curve (glucose 6-phosphate) is increased by ATP. From kinetic data the following conclusions could be drawn: (a) glucose 6-phosphate dehydrogenase is an allosteric enzyme which is composed of at least four subunits; ATP is a negative allosteric effector; (b) the binding sites of the subunits for the substrate show co-operative interactions; (c) the enzyme fits the "K-system", i.e. the allosteric effector changes the apparent K_m value without influencing the maximal velocity (V_{max}) of enzyme action; (d) the inhibition by $NADH_2$ is caused by competition between NADP and $NADH_2$ for the binding site of the coenzymes at the enzyme.

The enzyme is more or less strongly inhibited by nucleoside triphosphates (ATP, GTP, ITP, CTP, UTP) as well as by ADP. The inhibitory effect of AMP and other nucleoside monophosphates is negligible. A positive effector has not been found. Further data, however, suggest that the interaction between the nucleotide phosphates and the enzymes may be more complex than suggested in the foregoing. The inhibition caused by ATP can be completely relieved by magnesium sulphate if glucose 6-phosphate is present in quasisaturating concentrations. The inhibitory effect becomes minimal when the Mg^{2+}:ATP ratio exceeds 1·3.

With regard to the regulation of hexose degradation *via* the Embden-Meyerhof pathway, the sensitivity of glucose 6-phosphate dehydrogenase to adenylates is not surprising. Phosphofructokinase is the first enzyme which is specific for the glycolytic pathway and is allosterically inhibited by ATP and citrate and is activated by AMP and ADP. The result of this regulation is, as a rule, only recognized in intact cells if two processes of different energetic efficiency are compared, e.g. respiration and fermentation. In yeast, this type of regulation is referred to as the "Pasteur effect". It prevents the overproduction of metabolites and, therefore, makes sense. Generally the result of this allosteric regulation of a catabolic pathway remains unrecognized in intact cells. An allosteric catabolic control mechanism will, *in vivo*, easily be recognized only in those cells which are able to gain energy or reducing power in the absence of a carbon source. Possible candidates for observing inhibitory effects similar to those encountered in hydrogen bacteria are all facultative chemolithotrophs which can utilize hexoses. However, the authors are not aware of any bacterium of that kind which uses the Embden-Meyerhof degradative pathway. Therefore, the organism with which the inhibition of phosphofructokinase by high energy charge can be demonstrated *in vivo* has still to be discovered.

REGULATORY PHENOMENA IN METABOLISM OF KNALLGASBACTERIA 229

The discovery of other chemolithotrophic "Entner-Doudoroff organisms" containing an allosterically sensitive glucose 6-phosphate dehydrogenase and exhibiting an inhibition of hexose utilization by the inorganic energy source can be more hopefully envisaged. Related phenomena of enzyme repression have already been found in *Thiobacillus intermedius* (discussed by S. Rittenberg, 1969; Matin and Rittenberg, 1970a, b).

Glucose 6-phosphate dehydrogenases from *Pseudomonas aeruginosa* (Lessie and Neidhardt, 1967) and *Pseudomonas fluorescens* (Schindler and Schlegel, 1969) exhibit sigmoidal substrate-saturation curves (glucose 6-phosphate) and are inhibited by ATP. Thus, they are similar to the enzyme in *Hydrogenomonas* H 16. The enzymes obtained from *Escherichia coli, Acetobacter xylinum* and *A. suboxydans* are dissimilar to that of *Hydrogenomonas* H 16. The results support the idea of the glucose 6-phosphate dehydrogenases functioning only in a degradative pathway, being subject to adenylate control with ATP as a negative allosteric effector. The enzymes serving other functions are of the yeast type; they exhibit hyperbolic substrate-saturation curves and are not subject to adenylate control. These speculations deserve, however, further comparative investigations.

C. Formation and Function of the Tricarboxylic Acid Cycle Enzymes

Under chemolithotrophic conditions, hydrogen bacteria derive reducing power from the hydrogenase reaction, and the tricarboxylic acid cycle mainly serves the synthesis of metabolites. Theoretically, one enzyme (α-ketoglutarate dehydrogenase) is dispensable in this case (Smith *et al.*, 1967). The specific activity of the enzymes of the tricarboxylic acid cycle, and of those connected with the cycle, has been determined using a mutant of *Hydrogenomonas* H 16 which is able to utilize glucose (*Hydrogenomonas* strain H 16g$^+$).

The cells were grown on carbon dioxide plus hydrogen, fructose, acetate or glutamate. The differences in the specific activities of the enzymes engaged in the cycle itself were not enormous. Autotrophically-grown cells exhibited less activity than organotrophically-grown cells with regard to isocitrate dehydrogenase (NADP-dependent), fumarase and malate dehydrogenase. Acetate-grown cells were highest with respect to citrate synthase and fumarase (Trüper, 1965). The inhibition of the tricarboxylic acid-cycle enzymes during mixotrophic growth has not been tested so far. Some results are indicative of a complete suppression of the cycle by hydrogen. When [2-^{14}C]-acetate was incorporated into *Hydrogenomonas* H 16 cells in a hydrogen-oxygen

atmosphere as much as 96% of the radioactivity was found in C-2 and C-4 of the β-hydroxybutyric monomer, derived (for analysis) from the PHB formed during incubation. If a significant proportion of the acetate had been oxidized through the cycle, the [^{14}C]-carbon dioxide would have been re-incorporated by autotrophic fixation and would have been more widely distributed throughout the entire PHB molecule. The inhibition of the tricarboxylic acid cycle exerted by hydrogen may be caused by the high $NADH_2/NAD$ ratio and the inhibition of α-keto-glutarate dehydrogenase by $NADH_2$ (H. G. Schlegel, unpublished result); its sensitivity to $NADH_2$ is similar to that of the pyruvate dehydrogenase in *Escherichia coli K* 12 (Hansen and Henning, 1966).

The essential anaplerotic carbon dioxide fixation resulting in oxalo-acetate is catalysed in *Hydrogenomonas facilis* and in *Hydrogenomonas* H 16 by phosphoenolpyruvate carboxylase. In *Hydrogenomonas* H 16 this enzyme is stimulated by acetyl-coenzyme A and inhibited by aspartate (Frings, 1969).

D. Utilization of Nitrogenous Compounds

The nitrogen requirement of *Hydrogenomonas eutropha* strains and *H. facilis* can be satisfied by many compounds, e.g. ammonia, nitrate, urea, amino acids and purine derivatives. Some of the enzymes of nitrogen metabolism were found to be subjected to regulation by induction or repression-derepression mechanisms.

Hydrogenomonas H 16 contains, after growth on nitrate as the sole source of nitrogen, a membrane-bound assimilatory nitrate reductase (Pfitzner, 1969). Reduction of nitrate to ammonia seems to be a growth rate-limiting step, since autotrophic or heterotrophic growth with nitrate is somewhat slower than growth with ammonium chloride. The possible coupling of hydrogenase and nitrate reductase formation was mentioned in Section II C (p. 212). In *Micrococcus denitrificans*, the assimilatory nitrate reduction seems to be controlled separately from the dissimilatory nitrate reduction (Chang and Morris, 1962). The assimi-latory enzyme is repressed by ammonia or, according to Pichinoty (1970), not influenced by ammonia.

Hydrogenomonas eutropha and *Hydrogenomonas* H 16 contains two different glutamate dehydrogenases, one specific for NAD and the other for NADP (Krämer, 1970; Joseph and Wixom, 1970). For both strains, it was concluded that the NADP-dependent glutamate dehydrogenase is the enzyme involved in the biosynthesis of glutamate, whereas the NAD-dependent enzyme is operative in the degradation of glutamate. The relation of the levels of the two glutamate dehydrogenases was dependent on the growth conditions, thereby indicating unco-ordinated synthesis.

REGULATORY PHENOMENA IN METABOLISM OF KNALLGASBACTERIA 231

The differences in levels of activity, however, were not too pronounced. In *H. eutropha* (Joseph and Wixom, 1970) the activity of the NADP-dependent enzyme was almost independent of the nitrogen source used, i.e. ammonium chloride, L-glutamate or L-aspartate (variations between 5·2 and 7·4 nmoles NADP/min./mg. protein). The activity of the NAD-dependent enzyme was low (about 0·2 nmole NAD/min./mg. protein) with ammonium chloride and reached a maximum after growth on glutamate (0·9 nmole NAD/min./mg. protein). In *Hydrogenomonas* H 16 (Krämer, 1970) the activity of the NADP-dependent glutamate dehydrogenase was at its maximum with ammonium chloride as nitrogen source regardless of the simultaneous presence of glutamate. After exhaustion of ammonium chloride, or with glutamate as the sole source of nitrogen, the activity was half maximal. The NAD-dependent enzyme was found to have the same specific activity regardless of the nitrogen source used (i.e. ammonium chloride, ammonium chloride plus glutamate or glutamate alone). After exhaustion of the nitrogen source, however, the activity increased three-fold. Synthesis of the NAD-dependent glutamate dehydrogenase seems to be regulated in a co-ordinated manner with glutamine synthetase. The ratio between both enzyme activities was the same under all growth conditions tested. In this respect *Hydrogenomonas* H 16 resembles yeast (Kohlhaw *et al.*, 1965).

As investigated in *Hydrogenomonas* strain H 16, the utilization of urea depends on the presence of urease (König *et al.*, 1966). This enzyme is subject to a very efficient control mechanism. During cell growth in the presence of excess ammonia as the single nitrogen source, the specific activity of urease amounts to 7 units/g. protein. When urea is the only nitrogen source during exponential growth, the cells have a specific urease activity of 40–60 units/g. protein. The results may be interpreted as an "induction" of urease by urea. However, unexpected results have been obtained when the urease activity was determined after growth at the expense of ammonia became stationary due to complete consumption of ammonia; in this case, the specific activity of urease reached an upper level of 220–300 units/g. Finally, when the cells were grown at the expense of a limiting amount of urea as the only nitrogen source, an upper level of 600 units urease/g. protein was reached. Urease is not formed when both ammonia and urea are present in the growth medium. The results obtained indicate that formation of urease is controlled by repression (by ammonia) as well as by endogenous derepression caused by the absence of a nitrogen source. For the apparent "induction" of urease initiated by urea, conditions are required which promote endogenous derepression of urease formation.

Extending these experiments, it could be shown that, during growth of *Hydrogenomonas* strain H 16 in a fructose-containing medium with

9

urea as the single nitrogen source, oscillations of urease activity and ammonia concentration occur (König and Schlegel, 1967). When urease-poor cells, grown in the presence of ammonia, were inoculated into fructose-urea medium, the specific urease activity increased, this rise being followed by an increase in the concentration of ammonia in the medium. By hydrolysis of urea, more ammonia was produced than was used for amino-acid synthesis. Before the ammonia concentration reached its peak, a repression of urease formation commenced, and the specific activity of urease fell to a median level by "dilution". After utilization of the ammonium ions that accumulated as a product of urea hydrolysis, urease synthesis was re-initiated which, in turn, led to renewed production of ammonium ions. This alternating cycle of repression and derepression recurred three times. During continuous growth in the chemostat, by varying the ammonia concentration and the dilution rate, it was observed that urease formation is already completely repressed at a concentration of $1\cdot4$ mM NH_4^+.

As shown in a comparative study (Krämer *et al.*, 1967) control of urease formation by repression and endogenous derepression is not unique and not specific for *Hydrogenomonas* strain H 16. Strains H 1, H 2, H 20, 10/X, and 12/X as well as *Pseudomonas fluorescens*, *Ps. aeruginosa*, *Bacillus megaterium*, *Micrococcus denitrificans*, and *M. cerificans* react similarly. With respect to urease formation, this group can be differentiated from a group of bacteria producing urease constitutively (*Sporosarcina ureae*, *Proteus vulgaris*, *Bacillus pasteurii*, *Azotobacter vinelandii*) or not at all (*Escherichia coli* B, *Serratia marcescens*, *Bacillus polymyxa*).

In the stationary phase of growth, urea is possibly produced in the course of the degradation of purine derivatives formed during RNA breakdown. Thus, the apparent derepression of urease in the absence of nitrogen sources might be simulated by endogenously produced urea and thereby causing an induction of urease. Consequently, degradation of uric acid or allantoin were studied (Kaltwasser, 1968a, b, 1969). The experiments showed very clearly that urea is formed during uric acid degradation, but that this does not cause the formation of urease because of the simultaneous formation of ammonia. Urease was rather formed after exhaustion of uric acid and ammonia. Thus, it was confirmed that urease formation is initiated only by derepression in the absence of utilizable nitrogen sources. Presumably, with respect to the type of regulation, urease belongs to that group of enzymes (phosphatase, sulphatase) which are derepressed in the absence of the inorganic reaction product.

Hydrogenomonas H 16 and *H. facilis* are able to use purine derivatives as sources of carbon and nitrogen (Kaltwasser and Krämer, 1968).

Adenine, guanine, xanthine, and hypoxanthine are converted to uric acid, which, in turn, is transformed to allantoin by uricase. During degradation of allantoin, ammonia, urea, and glyoxylic acid are formed in the ratio $2:1:1$. This suggests degradation of allantoin *via* the intermediary products allantoic acid, ureidoglycine, and ureidoglycolic acid as in *Pseudomonas acidovorans* (Trijbels, 1967; Trijbels and Vogels, 1966a, b, 1967). D-Glycerate is formed from glyoxylic acid by the action of glyoxylate carboligase and D-glycerate dehydrogenase (Kaltwasser, 1968b, 1969). Regulation of uricase, glyoxylate carboligase, D-glycerate dehydrogenase and urease is shown in Figure 3. In the presence of uric

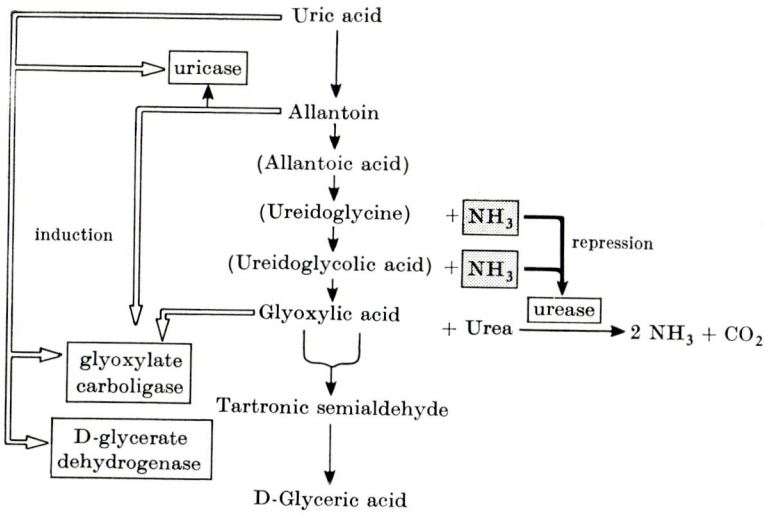

FIG. 3. Regulation of the formation of uricase, urease, glyoxylate carboligase, and D-glycerate dehydrogenase in *Hydrogenomonas* H 16 according to Kaltwasser (1968a, b, 1969). The possible induction of D-glycerate dehydrogenase by glyoxyate or allantoin was not tested.

acid as sole source of carbon and nitrogen, all of the above named enzymes, except for urease, were formed. After exhaustion of uric acid the specific activities of uricase and glyoxylate carboligase decrease, possibly because of degradation of the enzymes. Addition of ammonium chloride to cells using uric acid did not alter the increase and decrease in the specific activity. When uric acid and fructose were present simultaneously, *Hydrogenomonas* H 16 preferably used uric acid. Under these conditions the specific activities of uricase and glyoxylate carboligase increased faster than with uric acid as the sole substrate. After exhaustion of uric acid, fructose was utilized and the decrease in uricase and glyoxylate carboligase activity seemed to be more pronounced. In the presence of allantoin, glyoxylate carboligase was

induced, the specific activity being twice as high as with uric acid. Uricase was present in a small amount during the early phase of growth; later the activity decreased.

Glyoxylate carboligase also was formed in the presence of glyoxylate and ammonium chloride. Under these conditions fructose repressed the formation of glyoxylate carboligase; as mentioned above, with uric acid as a substrate, fructose did not affect the induction of glyoxylate carboligase (Kaltwasser, 1969).

Urease was not formed during the degradation of uric acid or allantoin. However, after transferring cells of *Hydrogenomonas* H 16 to a fructose-containing medium devoid of a nitrogen source, urease was formed. Under these conditions uricase and glyoxylate carboligase activity did not appear. Thus there is a clear difference in the regulation of the enzymes which participate in the degradation of purine derivatives.

As can be seen in Figure 3, synthesis of uricase and glyoxylate carboligase is not co-ordinate. Unfortunately, there are too few data to decide whether synthesis of D-glycerate dehydrogenase and of glyoxylate carboligase are coupled as in other bacteria (e.g. *Arthrobacter allantoicus* and *Streptococcus allantoicus*; cf. Vogels, 1963; Valentine *et al.*, 1964). The only pyrimidine derivative that can be utilized as a nitrogen source by *Hydrogenomonas* H 16 is cytosine (Kaltwasser and Krämer, 1968). However, only one of the three nitrogen atoms (the amino nitrogen) is incorporated into cell material, while uracil is accumulated in the medium. The deaminating enzyme, cytosine desaminase, apparently is induced by cytosine (Kaltwasser and Krämer, 1968).

Hydrogenomonas facilis is able to use thymine, uracil and cytosine as nitrogen sources, incorporating all of the nitrogen atoms of these compounds. Cytosine is deaminated to form uracil, which in turn is degraded reductively, *via* dihydrouracil and 3-ureidopropionate, to β-alanine, carbon dioxide and ammonia. While cytosine deaminase activity seemed to be rather independent of the nitrogen source (ammonium chloride, cytosine or uracil), the activities of dihydrouracil dehydrogenase and dihydrouracil hydrase were practically absent with ammonium chloride as a nitrogen source. Significant activity was found in cells grown on cytosine and the highest activity was found in cells grown on uracil. Urease, which has no relation to the degradation of pyrimidine derivatives, was not fully repressed by cytosine or uracil (Krämer and Kaltwasser, 1969).

IV. Biosynthesis of Amino Acids

The regulation patterns of biosynthetic pathways have become valuable characteristics for taxonomic investigations. While the basic

REGULATORY PHENOMENA IN METABOLISM OF KNALLGASBACTERIA   235

enzyme patterns are very similar within a group of organisms belonging
to the same metabolic type they may be different from those characteristic for other groups. Considering the metabolic differences between
autotrophic and heterotrophic carbon assimilation (the need for additional ATP and reducing power for autotrophic growth, and possible
differences in the pool sizes of metabolites related to autotrophic
carbon dioxide fixation) the elucidation of regulation patterns in chemolithotrophic organisms is of special interest. Since strain *Hydrogenomonas*
H 16 is readily subject to mutagenesis and to procedures resulting in the
enrichment of auxotrophic (Schlegel *et al.*, 1965) and deregulated
mutants, relevant studies have been initiated.

Already much attention has been paid to the regulation of the iso-
leucine-valine-leucine family of amino acids. The variation of the overall
control pattern among different groups of bacteria and of yeast is not
pronounced. In *Hydrogenomonas* strain H 16, the synthetic pathways
of these amino acids follow the general pattern (Reh and Schlegel, 1969).
Isoleucine and valine are synthesized *via* parallel reactions catalysed
by a set of four enzymes. The substrate for isoleucine synthesis is the
deamination product of threonine; α-ketoisovalerate is the common
precursor of valine and leucine. The metabolic flow is controlled by
repression as well as by feedback inhibition.

In the wild type during normal growth the enzymes are present at
median activity. Only when an isoleucine-valine auxotrophic mutant
(which is probably blocked in the dihydroxy acid dehydratase reaction)
was grown in a chemostat, and when growth was limited by the concentration of valine, was the formation of two representative enzymes
(threonine deaminase and acetohydroxy acid synthase) derepressed.
The activity of threonine deaminase was increased about 50% and that
of acetohydroxy acid synthase nearly five-fold. When growth was
limited by isoleucine, only threonine deaminase was derepressed; its
specific activity increased about 80%. The repressive effect of the end
product is not significant. When the wild-type strain is grown in the
presence of the corresponding end products (valine, isoleucine and
leucine) the activity of the enzymes is not diminished.

Allosteric interactions have been tested for both enzymes. Threonine
deaminase is inhibited by very low concentrations of isoleucine. At
0·3 mM L-isoleucine, the inhibition is already complete. The inhibition
is partially relieved by L-valine. The substrate-saturation curve (threonine) has a hyperbolic shape; in the presence of isoleucine (0.05 mM)
the curve is sigmoid. From an isoleucine-requiring mutant, defective
in threonine deaminase, a prototrophic revertant has been isolated.
The threonine deaminase of this revertant differs from the wild-type
enzyme by a diminished affinity for isoleucine; half maximal inhibition

occurs at $2 \cdot 6$ mM isoleucine, in contrast to $0 \cdot 2$ mM for the wild-type enzyme. As a result of this diminished allosteric sensitivity, the revertant excretes isoleucine.

The acetohydroxy acid synthase is inhibited only by L-valine. At its pH optimum ($9 \cdot 0$) the inhibition amounts only to 25%. At pH $7 \cdot 4$, the inhibition at $0 \cdot 1$ mM L-valine is 50%. By varying the Mg^{2+} concentration, an 80% inhibition was reached at $0 \cdot 2$ mM L-valine. The sensitivity of the enzyme is specific for L-valine; L-isoleucine and L-leucine neither decrease the activity nor act antagonistically. The inhibition of this enzyme is of more general interest in so far as a decrease of its activity not only impairs the biosynthesis of valine but that of isoleucine also. In *Escherichia coli* strain K 12, the acetohydroxy acid synthase is very sensitive to valine, even *in vivo*; $0 \cdot 1$ mM L-valine already prevents growth (Leavitt and Umbarger, 1962). In *Hydrogenomonas* H 16 there is neither a strong repression or a pronounced allosteric inhibition; growth is, therefore, not yet inhibited by 1 mM L-valine.

A type of regulation different from that in organisms investigated before was discovered when studying the leucine biosynthetic pathway in *Hydrogenomonas* H 16 (Hill and Schlegel, 1969a, b). This pathway starts from α-ketoisovalerate, which is the immediate precursor of valine. The first enzyme specific for leucine synthesis (α-isopropylmalate synthetase) catalyses the condensation of α-ketoisovalerate and acetyl-coenzyme A. Four further enzymes are involved in the conversion of α-isopropylmalate to α-ketoisocaproate and finally to leucine. This pathway, as investigated in *Salmonella typhimurium* and *Neurospora crassa* (Burns *et al.*, 1966; Webster and Gross, 1965), is subject to repression and to allosteric inhibition of α-isopropylmalate synthetase by L-leucine. When studying the effect of a structural analogue of L-leucine (DL-5′,5′,5′-trifluoroleucine) on *Hydrogenomonas* H 16, mutants resistant to the analogue were found which did not fit the usual enzyme control pattern.

Growth of *Hydrogenomonas* H 16 is completely suppressed by the leucine analogue at concentrations higher than $0 \cdot 05$ mM. After mutagenesis, mutants were selected which tolerated 250-fold that concentration. Four types of mutants were recognized: (a) constitutively derepressed for the formation of the enzyme α-isopropylmalate synthetase; (b) mutants containing an α-isopropylmalate synthetase insensitive to end product inhibition by L-leucine; (c) mutants constitutively derepressed for the formation of acetohydroxy acid synthase; and (d) a mutant containing an acetohydroxy acid synthase insensitive to end-product inhibition by L-valine. Furthermore, prototrophic revertants of isoleucine- or isoleucine-valine-requiring auxotrophic mutants of *Hydrogenomonas* H 16 which produce acetohydroxy acid synthase con-

REGULATORY PHENOMENA IN METABOLISM OF KNALLGASBACTERIA 237

stitutively, or which produce an enzyme with diminished valine sensitivity, proved to be resistant to 1 mM trifluoroleucine. While the types of mutants mentioned under (a) and (b) had to be anticipated among the mutants resistant to the leucine analogue, the cause of resistance of the other mutants could not be explained on the basis of information obtained so far. The isolation of mutants carrying regulatory defects in the regulation of the valine-isoleucine biosynthetic pathway, by means of trifluoroleucine, is attributable to an unusual property of the α-isopropylmalate synthetase of *Hydrogenomonas* H 16. L-Valine is not only related to the leucine pathway as a precursor, but acts as a positive effector for the first enzyme of the leucine pathway. In this way, overproduction of valine is followed by an overproduction of leucine, which finally confers resistance to the bacteriostatic effect of trifluoroleucine.

For the study of α-isopropylmalate synthetase, a crude extract of cells, which had been grown under conditions of derepression, was used. Maximal inhibition amounts to 93% at pH 7·4. The sensitivity to leucine is maximal at pH 7·2 and negligible at pH 8·4. The inhibition by trifluoroleucine amounts to 83%. Other structurally related amino acids have no inhibitory effect. However, they are able to relieve the inhibition by L-leucine when applied in 5–10-fold excess, and thus act antagonistically to L-leucine. Among the amino acids tested (L-valine, L-isoleucine, DL-α-aminobutyric acid, DL-norvaline, DL-norleucine, and cycloserine) valine is most effective. The antagonistic effect of valine depends on its concentration; at 0·5 mM L-leucine and varying concentrations of L-valine, the enzyme activity increases linearly up to 6 mM L-valine, and then levels off. Thus, valine confers a protection against L-leucine or trifluoroleucine as a result of the interaction between valine and leucine, or its structural analogue, with the enzyme α-isopropylmalate synthetase. While leucine acts as a negative effector in the leucine biosynthetic pathway, valine serves as a negative effector in its own pathway and additionally as a positive effector in the diverging pathway of leucine biosynthesis. The isolation of trifluoroleucine mutants, which are altered in the regulatory control of the valine pathway and which overproduce valine, supports the conclusion that the leucine-valine interactions found *in vitro* are effective *in vivo* as well.

Hydrogenomonas has been already included into the group of organisms in which the control of the aromatic amino-acid biosynthetic pathway has been studied (Jensen *et al.*, 1967). In the hydrogen-oxidizing bacteria assayed (*H. facilis*, *H. eutropha*, *H. pantotropha*), the first enzyme of the pathway is 3-deoxy-D-arabinoheptulosonic acid 7-phosphate synthetase (Gibson and Pittard, 1968).

V. Final Considerations

This review has demonstrated that important problems in the regulation of chemolithotrophic metabolism in hydrogen bacteria still are unsolved. More systematic studies of the key enzymes of chemolithotrophic growth and the control of their formation are urgently needed. Formation of these enzymes seems to be controlled by a great variety of factors including the presence or absence of hydrogen and organic substrates as well as by the growth rate, oxygen limitation and other environmental conditions. Consequently, measurement of enzyme formation or activity should be made under precisely defined conditions and must be accompanied by the determination of the energy charge and the $NADH_2/NAD$ ratio in the cell, as these have been shown to repress the formation and inhibit the activity of key enzymes (e.g. glucose 6-phosphate dehydrogenase). They may, therefore, determine the extent to which metabolic pathways will prevail during growth under prescribed cultural conditions. Furthermore, control of hydrogen activation and oxidation exerted by carbon dioxide still has to be elucidated.

As far as knallgasbacteria have been studied with respect to their organotrophic metabolism, they do not differ at all from heterotrophic bacteria. With regard to their basic metabolism, they are completely comparable to non-lithotrophic strains. The capability to gain energy and cell carbon from litho-autotrophic reactions will be recognized only if these bacteria are assayed for this property. Nevertheless, the possibility has to be kept in mind that, in acquiring the capability for litho-autotrophic growth, the cells had to adapt their regulation patterns of the basic metabolic pathways to those metabolic equilibria which are characteristic of autotrophic growth. In this context, lack of systematic studies with regard to the metabolite concentrations in the cells has to be emphasized. Preliminary results indicate that the intermediates of the pentose phosphate cycle, and of related pathways, during autotrophic carbon dioxide fixation, are present in higher concentrations than during hexose degradation and heterotrophic growth.

Interactions of the heterotrophic and autotrophic metabolism become evident if hydrogen and/or carbon dioxide and an organic compound are simultaneously present. One of these interactions is the repression of hydrogenase(s) by an organic substrate in the presence of hydrogen, i.e. a suppression of the chemolithotrophic metabolism by the heterotrophic metabolism. On the other hand, hydrogen inhibits the synthesis and the function of some catabolic enzymes thereby suppressing the heterotrophic reactions. These observations must be correlated in any uniform concept of the regulatory patterns of chemolithotrophic and organotrophic metabolism.

As outlined by Rittenberg (1969) the hydrogenomonads belong to that versatile group of chemolithotrophic bacteria that is able to grow mixotrophically, i.e. they do not have to choose between autotrophic or heterotrophic growth as alternatives. Rather, the assimilation of some organic compounds can be coupled with the oxidation of hydrogen (Gottschalk, 1964; DeCicco and Stukus, 1968; Rittenberg and Goodman, 1969; Stukus and DeCicco, 1970); conversely, organic compounds can serve as a source of energy and reducing power for the assimilation of carbon dioxide (Wilson *et al.*, 1953; Gottschalk *et al.*, 1964). However, the mixotrophic way of life has only been demonstrated in short-term experiments lasting one or a few generations. It is still questionable whether the regulatory mechanisms in fact allow continual mixotrophic growth. The apparently unco-ordinate regulation of the synthesis of hydrogenases and of the Calvin-cycle enzymes supports the concept that continual mixotrophic growth is possible.

Our knowledge of the physiology and biochemistry of hydrogen bacteria is restricted almost exclusively to the hydrogenomonads (*Pseudomonas* and *Alcaligenes*). An extensive examination of strains not belonging to this group may answer the question whether there are regulatory patterns common to all knallgasbacteria. The strains studied so far are very similar in that they grow easily under autotrophic conditions, though much slower than with most organic substrates. Considering the great taxonomic variety of hydrogen bacteria, it should not be surprising if there are strains which need organic compounds for good growth with hydrogen or need hydrogen for the utilization of organic compounds. The first example seems to be realized with certain strains of *Micrococcus denitrificans* (Banerjee and Schlegel, 1966) and *Hydrogenomonas* strain Z-56 (Zhilina, 1970). A hydrogen bacterium utilizing organic compounds only during hydrogen oxidation would be analogous to *Nitrobacter agilis* which is able to incorporate acetate only in the presence of nitrite (Smith and Hoare, 1968). In the three bacteria mentioned, regulation patterns different from those of *Hydrogenomonas* strain H 16 should be expected.

The study of hydrogen bacteria has still much to contribute to the elucidation of cellular control mechanisms and will be especially helpful in answering the question whether or not the control mechanisms found *in vitro* are of importance *in vivo*.

VI. Acknowledgements

The authors are very grateful to Dr. Carl Bovell for improving the manuscript and making many suggestions. The suggestions made by

W. Frings, N. Glaeser, H. Kühnemund, V. Oeding, and J. Pfitzner are greatly appreciated.

REFERENCES

Ahrens, J. (1966). Thesis: University of Göttingen.
Atkinson, D. E. and McFadden, B. (1954). *J. biol. Chem.* **210**, 885.
Banerjee, A. K. and H. G. Schlegel (1966). *Arch. Mikrobiol.* **53**, 132.
Bartha, R. (1962). *Arch. Mikrobiol.* **41**, 313.
Blackkolb, F. and Schlegel, H. G. (1968a). *Arch. Mikrobiol.* **62**, 129.
Blackkolb, F. and Schlegel, H. G. (1968b). *Arch. Mikrobiol.* **63**, 177.
Bock, E. and Engel, H. (1966). *Arch. Mikrobiol.* **54**, 191–198.
Bone, D. H., Bernstein, S. and Vishniac, W. (1963). *Biochim. biophys. Acta* **67**, 581–588.
Bongers, L. (1970). *J. Bact.* **104**, 145.
Bovell, C. (1957). Thesis: University of California, Davis, California, U.S.A.
Bowien, B. (1970). Thesis: University of Göttingen.
Burns, R. O., Calvo, I. M., Margolin, P. and Umbarger, H. E. (1966). *J. Bact.* **91**, 1570.
Chang, J. P. and Morris, J. G. (1962). *J. gen. Microbiol.* **29**, 301.
Cook, D. W., Tischer, R. G. and Brown, L. R. (1967). *Can. J. Microbiol.* **13**, 701.
Crough, D. J. and Ramsey, H. H. (1962). *J. Bact.* **84**, 1340.
Davis, D. H., Doudoroff, M. and Stanier, R. Y. (1969). *Int. J. syst. Bact.*, **19**, 375.
Davis, D. H., Stanier, R. Y., Doudoroff, M. and Mandel, M. (1970). *Arch. Mikrobiol.* **70**, 1.
De Cicco, B. T. and Stukus, P. E. (1968). *J. Bact.* **95**, 1469.
Eberhardt, U. (1965). Thesis: University of Göttingen.
Eberhardt, U. (1966a). *Arch. Mikrobiol.* **53**, 288.
Eberhardt, U. (1966b). *Arch. Mikrobiol.* **54**, 115.
Eberhardt, U. (1969). *Arch. Mikrobiol.* **66**, 91.
Fewson, C. A. and Nicholas, D. J. D. (1961). *Biochim. biophys. Acta* **48**, 208.
Frings, W. (1969). Thesis: University of Göttingen.
Gale, E. F. (1943). *Bact. Rev.* **7**, 139.
Gibson, F. and Pittard, J. (1968). *Bact. Rev.* **32**, 465.
Gottschalk, G. (1964). *Arch. Mikrobiol.* **49**, 96.
Gottschalk, G. (1965) *Biochem. Z.* **341**, 260.
Gottschalk, G., Eberhardt, U. and Schlegel, H. G. (1964). *Arch. Mikrobiol.* **48**, 95.
Hansen, R. G. and Henning, U. (1966). *Biochim. biophys. Acta* **122**, 355.
Hill, F. and Schlegel, H. G. (1969a). *Arch. Mikrobiol.* **68**, 1.
Hill, F. and Schlegel, H. G. (1969b). *Arch. Mikrobiol.* **68**, 18.
Hippe, H. (1967). *Arch. Mikrobiol.* **56**, 248.
Jensen, R. A., Nasser, D. S. and Nester, E. W. (1967). *J. Bact.* **94**, 1582.
Joseph, A. A. and Wixom, R. L. (1970). *Biochim. biophys. Acta* **201**, 295.
Kaltwasser, H. (1962). *Arch. Mikrobiol.* **41**, 282.
Kaltwasser, H. (1968a). *Arch. Mikrobiol.* **60**, 160.
Kaltwasser, H. (1968b). *Arch. Mikrobiol.* **64**, 71.
Kaltwasser, H. (1969). *Arch. Mikrobiol.* **65**, 288.
Kaltwasser, H. and Krämer, J. (1968). *Arch. Mikrobiol.* **60**, 172.

REGULATORY PHENOMENA IN METABOLISM OF KNALLGASBACTERIA 241

Kaserer, H. (1906). *Zentbl. Bakt. Parasitkde Abt.* (*II*). **16**, 681.
König, Chr., Kaltwasser, H. and Schlegel, H. G. (1966). *Arch. Mikrobiol.* **53**, 231.
König, Chr. and Schlegel, H. G. (1967). *Biochim. biophys. Acta* **139**, 182.
König, Chr., Sammler, I., Wilde, E. and Schlegel, H. G. (1969). *Arch. Mikrobiol.* **67**, 51.
Kohlhaw, G., Drägert, W. and Holzer, H. (1965). *Biochem. Z.* **341**, 224.
Krämer, J. (1970). *Arch. Mikrobiol.* **71**, 226.
Krämer, J., Kaltwasser, H. and Schlegel, H. G. (1967). *Zentbl. Bakt. Parasitkde* (*Abt. II*) **121**, 414.
Krämer, J. and Kaltwasser, H. (1969). *Arch. Mikrobiol.* **68**, 227.
Kuehn, G. D. and McFadden, B. A. (1968). *J. Bact.* **95**, 937.
Kuehn, G. D. and McFadden, B. A. (1969). *Biochemistry N.Y.* **8**, 2394.
Lascelles, J. (1960). *J. gen. Microbiol.* **23**, 499.
Leavitt, R. and Umbarger, H. E. (1962). *J. Bact.* **83**, 624.
Lebedeff, A. F. (1910). "Die Erforschung der Chemosynthese bei Bacillus hydrogenes" (translated from Russian), Odessa.
Lee, S. B. and Wilson, P. W. (1943). *J. biol. Chem.* **151**, 377.
Lessie, T. and Neidhardt, F. C. (1967). *J. Bact.* **93**, 1337.
Linday, E. M. and Syrett, P. J. (1958). *J. gen. Microbiol.* **19**, 223.
MacElroy, R. D., Johnson, E. J. and Johnson, M. K. (1968). *Biochem. biophys. Res. Comm.* **30**, 678.
MacElroy, R. D., Johnson, E. J. and Johnson, M. K. (1969). *Archs Biochem. Biophys.* **131**, 272.
Magasanik, B. (1961). *Cold Spring. Harb. Symp. quant. Biol.* **26**, 249.
Matin, A. and Rittenberg, S. (1970a). *J. Bact.* **104**, 234.
Matin, A. and Rittenberg, S. (1970b). *J. Bact.* **104**, 239.
McFadden, B. A. and Howes, W. V. (1962). *J. biol. Chem.* **237**, 1410.
McFadden, B. A. and Tu, Ch. L. (1965). *Biochem. biophys. Res. Comm.* **19**, 728.
McFadden, B. A., Kuehn, G. D. and Homann, H. R. (1967). *J. Bact.* **93**, 879.
McFadden, B. A. and Tu, Ch. L. (1967). *J. Bact.* **93**, 886.
Packer, L. and Vishniac, W. (1955). *J. Bact.* **70**, 216.
Pardee, A. D. and Beckwith, J. R. (1963). In "Informational Macromolecules". (H. J. Vogel, V. Bryson and J. O. Lampen, eds.), pp. 255–269. Academic Press, New York.
Peck jr., H. D. (1968). *A. Rev. Microbiol.* **22**, 489.
Pfitzner, J. (1969). Thesis: University of Göttingen.
Pfitzner, J., Linke, H. A. B. and Schlegel, H. G. (1970). *Arch. Mikrobiol.* **71**, 67.
Pichinoty, F. (1970). *Arch. Mikrobiol.* **71**, 116.
Ramsey, H. H. (1968). *Antonie van Leeuwenhoek* **34**, 71.
Reh, M. and Schlegel, H. G. (1969). *Arch. Mikrobiol.* **67**, 110.
Repaske, R. (1966). *Biotechnol. Bioengng* **8**, 217.
Repaske, R. and Dans, C. L. (1968). *Biochem. biophys. Res. Comm.* **30**, 136.
Rittenberg, S. C. (1969). *Adv. microbial Physiol.* **3**, 159.
Rittenberg, S. C. and Goodman, N. S. (1969). *J. Bact.* **98**, 617.
Ruhland, W. (1924). *Jb. wiss. Bot.* **63**, 321.
Schindler, J. (1964). *Arch. Mikrobiol.* **49**, 236.
Schindler, J. and Schlegel, H. G. (1969). *Arch. Mikrobiol.* **66**, 69.
Schlegel, H. G. (1955). *Arch. Mikrobiol.* **23**, 195.

Schlegel, H. G. (1966). *Adv. comp. Physiol. Biochem.* **2**, 185.

Schlegel, H. G., Gottschalk, G. and Bartha, R. (1961). *Nature, Lond.* **191**, 463.

Schlegel, H. G. and Gottschalk, G. (1965). *Biochem. Z.* **341**, 249.

Schlegel, H. G., Schuster, E., Reh, M. and Metz, H. (1965). *Zentbl. Bakt. Parasitkde (Abt. II)* **119**, 225.

Schlegel, H. G. and Trüper, H. G. (1966). *Antonie van Leeuwenhoek* **32**, 277.

Schlegel, H. G. and Blackkolb, F (1967). *Nachr. Akad. Wiss. Göttingen, II. Kl.*, 51.

Schlegel, H. G., Lafferty, R. and Krauss, I. (1970). *Arch. Mikrobiol.* **71**, 283.

Schuster, E. and Schlegel, H. G. (1967). *Arch. Mikrobiol.* **58**, 380.

Smith, A. J., London, J. and Stanier, R. Y. (1967). *J. Bact.* **94**, 972.

Smith, A. J. and Hoare, D. S. (1968). *J. Bact.* **95**, 844.

Stukus, P. E. and De Cicco, B. T. (1970). *J. Bact.* **101**, 339.

Trijbels, F. (1967). Thesis: University of Nijmegen.

Trijbels, F. and Vogels, G. D. (1966a). *Biochim. biophys. Acta* **113**, 292.

Trijbels, F. and Vogels, G. D. (1966b). *Biochim. biophys. Acta* **118**, 387.

Trijbels, F. and Vogels, G. D. (1967). *Biochim. biophys. Acta* **132**, 115.

Trüper, H. G. (1965). *Biochim. biophys. Acta* **111**, 565.

Valentine, R. C., Drucker, H. and Wolfe, R. S. (1964). *J. Bact.* **87**, 241.

Vishniac, W. and Trudinger, P. A. (1962). *Bact. Rev.* **26**, 168.

Vogels, G. D. (1963). Thesis: University of Delft.

Webster, R. E. and Gross, S. R. (1965). *Biochemistry N.Y.* **4**, 2309.

Wilde, E. (1962). *Arch. Mikrobiol.* **43**, 109.

Wilson, E., Stout, H. A., Powelson, D. and Koffler, H. (1953). *J. Bact.* **65**, 283.

Wittenberger, C. L. and Repaske, R. (1961). *Biochim. biophys. Acta* **47**, 542.

Zhilina, T. N. (1970). *Mikrobiologija* **39**, 812.

Energy Conversion and Generation of Reducing Power in Bacterial Photosynthesis

HOWARD GEST

Department of Microbiology, Indiana University,
Bloomington, Indiana 47401, U.S.A.

If we believe that the initiative for scientific research lies in observation, that scientific knowledge grows out of the evidence of the senses, then our first duty as scientists must be to observe nature faithfully, intently, and without misleading preconceptions.

P. B. Medawar (1969).

I. Introduction	243
II. On the "Eras" of Photosynthesis; Primary versus Secondary Processes	245
III. Photophosphorylation	248
IV. Generation of Net Reducing Power	251
A. Molecular Hydrogen as an Accessory Electron Donor . .	253
B. Reduction of NADP by $NADH_2$ (Transhydrogenase) . .	258
C. Succinate as a Hydrogen Donor for Photoreduction of NAD .	260
D. Another Way in Which ATP May "Drive" Nicotinamide Nucleotide Reduction	263
E. Photoproduction of Molecular Hydrogen	264
F. General Comments on Proposed Non-Cyclic Electron-Flow Mechanisms	268
V. A Comparison of Energy Metabolism and Electron-Transfer Patterns in Photosynthetic Bacteria and Clostridia	269
VI. Regulatory Mechanisms	273
VII. Epilogue	277
VIII. Acknowledgements	278
References	278

I. Introduction

The outstanding common physiological property of the photosynthetic bacteria is the ability to grow rapidly under strictly anaerobic conditions using light as the ultimate energy source.* This property makes the photosynthetic bacteria virtually unique in the biological world.

* Uffen and Wolfe (1970) have recently shown that, if very strict anaerobic procedures (Hungate technique) are used, several non-sulphur purple bacteria can grow anaerobically in darkness, at very slow rates, in certain media.

244   HOWARD GEST

Numerous investigators have sought, over many decades, the basic reasons and meaning of this apparent uniqueness, and this essay attempts an assessment of the progress made.

As a group, the photosynthetic bacteria display a remarkable metabolic diversity and versatility, and this has contributed to difficulty in defining the basic issues relating to the special aspects of their biochemistry. In this connexion, we may briefly consider carbon nutrition, one of the classical bases used in categorizing bacteria. A number of photosynthetic bacteria can grow on carbon dioxide as the sole carbon source, in purely mineral media, when particular "accessory" electron donors such as hydrogen or hydrogen sulphide are made available. On the other hand, with many of these organisms, a single organic compound such as malate or glucose can serve as both carbon and ostensible "reducing power" source for photoheterotrophic growth (i.e. light is still required). Some species have the capacity to use either carbon dioxide, under the conditions noted, or organic substrates. At this writing, it seems quite unlikely that any of the major carbon-assimilation pathways involved will prove to be found only in photosynthetic bacteria. Thus, when carbon dioxide is the carbon source, the reductive pentose cycle appears to predominate, as in most other types of autotrophs. With organic substrates, carbon dioxide reduction is not necessarily a quantitatively major process, and a multitude of "heterotrophic" carbon-assimilation pathways are possible, depending on the particular circumstances. Clearly, a number of variations and combinations of known reaction sequences may be expected in the numerous species that have evolved, and these are being gradually uncovered [see recent review by Fuller (1969), and Sirevåg and Ormerod (1970)].

Similarly, there are no compelling reasons to pin-point aspects of nitrogen metabolism as being particularly unique for photosynthetic bacteria. The assimilation of NH_4^+, molecular nitrogen and amino acids by these bacteria seems to occur through mechanisms of the kinds observed in other types of micro-organisms. These remarks on the pathways of carbon and nitrogen flow in biosynthetic metabolism are intended in an overall sense; obviously, special reactions or sequences may be encountered in the fabrication of substances found only in photosynthetic bacteria, such as bacteriochlorophyll (BChl).

Generally speaking, in the anabolic phase of cell growth, appropriate organic intermediates and "monomers" are produced from the carbon and nitrogen sources of the external medium, and these are utilized for the synthesis of proteins, nucleic acids, lipids, carbohydrates and various other complex cellular substances. Synthesis of the intermediates and monomers commonly requires energy in the form of ATP and "reducing power" (primarily as reduced nicotinamide nucleotide). If uniqueness

is to be sought in the metabolic machinery of photosynthetic bacteria, the field becomes narrowed to the mechanisms these organisms employ to regenerate ATP and reducing power under anaerobic conditions, using radiant energy as the driving force. It should be kept in mind, however, that other factors may also contribute to the physiological pattern that distinguishes these organisms (indeed, any other "physiological group"), such as: (a) particular constellations of "ordinary" nutritional and biochemical features, in combinations that lead to metabolic capacities under unusual circumstances; and (b) specialized regulatory devices, for example, in connexion with control of energy metabolism and reducing-power generation. The knowledgeable microbial physiologist will appreciate that yeast cells constitute a special, and outstanding, metabolic class for what might be considered rather trivial reasons; basically, the presence of pyruvate decarboxylase and an associated reaction for regenerating oxidized nicotinamide nucleotide through reduction of acetaldehyde to ethanol. In connexion with (b) above, note should be made of Peck's (1968) suggestion that obligate (chemosynthetic) autotrophy "may simply be the outward manifestation of control and regulatory mechanisms essential for the conservation of ATP".

II. On the "Eras" of Photosynthesis; Primary versus Secondary Processes

Kamen (1963) has discussed the time sequence of events in photosynthesis in terms of "eras", as shown in Fig. 1. The time scale is in units of pt_s, an abbreviation for "logarithm of the reciprocal of time, expressed in seconds" (thus, a unit analogous to pH and pK). Looked at in this way, the range of the time scale encompasses an impressive number of orders of magnitude; approximately 19. From Kamen's discussion, the impression emerges that the eras are quite separate in the sense that one ends and then the next begins. To be sure, time overlaps are indicated, but precious little between "photochemistry" and "biochemistry". Communication overlaps among investigators who study the different eras also tend to be minimal—in Kamen's words: "The passage through all these eras requires learning the languages of radiation physics, solid state physics, photochemistry of condensed systems, quantum chemistry, biochemistry, enzymology, plant physiology and descriptive biology. It is hardly surprising that lack of communication often exists among different investigators in the many areas of photosynthetic research". This state of affairs, of course, is not restricted to the field of photosynthesis. Linguistic problems are becoming acute within other research areas, notably in what might be called

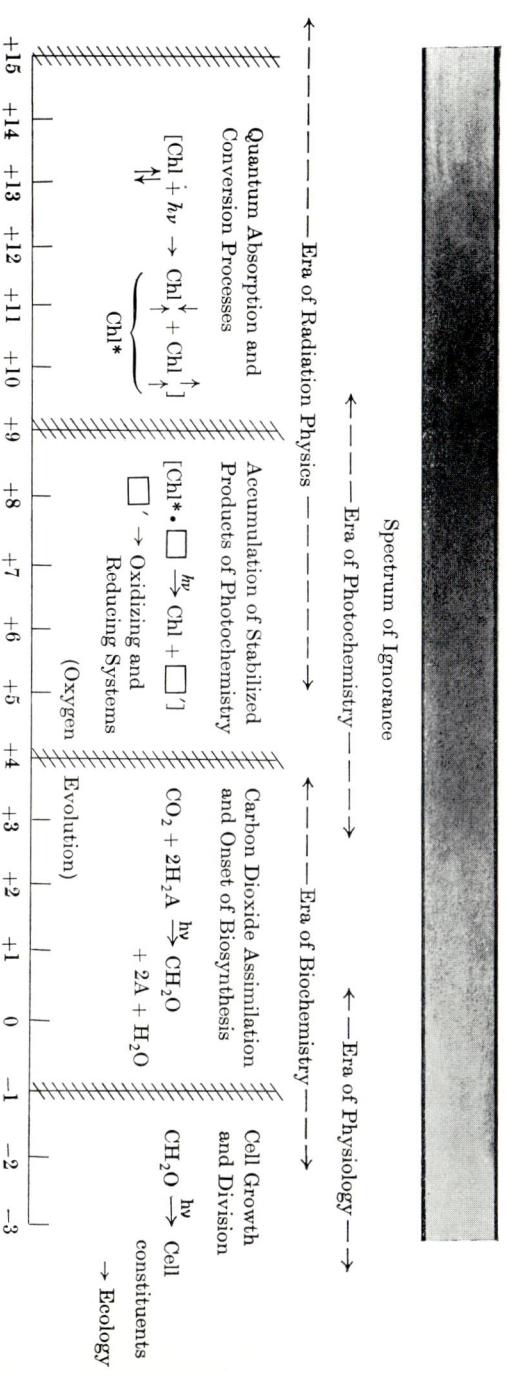

FIG. 1. The eras of photosynthesis according to Kamen (1963). Spectrum of ignorance: density at each time point is proportional to ignorance about the nature of the process at that point. Chl indicates chlorophyll; pt_s is defined in the text.

"genetics". Photosynthesis is one of the oldest problems of biochemistry and its communication difficulties started to develop well before the age of molecular biology began to flower. One apparent consequence of the language barrier is that genetic approaches to elucidating chemical aspects of photosynthesis have been seriously undertaken only quite recently.

An important question not considered by Kamen is now posed. Can events in the "slow" biochemical era influence events in the "fast" eras of radiation physics and photochemistry? This must be possible—not necessarily in the sense that the elementary physical and photochemical processes themselves are directly affected, but at least in that "biochemistry" has the capacity to alter the composition of the energy transduction system and, thereby, its overall function. The transfer of electronic excitation energy within the pigment system will certainly depend somehow on the relative concentrations of the several components involved and, in principle, maximal efficiency should require quite specific structural associations. It is well known (e.g. see Cohen-Bazire et al., 1957; Fuller et al., 1963) that changes in nutritional conditions can markedly affect the pigment composition of photosynthetic bacteria, and it seems that the chemical structure of the energy-transducing membrane complex changes in response to the prevailing biochemical "economy" (Ketchum and Holt, 1970; Steiner et al., 1970). An extreme example, which amounts to an "on-off" situation, is encountered with non-sulphur purple bacteria. With many such organisms, the presence of a high oxygen tension during growth leads to severe inhibition of synthesis of bacteriochlorophyll (Cohen-Bazire et al., 1957). Thus, "biochemistry" can tell the cell not to bother about radiation physics at all. Less extreme situations can be easily imagined, in which compositional changes (e.g. in respect to reaction-centre chlorophyll levels), dictated by "biochemistry", might lead to a variety of "new" biophysical interrelationships. It is also conceivable, if not likely, that feedback control systems exist through which chemical signals can affect the course of events in the early eras.

The term "primary" is frequently used to convey the spirit of "most important", as well as first in time. From the vantage point of the biosynthetic machinery of the growing cell, the fact that "primary" reactions may be concluded within, say, 10^{-4} seconds is doubtless of little relevance; the machinery awaits supplies of intermediates, monomers, reducing power and ATP, and might be said to care as little about the time constants of excitation energy transfer as the average consumer of electricity in a large city is concerned about whether the current is being generated by an atomic power plant or by turbines driven by falling water. Real concern is likely to arise only when there is a power failure.

Since there is virtually no power storage in photosynthetic bacteria (Sojka *et al.*, 1967), the rationale for design of a system requiring chemical coupling between very fast photophysical processes and slow enzymic reactions deserves comment. It may simply be that the processes of radiation physics and photochemistry are inherently extremely rapid and if evolving cells were to take advantage of light energy, there was no alternative but to invent means of bridging the gap. Remarks by Duysens (1964) on evolution of the photosynthetic apparatus gives food for thought in this connexion: "Photosynthesis may have started when a cell acquired the ability to synthesize a pigment able to function as a light driven oxidation reduction catalyst. . . . Subsequently, other pigments were evolved in the primitive photosynthetic cells, which were able to transfer excitation energy by induced resonance to the photochemically active pigment, which thus became an energy trap or reaction centre."

It may come as a surprise to some that, even now, there is not general agreement on the *direct biochemical sequelae* of the early eras of photosynthesis (before pt_s 4) in the bacteria. Succinctly stated, the question is: does the photochemical apparatus directly regenerate both reducing power and ATP for biosynthesis, as in green plants, or only ATP? The answer will aid us greatly in understanding the physiology of photosynthetic bacteria, and has far-reaching implications for biochemical evolution as well as comparative biochemistry. The remainder of this paper is largely devoted to reviewing experimental evidence relating to the specific question at hand, and to its various ramifications. For detailed consideration of photosynthesis events prior to pt_s 4, the reader is referred to Kamen (1963), Duysens (1964), Clayton (1965) and Frenkel (1970).

III. Photophosphorylation

$$(ADP + P_i \xrightarrow[Mg^{2+}]{light} ATP)$$

The rather simple concept that radiant energy can be converted to ATP energy by a relatively direct mechanism in photosynthetic organisms had a surprisingly long incubation period [extended, I think, by theoretical objections to the idea that the energy of a quantum (e.g. 40 kcal) can somehow be parcelled into smaller "packets" (e.g. 10 kcal)]. Following a period of equivocal experimentation, the application of tracer techniques (Gest and Kamen, 1948) showed that illumination causes a rapid turnover of ^{32}P between soluble and insoluble cell fractions of intact purple bacteria and algae; thus, the probability of a "light-induced" phosphorylation seemed to increase. Frenkel's (1954, 1956)

discovery that pigmented particles obtained by disruption of *Rhodo-spirillum rubrum* could rapidly phosphorylate ADP when illuminated anaerobically in the *absence* of appreciable quantities of electron donors or acceptors was a key finding, with profound implications for the mechanism of bacterial photosynthesis.

Detailed review of the numerous published studies on bacterial photophosphorylation, which would be a formidable undertaking, is not attempted here. Rather, a distillation of salient facts and current general interpretations is presented. In fact, the actual mechanism of the coupling between inorganic phosphate and ADP is still unknown, as is the case for oxidative phosphorylation, and several distinct alternatives are being vigorously debated.

The photosynthetic phosphorylation complex (comprised of bacteriochlorophyll carotenoids, cytochromes, and other electron carriers, etc.) is localized in the cytoplasmic membrane and its invaginations (Tuttle and Gest, 1959).* Disruption of the membrane by vigorous procedures such as sonication yields fragments which appear to seal spontaneously to yield spherical vesicles (formerly called "chromatophores"); these seem to have an inverse polarity as compared with the intact cell (Scholes *et al.*, 1969). The vesicles phosphorylate ADP in association with a *cyclic* flow of electrons, induced by illumination (in the absence of added ADP, light-dependent formation of inorganic pyrophosphate from inorganic phosphate has been observed (Baltscheffsky *et al.*, 1966; Baltscheffsky *et al.*, 1971)). Results of investigations using various approaches (spectrophotometric, inhibitor studies) suggest a sequence of electron carriers such as indicated in Fig. 2.

At least two species of bacteriochlorophyll are involved; a "bulk" light-harvesting variety absorbs and transmits radiant energy to "reaction centre bacteriochlorophyll," commonly described as bacteriochlorophyll in a special environment. The cyclic electron flow begins with transfer of an electron from bacteriochlorophyll to "X", thereby creating X.e—which might be called the primary reductant—and $BChl^+$, the final electron acceptor in the cycle. Evidence has been obtained for two phosphorylation sites (Baltscheffsky and Arwidsson, 1962; Baltscheffsky *et al.*, 1971), one of which is between cytochromes *b* and *c* (Baltscheffsky, 1967b; Baltscheffsky and Baltscheffsky, 1971).

For optimal phosphorylation activity *in vitro*, electron carriers concerned must be in suitable redox states; an "overoxidizing" or "overreducing" environment causes marked inhibition of phosphorylation activity (Horio and Kamen, 1962; Bose and Gest, 1963a) [with

---

* It has been suggested that in green photosynthetic bacteria, the photosynthetic functions occur in peripheral vesicles which seem not to be connected with the cytoplasmic membrane (Cruden and Stanier, 1970).

Chromatium particles, optimal photophosphorylation is observed when the apparent redox potential of the system is between +50 and 100 mV (Cusanovich and Kamen, 1968)]. As in other electron-transport phosphorylation systems, a protein "coupling factor" is required for the terminal stages of bacterial photophosphorylation (Baccarini-Melandri *et al.*, 1970). A recent study (Melandri *et al.*, 1971) clearly indicates that "photophosphorylation" coupling factor preparations (from *Rhodopseudomonas capsulata*) can also function as coupling factor for dark

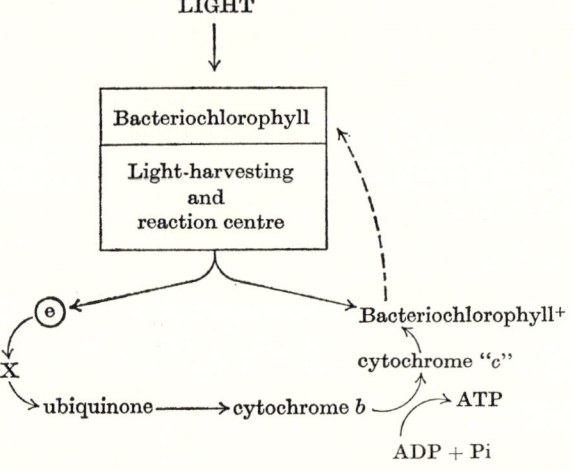

FIG. 2. General outline of a scheme for light-dependent cyclic electron transfer and phosphorylation in purple bacteria, based primarily on studies with non-sulphur purples. See Horio and Kamen (1970) for a review of cytochrome function in bacterial photosynthesis, Yamamoto *et al.* (1970) in respect to ubiquinone, and Vernon (1968) for other details. It is believed that there are two phosphorylation sites, but only one has been located with reasonable certainty. Fuller and Nugent (1969) have suggested that "X" may be a low redox potential pteridine.

aerobic oxidative phosphorylation. Pigmented vesicles from purple bacteria show a light-stimulated H^+ uptake (von Stedingk and Baltscheffsky, 1966). Removal of coupling factor from *Rps. capsulata* particles does not appreciably change the extent of proton translocation observed in the absence of phosphorylation substrates $(ADP + AsO_4^{3-} + Mg^{2+})$, but eliminates the H^+ uptake stimulation otherwise seen when the substrates are added (Melandri *et al.*, 1970).

Note should be made of earlier experiments (Nozaki *et al.*, 1961) purported to demonstrate a "non-cyclic" photophosphorylation, i.e. phosphorylation supposedly associated with a non-cyclic electron flow in which bacteriochlorophyll is presumed to act as an essential redox catalyst (as in chloroplasts). The interpretation of these experiments was

BACTERIAL PHOTOSYNTHESIS 251

based on the assumption that cyclic electron flow and phosphorylation were completely inhibited by the presence of antimycin A, but has been criticized (Bose and Gest, 1963a) on the grounds that this was not the case in the particular circumstances employed. More definitive experiments to counter the objections have not been forthcoming, and the existence of a light-dependent non-cyclic phosphorylation process in photosynthetic bacteria is therefore considered very dubious (Gest, 1966).

For convenience in the following discussion, phosphorylation is assumed to occur through the "chemical coupling" mechanism. In any event, alternative postulations or descriptions would not materially alter the biochemical physiology considered.

IV. Generation of Net Reducing Power

According to van Niel's (1941) water-cleavage model of bacterial photosynthesis, and its subsequent variations, net reducing power is obtained from a non-cyclic hydrogen (or electron) flow, in which the ultimate donor is an oxidizable substance provided in the growth medium (e.g. hydrogen, hydrogen sulphide, organic compounds). In its more modern form, non-cyclic flow entails light-activated transfer of electrons from BChl to NAD, yielding $BChl^+$ and $NADH_2$. Electron flow from the substrate ("accessory" donor) to $BChl^+$ presumably then regenerates BChl, and the process may now be repeated. This formulation specifies BChl as a "photoredox" catalyst in the transfer of electrons from the accessory donor to NAD.

As of the early 1960s, an alternative concept could be seriously entertained. Namely, that light was directly required for photophosphorylation, but only indirectly involved in regeneration of reducing power. With certain accessory donors, ordinary dark reduction of nicotinamide nucleotide could be expected. With donors of high redox potential, there was the possibility of "energy-linked electron transfer", i.e. electron transfer against the thermodynamic gradient, facilitated by ATP or its immediate precursors. This latter kind of electron transfer was described by Chance, starting in about 1957, particularly in connexion with the mitochondrial reduction of NAD by succinate. Despite the fact that the reaction, as written below, is thermodynamically reasonable, biochemists were for some time rather reluctant (curiously) to accept experimental data in its support (for a review of energy-linked electron-transfer reactions in mitochondria, see Ernster and Lee, 1964).

$$\text{Succinate} + \text{NAD} \xrightarrow{\text{energy*}} \text{fumarate} + NADH_2$$

* From ATP or "oxidative phosphorylation".

252 HOWARD GEST

Although Frenkel (1958, 1959) had demonstrated a very similar *light-dependent* oxido-reduction reaction between succinate and NAD, catalysed by pigmented particles from *R. rubrum*, the mechanism was ordinarily interpreted in terms of photochemical non-cyclic electron flow —that is, it was believed that light caused transfer of electrons from the pigment system to NAD and that these, in turn, were replaced by electrons from succinate.

$$\text{Succinate} + \text{NAD} \xrightarrow[\text{BChl}]{\text{light}} \text{fumarate} + \text{NADH}_2$$

Chance and his colleagues (Chance and Olson, 1960; Chance and Nishimura, 1960) suggested that energy-linked nicotinamide nucleotide reduction might occur in bacterial photosynthesis, on the basis of the

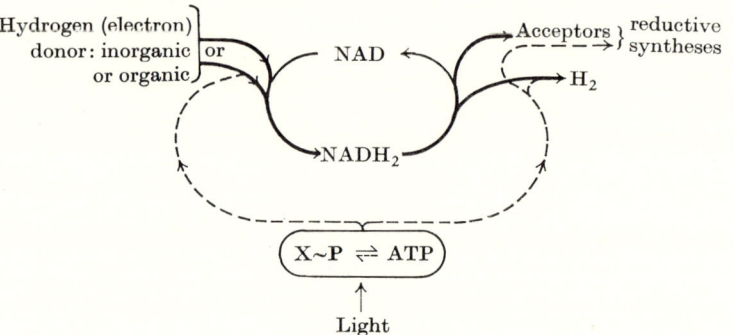

FIG. 3. Scheme for hydrogen (electron) flow from donors to acceptors, and photoproduction of molecular hydrogen in bacterial photosynthesis (Gest, 1963).

kinetics of spectroscopic changes observed upon illumination of intact cells.* Their various findings and other stimulating discussions in the literature (Wassink, 1947; Hill and Bendall, 1960) prompted experimental studies by Bose and Gest (1962, 1963b) of the possibility that several net oxido-reduction reactions dependent on, or promoted by, light in photosynthetic bacteria were indeed examples of energy-linked electron transfer. These investigations (see also Gest, 1963) indicated that such reactions could be explained plausibly on the basis of energy-dependent "dark" electron flow and, accordingly, the generalized scheme for bacterial photosynthesis shown in Fig. 3 was advanced.

* Jackson and Crofts (1968) have recently confirmed that "photoreduction" of nicotinamide nucleotide (with endogenous hydrogen donors) in intact cells lags behind light-induced oxidation of cytochrome; they also reported that certain inhibitors of cyclic photophosphorylation cause severe, or complete, inhibition of the nicotinamide nucleotide reduction process.

A. Molecular Hydrogen as an Accessory Electron Donor

1. Reduction of Nicotinamide Nucleotides

A number of species of photosynthetic bacteria can grow using carbon dioxide as the sole (or primary) source of carbon with molecular hydrogen serving as the accessory electron donor. The H_2/H^+ redox couple has a low potential ($E_0' = -0.42$ V) and, in principle, is an excellent electron donor for reduction of nicotinamide nucleotides ($E_0' = -0.32$ V) via dark metabolism. Pigmented particles from R. rubrum grown photoheterotrophically with glutamate as the nitrogen source (i.e. cells with a high capacity for photoproduction of molecular hydrogen) contain active hydrogenase. The particles, however, do not reduce NAD with molecular hydrogen unless treated with deoxycholate (or sodium lauryl sulphate) and supplemented with flavine mononucleotide (FMN) (Bose and Gest, 1962). Experiments in which a trapping system of pyruvate plus lactate dehydrogenase was employed, to re-oxidize the $NADH_2$ formed, indicated the (dark) hydrogen transfer sequence:

$$H_2 \rightarrow FMNH_2 \rightarrow NADH_2$$

Although the standard potential of the FMN redox couple ($E_0' = -0.22$ V) is *higher* than that of the NAD couple, it is evident that the potentials of the two systems will overlap (thus facilitating reaction) if FMN is maintained primarily in the reduced state by hydrogenase action, and the nicotinamide nucleotide in the oxidized form by the trapping system. Frenkel (1958, 1959) observed that light was required for reduction of NAD when *chemically* reduced FMN was the hydrogen donor. However, he neither provided for continual regeneration of $FMNH_2$ nor used a trapping system for re-oxidation of $NADH_2$, and it has been suggested (Bose and Gest, 1962) that light was necessary to provide energy for overcoming an unfavourable (though relatively small) potential difference between the two redox couples.

There is a strong impression in the literature that all photosynthetic organisms contain low-potential ferredoxins which, of course, would be expected to be capable of mediating electron transfer from H_2 to NAD, or NADP, in darkness. Thus far, however, it seems that there has been no published account that convincingly documents such proteins from non-sulphur purple bacteria. The presence of a ferredoxin (Fd) in R. rubrum has been briefly noted in papers from Arnon's laboratory (Tagawa and Arnon, 1962; Buchanan et al., 1967). On the other hand, von Stedingk (1964) attempted to isolate Fd from N_2-grown cells of R. rubrum by procedures successful with Clostridium pasteurianum, and found that the specific activity of the "R. rubrum fraction" was less than 1% of that

of clostridial Fd in catalysing either the phosphoroclastic split of pyruvate to acetyl phosphate or the formation of molecular hydrogen with dithionite as electron source (with *C. pasteurianum* preparations). Yamanaka and Kamen (1965, 1967) have described a non-haem iron protein from *Rhodopseudomonas palustris* that resembles Fd in certain respects. Its midpoint redox potential, however, is rather high (estimated to be between $+0.05$ and $+0.24$ V), and as yet there are no clear-cut indications of a possible physiological function. According to a personal communication from Dr. R. G. Bartsch and Dr. T. Horio, they have been successful in isolating low potential ferredoxins from certain non-sulphur purple bacteria. I am also aware of an undercurrent of unpublished negative attempts, which are frequently rationalized as being due to extreme lability (presumably to molecular oxygen) of these particular proteins. It is also possible that because of differences in specificity, the Fd assay procedures ordinarily used may not suffice. Obviously, this important matter remains to be clarified.

It should be noted that a role for (reduced) Fd in *R. rubrum* metabolism, not directly related to nicotinamide nucleotide reduction, has been proposed—namely, as the reductant for pyruvate and α-ketoglutarate synthases, which are presumed to participate in a so-called reductive carboxylic acid cycle. Buchanan *et al.* (1967) suggest that this biosynthetic cycle may function in *R. rubrum*, especially during growth on hydrogen plus carbon dioxide. In their experiments, the synthase activities in a soluble cell fraction were detected using *C. pasteurianum* ferredoxin (only one experiment is reported with *R. rubrum* "ferredoxin", reduced photochemically by heated chloroplasts).

Klemme and Schlegel (1967a, b) studied reduction of nicotinamide nucleotides by H_2 with particles from *Rhodopseudomonas capsulata*, and found that the process was *light dependent*. The activity was observed in preparations from cells grown photoheterotrophically, but considerably higher rates were obtained with particles from cells cultivated on hydrogen plus carbon dioxide (anaerobically in the light). Nicotinamide adenine dinucleotide functions specifically as the acceptor; NADP is "photoreduced" only in the presence of small amounts of NAD. The mechanism of the NAD photoreduction was later investigated in more detail by Klemme (1969a, b), whose results are for the most part consistent with the view that the reduction, with either molecular hydrogen or succinate, occurs through an energy-linked "dark" reaction. As in a number of other similar studies, this conclusion was based on the following criteria: (1) inhibition of the light-driven oxidation-reduction reaction by various inhibitors and uncouplers of photophosphorylation [e.g. antimycin A and *m*-chlorocarbonylcyanide phenylhydrazone (CCCP)]; (2) inhibition of the oxidation-reduction reaction by the phos-

phorylation substrates, ADP + Pi + Mg^{2+} (indicating a competition between the oxidation-reduction reaction and the photophosphorylation system for a common "intermediate"); and (3) success in demonstrating that the oxidation-reduction reaction proceeds *in darkness* when ATP or an ATP-generating system is provided (in Klemme's experiments, the ATP-driven dark reaction rates were of the order of 20–30% of the light-dependent reaction rate).

If the light dependency is in fact due to production of an "energy-rich precursor" of ATP (also derivable from ATP) that drives the oxidation-reduction reaction, oligomycin would be expected to stimulate the light-driven process and inhibit the ATP-driven (dark) reaction, both of which results were observed with the *Rps. capsulata* particles. Except for differences in detail, the foregoing illustrates the approach used by various investigators with different organisms and electron donors. Klemme (1969a) points out certain (inevitable!) difficulties in interpretation of the results of inhibitor experiments, and concludes that the succinate plus NAD reaction occurs by energy-linked electron transfer, while the H$_2$ plus NAD reaction may involve two mechanisms, viz. energy-linked transfer *and* a non-cyclic electron flow that is insensitive to certain inhibitors. Klemme also proposes a non-cyclic flow when the donor system is ascorbate + dichlorophenolindophenol (DCPIP), while from more recent studies with a similar artificial donor system [ascorbate + N,N,N′,N′-tetramethyl-*p*-phenylenediamine (TMPD)] Isaev *et al.* (1970) concluded that their data "show unambiguously that photo-reduction of NAD$^+$ in *R. rubrum* chromatophores occurs *via* an energy-dependent reversed electron transfer pathway".

The light dependency of the H$_2$ plus NAD reaction in *Rps. capsulata* particles is of special interest in several connexions. In the first instance, the light requirement would not be predicted on the basis of the redox potentials of the reactants. This suggests that the immediate electron acceptor of the hydrogenase has a rather high redox potential, and that light energy is required to now move electrons against the thermodynamic gradient to NAD. The logic behind this kind of scheme is at present best known to the organism, but may well have a completely reasonable explanation. Secondly, the light dependency in *Rps. capsulata* makes it seem unlikely that a Fd is involved in this organism. In photosynthetic bacteria known to contain significant levels of Fd, it has been possible to demonstrate ferredoxin-dependent reduction of nicotinamide nucleotides by molecular hydrogen *in darkness* (as would be expected from earlier observations with clostridial preparations). In experiments with *Chromatium* extracts, Weaver *et al.* (1965) found that NAD was reduced much more rapidly than NADP. They also reported similar observations with *Chlorobium thiosulfatophilum* extracts, but these had

low hydrogenase activities, which posed complications. A dark Fd-dependent reaction in *Chromatium* extracts was also described by Buchanan and Bachofen (1968), who noted in addition: "So far it has not been possible with these preparations to reduce NAD^+ or $NADP^+$ photochemically or to replace H_2 with other reductants". The photo-reduction of NAD with molecular hydrogen by *Rps. capsulata* particles is, in fact, not stimulated by bacterial ferredoxins (Klemme, 1969b).

2. Reduction of Fumarate

An interesting model system has been described (Bose and Gest, 1962, 1963b; Gest, 1966) in which the velocity of the following reaction, catalysed by *R. rubrum* particles, can be influenced by light, depending on the potential of the redox mediator:

$$H_2 + \text{fumarate} \xrightarrow[\text{mediator}]{\text{redox}} \text{succinate}$$

The particles (derived from photoheterotrophically grown cells with a high capacity for photoproduction of molecular hydrogen) do not oxidize molecular hydrogen with fumarate unless supplemented with an electron carrier. As shown in Table 1, mediators effective in the dark have standard potentials between, or overlapping, those of the H^+/H_2 couple ($E_0' = -0.42$ V) and the fumarate/succinate couple ($E_0' \cong 0$ V).

TABLE 1. Effect of light on reduction of fumarate with hydrogen by *Rhodospirillum rubrum* particles; relationship with E_0' value of the redox mediator. After Bose and Gest (1963b)

Mediator	E_0' (V)	Light effect (requirements)
Benzyl viologen	−0.359	None
Janus green B	−0.256	None
Flavin mononucleotide[a]	−0.219	None
Nile blue A	−0.119	None
Menadione (vitamin K_3)	−0.044	None
Methylene blue	+0.011	Stimulatory
Thionine	+0.063	Stimulatory
N-Methylphenazonium methosulphate	+0.080	Stimulatory
Coenzyme Q_{10} (*R. rubrum*)	+0.089	Stimulatory
Toluylene blue	+0.115	Stimulatory
2,6-Dichlorophenolindophenol	+0.217	Obligatory
Tetramethyl-*p*-phenylenediamine	+0.260	Obligatory
Cytochrome *c* (mammalian)	+0.262	Obligatory
p-Benzoquinone	+0.280	Obligatory
Cytochrome c_2 (*R. rubrum*)	+0.330	Obligatory

[a] Mediator function not observed in the absence of a detergent such as deoxycholate.

With mediators of somewhat higher potential, in the range of 0 to 0·115 V (e.g. CoQ_{10}), the overall reaction is stimulated by light. A light requirement becomes apparent with redox mediators of still higher potential, such as cytochrome c or c_2. All of the mediators are reduced by molecular hydrogen in darkness, and it is therefore evident that the light effect is concerned with electron transfer between the mediator and the fumarate/succinate couples (Fig. 4).

The "light-dependent" reaction with tetramethyl-p-phenylenediamine (TMPD) serving as redox mediator was particularly studied in respect to effects of phosphorylation inhibitors and uncouplers. In contrast to the dark process catalysed by a low potential dye, the light-dependent TMPD reaction was severely inhibited by agents such as CCCP and pentachlorophenol. These and related experiments led to the conclusion that the action of light with carriers such as cytochrome c

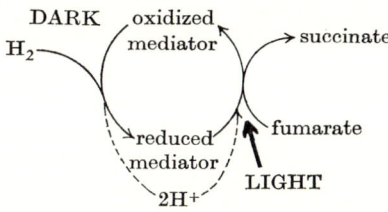

Fig. 4. Hydrogen (electron) flow in the "light-dependent" reduction of fumarate with molecular hydrogen. Dashed line, to show hydrogen balance when the mediator carries electrons only.

was concerned with providing energy for driving "dark" hydrogen (electron)-transfer against the unfavourable thermodynamic gradient between the reduced mediator and fumarate, rather than with a hypothetical non-cyclic electron flow.

This system, which can be studied with physiological substrates and redox mediators (molecular hydrogen, fumarate and, for example, cytochrome c_2) has not been exploited to the full, and takes on added significance in view of more recent work. Figure 5 gives a possible scheme for electron flow to fumarate, and NAD, when molecular hydrogen is the accessory hydrogen donor.

The energy-dependent reversal of electron flow at the cytochrome level has been especially studied by Dr. M. Baltscheffsky with pigmented particles from *R. rubrum*. Her researches (Baltscheffsky, 1967a, 1969) showed that endogenous cytochrome b in the particles becomes reduced in darkness upon addition of either ATP or inorganic pyrophosphate. Uncouplers such as gramicidin and desaspidin inhibit the process; oligomycin inhibits with ATP, but not the inorganic pyrophosphate-driven

reaction. Simultaneous with the pyrophosphate-induced reduction of cytochrome b, c-type cytochromes are oxidized. It is implied that there is an energy-dependent electron flow from cytochrome $c . e$ to cytochrome b, as indicated in Fig. 5.

Several kinds of chemosynthetic autotrophs are obliged to generate reducing power, for carbon dioxide reduction, using inorganic reductants of high redox potential, e.g. in *Nitrobacter*, the ultimate electron source is NO_2^-, which is oxidized to NO_3^- ($E_0' \geqslant +0.35$ V). It has been evident for some time that such organisms must utilize energy-driven electron flow to create the needed reducing power, and there is now considerable

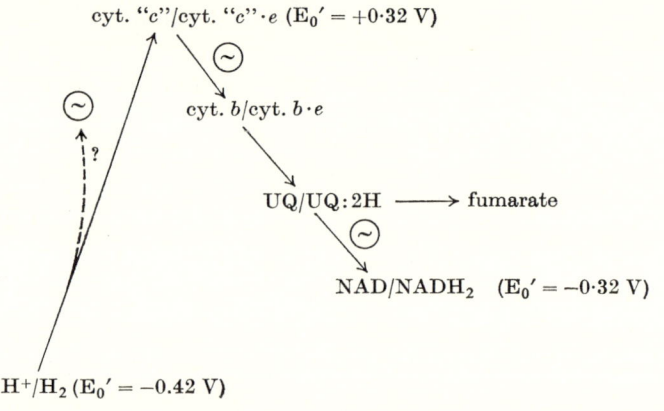

FIG. 5. Scheme for net "photoreduction" of NAD and fumarate with molecular hydrogen in purple bacteria (cf. Klemme, 1969a). Solid arrows signify directions of hydrogen(e) transfer. ⊝ = ATP or some related "energy-rich" compound produced by the photophosphorylation system; cyt., cytochrome; UQ, ubiquinone. Dark reduction of cytochrome c by hydrogen (*Rhodospirillum rubrum* particles) has been demonstrated by Bose and Gest (1962) and Frenkel and Cost (1962). Section IV, C (p. 260) considers succinate as a donor for "photoreduction" of NAD.

experimental evidence for the occurrence of ATP-dependent reduction of nicotinamide nucleotide with reduced cytochrome serving as electron donor (see review by Peck, 1968). Although Peck (1968) discusses possible complications in interpretation of certain published data in this connexion, it seems quite certain that some chemosynthetic autotrophs, at least, closely resemble purple bacteria in employing energy-driven electron transfer reactions to regenerate reducing power.

B. REDUCTION OF NADP BY $NADH_2$ (TRANSHYDROGENASE)

Energy-linked transhydrogenase activity, first observed in submitochondrial preparations (Danielson and Ernster, 1963), can be represented by the equation:

$$NADH_2 + NADP + {\sim}X \rightarrow NAD + NADPH_2 + X$$

where ${\sim}X$ is a "high energy intermediate" in terms of the chemical coupling theory of phosphorylation. This activity was first observed in *Rps. spheroides* particles by Orlando *et al.* (1966) and in *R. rubrum* membrane fragments by Keister and Yike (1966). The reaction in the *Rps. spheroides* system is driven either by light, or by ATP in darkness. As expected for the mechanism suggested by the foregoing equation, the light-dependent reaction is unaffected by oligomycin, while the dark ATP-driven reaction is completely inhibited. Orlando *et al.* (1966) also found that the photoreaction was inhibited by phosphorylation substrates and certain phosphorylation inhibitors (e.g. antimycin A). Similar observations with *R. rubrum* particles were made by Keister and Yike (1966, 1967b); in addition, the capacity of inorganic pyrophosphate to drive the reaction in darkness was reported.

Although the *R. rubrum* transhydrogenase system is tightly bound to the energy-converting membranes (Keister and Yike, 1966), extensive washing of the particles leads to simultaneous loss of the light-, ATP-, and pyrophosphate-driven activities (Fisher and Guillory, 1969a). This is due to loss of a protein, "transhydrogenase factor", which can restore activity to depleted particles (Fisher and Guillory, 1969a, b). The factor is evidently not a phosphorylation "coupling factor", since it does not significantly affect photophosphorylation rate (Fisher and Guillory, 1969a).

In the dark ATP-driven reaction (*R. rubrum*), approximately one ATP is consumed per $NADPH_2$ produced (Keister and Yike, 1967b). It is of interest that in the light-driven reaction, the intensity required for half-saturation of the transhydrogenase activity rate is considerably lower than for light-driven NAD reduction with succinate, or for photophosphorylation (Keister and Minton, 1969a). Thus, the transhydrogenase reaction could be said to be very sensitive to the "energy level" of the system. In this connexion, it is relevant to consider the equilibrium constants for transhydrogenase activities. The non-energy-linked reaction has a constant of the order of one, while in the energy-linked system the constant $[(NADPH_2)(NAD)/(NADH_2)(NADP)]$ is very much higher, e.g. Keister and Yike (1967b) report $K = {\sim}28$ for *R. rubrum* particles. A displacement in this direction makes good sense physiologically, at least in actively growing cells which require a constant replenishment of $NADPH_2$ for reductive syntheses. Although alternative means of accomplishing this may exist (see p. 263) a *direct* promotion of $NADPH_2$ regeneration, via transhydrogenase action, by the energy-converting machinery seems a plausible device for "pushing" biosynthetic metabolism.

C. Succinate as a Hydrogen Donor for Photoreduction of NAD

Succinate can be readily used as both the sole carbon source and ultimate electron donor for anaerobic photoheterotrophic growth of various photosynthetic bacteria.* The capacity to use succinate anaerobically for these purposes is an unusual metabolic characteristic, and suggests that many photosynthetic bacteria have a facile means of generating net reducing power, at the nicotinamide nucleotide level, from the rather high potential succinate/fumarate system. Early experiments of Ormerod (1956) with cell suspensions of *R. rubrum* showed that during photometabolism of succinate, which results in net carbon dioxide production, a substantial fixation of carbon dioxide also occurs. His results implied that the pair of hydrogen atoms "released" by the oxidation of succinate to fumarate is somehow employed for reduction of nicotinamide nucleotide which, in turn, is used for the reductive phosphoglyceric acid to triosephosphate step of the familiar carbon dioxide-fixation cycle. At the sub-cellular level, the light-dependent oxidation of succinate with NAD by pigmented particles derived from the same organism was soon thereafter demonstrated by Frenkel (1958). It was noted above that initial interpretations of the mechanism were invariably along the lines of non-cyclic electron flow. An important experimental step forward in the "energy-linked" versus "non-cyclic" flow controversy was the report of Löw and Alm (1964) that the reaction proceeded in darkness when ATP was provided and, moreover, that the dark reaction was inhibited when the "energy supply" from ATP was cut off by addition of oligomycin. These important observations have since been confirmed and extended by others, notably in investigations by Keister with *R. rubrum* particles.

The experiments of Keister and Yike (1967a) with inhibitors and uncouplers of photophosphorylation showed that uncouplers inhibited both the light-driven and the dark ATP-driven reactions (note: inorganic pyrophosphate was also observed to drive the reaction in darkness, at about 30–40% of the rate obtained with ATP). On the other hand, inhibitors of electron transport associated with photophosphorylation interfered with the light-driven process, but had no effect when ATP was the energy source. Keister and Yike (see also Keister and Minton, 1969b) suggested that electron transport in the energy-linked reduction of NAD by succinate occurs via a segment of the "aerobic" electron transfer chain (but, with electrons moving in the "reverse" direction). Others have made similar suggestions for the non-sulphur purple bacteria,

* The nonsulphur purple bacteria (and certain sulphur purple bacteria) require trace quantities of organic growth factors, which vary depending on the species.

and it seems that in such organisms a "respiratory" chain may always be present even in anaerobically-grown cells (Geller, 1962).* Whether or not there is a common electron-transfer carrier sequence in "cyclic" and "respiratory" electron flow is still a moot point. In any event, it seems premature to identify the energy-linked electron flow from succinate to NAD with a "respiratory" chain of carriers. Even if a separate carrier chain is involved, distinct from the cyclic electron transfer pathway, there is no *a priori* reason to believe that it would necessarily serve a "respiratory" role as its prime function. From additional experiments with inhibitors and uncouplers, Keister and Minton (1969a) conclude that "all of the NAD^+ reduction observed in *R. rubrum* is by an energy-linked process" (i.e. electron transfer promoted by a "high-energy intermediate" of phosphorylation). They also point out certain complexities that arise in experiments with inhibitors and uncouplers, e.g. the degree of inhibition is sometimes influenced by the concentration of the electron donor or by the absolute activity level of the system.

Similar studies by Jones and Vernon (1969) with *R. rubrum* particles led to the same general conclusions. They report a stoichiometry of ~1·8 ATP used per $NADH_2$ formed for the dark ATP-driven process. Several other aspects of the Jones and Vernon study are noteworthy. Inhibition of the O/R reaction by the phosphorylation substrates was again confirmed, as it has been by a number of others (e.g. Horio *et al.*, 1963; Keister and Yike, 1967a) since Frenkel's (1959) first demonstration of this effect. Results of experiments on light-dependent reduction of NAD with the donor system of ascorbate plus tetramethyl-*p*-phenylene-diamine suggested an energy-dependent electron transfer mechanism in this case also (as noted earlier, Isaev *et al.* (1970) reached the same conclusion).

Comparable investigations with *Rps. capsulata* (Klemme, 1969a, b) have given a very similar spectrum of results. Thus, at minimum, the three "criteria" stipulated on pp. 254–255 have been met in *R. rubrum* and *Rps. capsulata*. The picture with other kinds of photosynthetic bacteria is not nearly as clear; contradictory experimental results have been published, and the general paucity of information does not permit unambiguous interpretations in terms of any kind of mechanism. Hood (1964) noted conflicting reports on the ability of *Chromatium* preparations to photoreduce nicotinamide nucleotides, and described a slow light- and succinate-dependent NAD reduction by a "small particle" fraction. Hinkson (1965) also observed the reaction with *Chromatium*

---

* *Rhodopseudomonas molischianum*, however, and certain other Athiorhodaceae are reported to be unable to grow aerobically in darkness at normal atmospheric oxygen tensions (Sistrom, 1965); it would be of considerable interest to examine such bacteria in connexion with the mechanism of photoreduction of NAD.

262 HOWARD GEST

particles, but Buchanan and Bachofen (1968) apparently were unable to do so. Two reports (Buchanan and Evans, 1969; Evans, 1969) claim that *Chlorobium thiosulfatophilum* particles, unable to photophosphorylate at a significant rate, catalyse a photoreduction of NAD (NADP more slowly) dependent on: Fd, a soluble protein factor (replaceable by the Fd-NADP reductase of spinach), and an electron donor which can be succinate, sodium sulphide, or 2-mercaptoethanol. It is maintained (Evans, 1969) that the photoreduction is not inhibited by inhibitors and uncouplers of phosphorylation. The overall reaction seems to be very slow (90-minute incubations were usually employed), and clearly requires more study.

In reviewing papers for this essay, I have been struck by the fact that the composition of the medium used for growing cells, either for intact cell experiments or as a source of particles, is frequently treated as a well-guarded secret. Similarly, the time of harvest and other details of culture conditions may not be given or are shrouded in the mystery of "as described previously" (which, more often than not, leads to a wild goose-chase ending in ambiguity). Obviously, these factors are sometimes of decisive importance in either observing a phenomenon, or in reproducing results in the literature. Most microbiologists need not be lectured about this, but biochemists and others who work with photosynthetic

TABLE 2. Rates of light-induced and dark redox reactions catalysed by chromatophores of *Rhodopseudomonas capsulata*. After Klemme (1969b)

Reaction	Reaction rate (μmoles/mg. bacteriochlorophyll per hour) Chromatophores from cells grown	
	photoauto-trophically	photohetero-trophically
NAD photoreduction:		
with succinate	21	1·5
with hydrogen	13	—
NADH$_2$ dehydrogenase:		
cytochrome c-linked	86	46
dichlorophenolindophenol-linked	83	45
NADH$_2$ oxidase	130	41
Hydrogenase:		
(cytochrome c-linked)	109	20
Succinate dehydrogenase:		
(dichlorophenolindophenol-linked)	62	40

bacteria seem to need repeated admonitions. Examples of considerable variation in rates of oxidation-reduction photoreactions (catalysed by pigmented particles) due to differences in growth conditions are given by Orlando *et al.* (1966) and Klemme (1969a, b; see Table 2).

It is interesting that active photoreduction of NAD with succinate (Jones and Vernon, 1969) or energy-linked transhydrogenase activity (Keister and Yike, 1967b) does not appreciably affect the velocity of photophosphorylation. This may have a bearing on one of the controls of light-induced cyclic electron transfer; it is conceivable that an increased drain of "energized intermediates" due to simultaneous ATP synthesis and energy-driven electron flow to nicotinamide nucleotides causes stimulation of cyclic electron transfer and, thereby, an acceleration of energy conversion.

D. Another Way in Which ATP May "Drive" Nicotinamide Nucleotide Reduction

Krebs and Veech (1969, 1970) have developed a most interesting analysis of the factors responsible for regulation of the nicotinamide nucleotide redox states in rat liver cells. It seems to me that their general conclusions have significant implications for understanding how reducing power levels are controlled in all kinds of cells, including the photosynthetic bacteria. In brief, the new experimental findings and calculations of Krebs and Veech indicate that the degree of phosphorylation of the adenylate nucleotide system ultimately regulates a network of "near-equilibria" (reactions of various dehydrogenases, transaminases, etc.) which directly influences the steady states of the nicotinamide nucleotide redox couples. The network is composed of a number of reactions linked through the involvement of common substrates, which include the nicotinamide nucleotides, ATP and ADP. It is of particular interest that the integrated series of reactions results in a large difference (approximately 10^5) in the cytoplasmic ratios of $NAD/NADH_2$ and $NADP/NADPH_2$, which are about 1000 and 0·01, respectively. As noted by Krebs and Veech (1970): "It is remarkable that two very similar redox couples like the NAD- and NADP-systems, with almost identical physical and chemical properties and about the same standard redox potential, should show large differences in their *actual* redox potential in the liver cell, that of the NADP-couple being about 0·150 volt more negative".

The low actual potential of the NADP couple in the cytoplasm is, of course, admirably suited to its supposed primary physiological function, viz. as a hydrogen donor system for reductive biosynthesis. On the other hand, the potential of the cytoplasmic NAD couple is "set" in a range optimal for *its* particular functions (i.e. in glycolysis and gluconeogenesis

264   HOWARD GEST

in the liver cell). This analysis suggests that the supply of ATP can regulate the redox states of the nicotinamide nucleotide couples in mammalian cells without resort to special "energy-driven" reactions of the kind discussed earlier. It would not be surprising to find that photosynthetic bacteria (whose dark fermentative systems are extremely limited) use an interplay of both kinds of schemes—Krebs-Veech and energy-linked electron transfer—to surmount the difficulty of generating an adequate supply of net reducing power for biosynthesis under anaerobic conditions using light energy as the ultimate driving force.

E. Photoproduction of Molecular Hydrogen

1. By Photosynthetic Bacteria

Light-dependent formation of molecular hydrogen by photosynthetic bacteria can be construed as energy-promoted generation of net reducing power from relatively oxidized substrates. This process, first observed in *R. rubrum* (Gest and Kamen, 1949), is inhibited by molecular nitrogen and this effect led to the discovery that photosynthetic bacteria fix molecular nitrogen (Kamen and Gest, 1949; Gest *et al.*, 1950; see also Gest, 1951; and Gest and Kamen, 1960). In the Athiorhodaceae, production of molecular hydrogen is dependent on the presence of oxidizable organic substances such as citric acid cycle intermediates.* The purple sulphur bacterium *Chromatium* shows similar properties and, in addition, manifests a slow light-dependent evolution of hydrogen with thiosulphate as electron donor (Losada *et al.*, 1961; Ormerod *et al.*, 1961). Our understanding of the mechanism and significance of light-dependent hydrogen formation stems primarily from studies on *R. rubrum*, and the discussion will be restricted to this particular organism; it seems quite likely that the general conclusions apply to other photosynthetic bacteria as well. The reader is referred to the following papers for the major part of the experimental results which form the basis of the conclusions outlined below: Gest *et al.*, 1950; Ormerod *et al.*, 1961; Gest *et al.*, 1962; Ormerod and Gest, 1962; Bose and Gest, 1963b.

Hydrogen is produced by cultures of *R. rubrum* growing photoheterotrophically on substrates such as malate or succinate when certain amino acids are used as nitrogen sources (e.g. glutamate). Synthesis of one or more components of the hydrogen-producing system is repressed, on the other hand, when the nitrogen source is an ammonium salt; if the latter is made limiting, hydrogen is evolved sometime after the NH_4^+ has been

---

* Only light-dependent production of hydrogen will be considered; some photosynthetic micro-organisms can also produce molecular hydrogen in darkness from certain organic compounds, apparently by typical "heterotrophic" mechanisms (Gest, 1951, 1954; Gest and Kamen, 1960).

exhausted. Under appropriate experimental conditions, resting cells of
R. rubrum (from glutamate cultures) photometabolize organic substrates
with the formation of large quantities of both hydrogen and carbon
dioxide. The following (approximate) stoichiometries have been demonstrated.

$$
\begin{array}{ll}
\text{Acetate:} & C_2H_4O_2 + 2H_2O \rightarrow 2CO_2 + 4H_2 \\
\text{Succinate:} & C_4H_6O_4 + 4H_2O \rightarrow 4CO_2 + 7H_2 \\
\text{Fumarate:} & C_4H_4O_4 + 4H_2O \rightarrow 4CO_2 + 6H_2 \\
\text{L-Malate:} & C_4H_6O_5 + 3H_2O \rightarrow 4CO_2 + 6H_2
\end{array}
$$

That is, complete dissimilation to carbon dioxide and hydrogen. The
effects of fluoroacetate, plus other considerations, clearly indicate that
these conversions occur by means of an anaerobic citric-acid carbon
cycle, coupled with a light-dependent process that effects the re-oxidation
of reduced nicotinamide nucleotides (generated in oxidative steps of the
cycle) by liberation of molecular hydrogen (Gest *et al.*, 1962; Ormerod
and Gest, 1962).

Hydrogen production is greatly, or completely, suppressed by inhibitors and uncouplers of photophosphorylation (antimycin A, pentachlorophenol, CCCP; also by reduced dyes that inhibit phosphorylation
activity by "overreducing" natural electron carriers concerned). Photoproduction of molecular hydrogen by resting cell suspensions is also
inhibited by addition of NH_4^+ or molecular nitrogen, both of which are
readily used for reductive biosyntheses of amino acids and other cellular
constituents, i.e. with both nitrogen sources, reducing power is directed
toward synthesis rather than being "wasted" in the form of hydrogen.
Molecular nitrogen seems to inhibit hydrogen formation by an additional, "catalytic", mechanism (Bregoff and Kamen, 1952) and it
has been suggested (Gest, 1963) that this may be a specific allosteric
inhibition of the hydrogen-evolving complex by molecular nitrogen. In
this case, the allosteric "signal" for cessation of energy-dependent
hydrogen production is the appearance of a readily utilizable nitrogen
source that completes the requirements for immediate endergonic synthesis of proteins and other macromolecules (Gray and Gest, 1965).

The salient facts summarized above are interpreted as indicating that
photoproduction of hydrogen is in essence a "dark" energy-driven process that fulfils a regulatory function. As represented in Fig. 6, the
precursor of molecular hydrogen is believed to be reduced nicotinamide
nucleotide, generated by citric-acid cycle oxidations or, in some circumstances, by other reactions. Maximal photoproduction of hydrogen is
observed under conditions (resting cells plus a carbon source such as
malate or succinate plus light) in which the cells must be regenerating
ATP and reduced nicotinamide nucleotides in great excess of the demands
from the biosynthetic apparatus. In these circumstances, it would be

expected that a high ratio of reduced/oxidized nicotinamide nucleotide would tend to build up, the "reducing potential" presumably being augmented by energy-driven transhydrogenase activity. There is no *theoretical* reason why molecular hydrogen could not be produced from $NADPH_2$ (or $NADH_2$) without further input of energy, if a sufficiently high ratio of reduced/oxidized nicotinamide nucleotide is attained, and this seems to be possible in some kinds of anaerobes (see Section V, p. 270). *Rhodospirillum rubrum* and similar non-sulphur purple bacteria, however, do not have a strongly "reducing" metabolic pattern (e.g. in comparison with clostridia) and on this account a further energy contribution may be necessary for hydrogen formation.

The energy-dependent production of molecular hydrogen can be viewed as a "valve" that aids in regulating the flux of reduced nicotinamide nucleotides and ATP into biosynthetic pathways, to values ap-

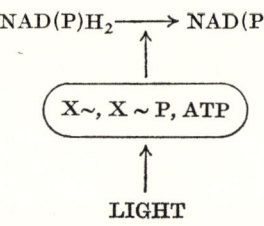

NAD(P)H$_2$ ⟶ NAD(P)

X~, X ~ P, ATP

LIGHT

FIG. 6. Proposed molecular mechanism for "photoproduction" of molecular hydrogen by photosynthetic bacteria. $NAD(P)H_2$ and $NAD(P)$, indicate reduced and oxidized nicotinamide nucleotide, respectively; X~ and X ~ P, "energy-rich" compounds produced by the photophosphorylation system.

propriate to the overall biosynthetic activity level dictated by the nutritional circumstances. It may be remarked that, aside from various other difficulties, the outmoded explanation of hydrogen photoproduction in terms of a non-cyclic electron flow from substrate to hydrogenase via bacteriochlorophyll also suffered from complete lack of a physiological rationale.

In connexion with succinate as a substrate, it should be noted that the stoichiometry of hydrogen and carbon dioxide formation with resting cells indicates that the pair of hydrogen atoms (electrons) removed in the oxidation of succinate to fumarate can be converted to molecular hydrogen. As far as I know, this potentiality is unique to photosynthetic bacteria; succinate is ordinarily a "dead-end" in anaerobic metabolism (except when it can be oxidized to fumarate by means of energy-dependent electron flow (to NAD), or in certain instances where an exogenous inorganic oxidant such as nitrate can function as electron acceptor). Formation of molecular hydrogen from succinate requires

electron transport against the thermodynamic gradient over a considerable range (approximately 400 mV, based on midpoint potentials). The possibility of "pushing" electron transfer from succinate to nicotinamide nucleotide with the aid of ATP energy makes the formation of molecular hydrogen feasible in organisms with a potent cyclic photophosphorylation system, and as envisaged in Fig. 7 this feat would occur in several stages.

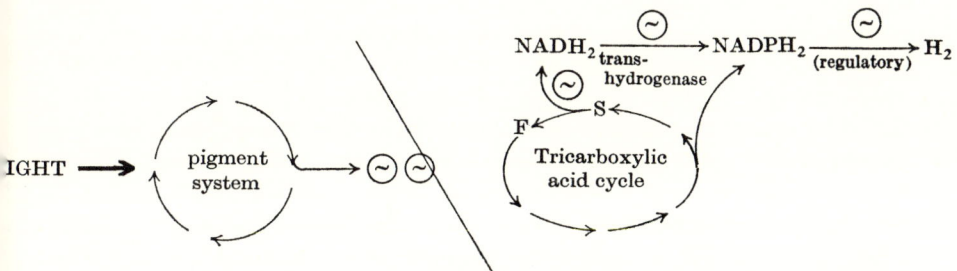

FIG. 7. A scheme for generation of net reducing power in bacterial photosynthesis. S indicates succinate; F, fumarate. In addition to tricarboxylic acid cycle substrates (not all the oxidative steps of the cycle are indicated), a variety of other substances may serve as hydrogen(e) donors for dark or energy-driven nicotinamide nucleotide reduction. During growth with molecular hydrogen as the accessory hydrogen(e) donor, net "photoproduction" of molecular hydrogen would not be expected. Adenosine triphosphate may also drive electron transfer from substrates to nicotinamide nucleotides by the Krebs-Veech mechanism.

2. By Algae

Photoproduction of molecular hydrogen by anaerobically adapted algae, first observed by Gaffron and Rubin (1942), has generally been interpreted in terms of non-cyclic electron flow schemes (e.g. see Gaffron, 1944). Various observations, which will not be detailed here, have suggested that the mechanisms in algae and photosynthetic bacteria may differ significantly. However, Abeles (1964) demonstrated light-dependent evolution of hydrogen from $NADH_2$ and $NADPH_2$ by cell-free extracts obtained from *Chlamydomonas eugametos*. More recently, Healey (1970) investigated the process in intact cells of *Chlamydomonas moewusii*, and his findings indicate that photoproduction of molecular hydrogen by algae and photosynthetic bacteria may be, in some respects, more similar than was previously believed. Evidence was obtained that "oxidative" carbon metabolism is a possible source of reductant for light-dependent hydrogen formation. Healey suggests that $NADH_2$ generated by the citric-acid cycle can provide electrons to (green plant) System I, which in turn "upgrades" the electrons to the ferredoxin redox potential level through the action of light (reduced ferredoxin than is

presumed to be the immediate electron donor for hydrogenase). In contrast to observations with photosynthetic bacteria, the uncoupler CCCP was found to stimulate photoproduction of molecular hydrogen by *C. moewusii* under certain conditions. The basis of this stimulation, however, is not clear, especially since the gas exchange is complex; photoreduction of carbon dioxide with hydrogen also occurs, and inhibition of this process by CCCP could cause an apparent stimulation of net hydrogen production.

It seems possible to me that photoevolution of hydrogen by anaerobically adapted algae could be explained as follows: ferredoxin (Fd), photoreduced by the "usual" green plant process, is the electron donor for the hydrogen-evolving hydrogenase that develops during anaerobic incubation. Reduced nicotinamide nucleotide may also funnel electrons to ferredoxin, if a high ratio of reduced/oxidized nicotinamide nucleotide is created: with $NADPH_2$, via a Fd-NADP reductase; with $NADH_2$ (generated by citric-acid cycle oxidations), via Healey's suggested mechanism and/or by way of the $NADPH_2$ route, facilitated by energy-linked transhydrogenase activity.

F. GENERAL COMMENTS ON PROPOSED NON-CYCLIC ELECTRON-FLOW MECHANISMS

The steady accumulation of positive results indicating that photoreduction of nicotinamide nucleotide occurs via energy-linked (reverse) electron flow has shifted the burden of proof to those who prefer, for whatever reason, a "non-cyclic" explanation for generation of net reducing power in bacterial photosynthesis. A non-cyclic process is broadly defined as:

$$\text{exogenous substrate} \longrightarrow e \begin{array}{c} \nearrow \text{BChl} \\ \searrow \text{BChl}^+ \end{array} \overset{\text{light}}{\underset{}{\Big)}} e \overset{H^+}{\longrightarrow} \begin{array}{c} \nearrow \text{PN nicotinamide nucleotide} \\ \searrow \text{PNH}_2 \text{ reduced nicotinamide nucleotide} \end{array}$$

In recent years, the possibility of a non-cyclic mechanism has been suggested in a number of studies concerned with spectroscopically detectable light-induced changes in pigments and electron carriers present in "chromatophores" (membrane fragments). This is a currently popular experimental approach and an almost bewildering assortment of findings has been published on kinetic changes in chromatophores of various kinds incubated under different conditions (of redox potential, availability of hydrogen donors and phosphorylation substrates, presence of "differential" inhibitors). In general, it can be said that complicated kinetics are invariably observed, which is not surprising considering the inherent chemical and physical complexity of the energy-converting membrane

structure and the manifold possibilities for interactions among the electromotively active components. In these circumstances, numerous problems in interpretation can be expected. For example, because of overlapping spectral characteristics, the identity of a "component" being followed at a particular wavelength is not always as definite as might be hoped. Also, the relationship between certain changes—as to whether they occur "in parallel" with one another or "in series"—is frequently very difficult to assess with confidence.

In relation to the question at hand, the greatest failing of typical spectroscopic investigations is that, ordinarily, no measurements are made of the net electron flow, i.e. of reduced nicotinamide nucleotide formation. This is a severe limitation that compounds the difficulty of sorting out the physiological electron-transfer pathways from the background of light-induced effects detectable by sensitive spectrophotometric techniques. Although many studies of this nature are quite ingenious, it is becoming clear that the spectroscopy will have to be augmented with appropriate biochemical determinations.

In the meantime, there is no shortage of hypothetical schemes for rationalizing available results. Mention may be made of a model proposed by Cusanovich *et al.* (1968) for *Chromatium*. They envisage *two* light-activated electron transfer chains: (1) a cyclic variety, whose function is generation of ATP, and (2) a non-cyclic sequence, responsible for production of reduced nicotinamide nucleotides, with sulphur-containing compounds serving as the ultimate electron donors. It is suggested that a single photosystem is involved, which can exist in two alternative functional states [for (1) and (2)]; interconversion of the states presumably is dependent on the ambient redox potential dictated by nutritional conditions; autotrophically, a relatively low potential would shift the balance in favour of (2), while photoheterotrophically (at higher potentials), function (1) would be dominant. Fowler and Sybesma (1970) suggest a basically similar model (for *R. rubrum*), but with separate light reactions for the two functions.

V. A Comparison of Energy Metabolism and Electron-Transfer Patterns in Photosynthetic Bacteria and Clostridia

Photosynthetic bacteria and clostridia present us with two interesting metabolic assemblages which show similarities and antipodal characteristics, and it still seems profitable to explore their meanings.

Certain photosynthetic bacteria [e.g. *Chromatium* (Bachofen and Arnon, 1966) and *Chlorobium thiosulfatophilum* (Buchanan *et al.*, 1969)] resemble clostridia in having a substantial content of ferredoxin. Moreover, there is evidence that *Chromatium* (strain D) possesses a typical

clostridial-type phosphoroclastic system in which ferredoxin mediates hydrogen formation in darkness (Bennett et al., 1964; note that photosynthetically grown cells of R. rubrum ferment pyruvate in the dark, but molecular hydrogen is not produced (Kohlmiller and Gest, 1951); see also relevant experiments with other nonsulphur purple bacteria by Qadri and Hoare (1967)). Bennett et al. (1964) believe that ferredoxin in Chromatium may be more closely connected with the nitrogen-fixation process than with photometabolism per se, and this notion is attractive for a number of reasons. In fact, the data available in support of the idea that photoreduction of ferredoxin might be a "primary" process in bacterial photosynthesis are not very convincing; there is only indirect evidence with particles from green bacteria; reported rates of ferredoxin reduction appear to be quite slow (see Evans and Buchanan, 1965; Buchanan and Evans, 1969). Results of recent experiments with Clostridium kluyveri (discussed below) suggest the possibility that "photoreduction" of ferredoxin in organisms such as Chlorobium thiosulfatophilum may be a secondary process that actually occurs via "dark" reduction of ferredoxin by "photoreduced" nicotinamide nucleotide.

As described in the foregoing, light-dependent production of hydrogen by photosynthetic bacteria is interpreted as an energy-driven anaerobic oxidation of $NAD(P)H_2$. Thus far, however, a reaction of this kind has not been demonstrable with cell-free extracts of photosynthetic bacteria (but see comments on algal production of hydrogen on p. 267). Burns and Bulen (1966) have observed dark ATP-dependent evolution of hydrogen from dithionite by extracts of R. rubrum, but lability of the system has apparently thwarted further work. Adenosine triphosphate-promoted hydrogen production by clostridial preparations, with dithionite as electron source, is now being studied in a number of laboratories, primarily because of its connexion with nitrogen-fixation enzymes. It seems likely that recent observations on hydrogen production by extracts of Cl. kluyveri may help in finding a common denominator in photo- and fermentative hydrogen formation. Clostridium kluyveri extracts produce molecular hydrogen from $NADPH_2$ by a ferredoxin-dependent reaction, if the nicotinamide nucleotide is maintained in the reduced state by an $NADPH_2$-regenerating system and if NAD is also provided (and kept predominantly in the oxidized form) (Jungermann et al., 1969). If the NAD is kept reduced by an $NADH_2$-regenerating system, molecular hydrogen is not evolved from $NADPH_2$. Molecular hydrogen can also be formed from $NADH_2$, via ferredoxin, but only in the presence of acetyl-CoA, which may function as an allosteric effector (Thauer et al., 1969). The possible alternatives in Cl. kluyveri extracts in respect to hydrogen production are summarized in Fig. 8.

Thus, molecular hydrogen can be produced from reduced nicotinamide

nucleotides despite the fact that the standard potentials of the nucleotide couples are higher than that of the hydrogen electrode. Clearly, it is not the standard potentials that are decisive, but rather the *actual* potentials that can be achieved by increase of the ratio of reduced/oxidized nicotinamide nucleotide. A very similar situation may obtain in *in vivo* photoproduction of hydrogen—that is, the ratio of reduced to oxidized nucleotide may attain very high values (especially under conditions where biosynthesis is limiting) with the aid of a strong reduction pressure exerted by the photosystem—in photosynthetic bacteria, through citric-acid cycle dehydrogenase activity augmented by energy-linked electron transfer, and in algae by the mechanisms already noted.

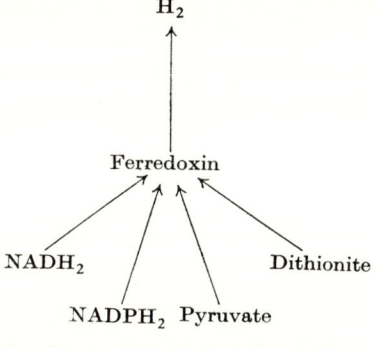

FIG. 8. Hydrogen production reactions in *Clostridium kluyveri* extracts, after Thauer *et al.* (1969). In connexion with reduction of ferredoxin by reduced nicotin-amide nucleotides, it is of interest that Gottschalk and Chowdhury (1969) have demonstrated (with *Cl. kluyveri* extracts) a ferredoxin-dependent synthesis of pyruvate from acetyl-coenzyme A and carbon dioxide with reducing power provided by $NADH_2$ or $NADPH_2$.

Several instances have been cited above of oxidation-reduction re-actions which *seem* to represent electron flow against the thermodynamic gradient, judging from standard redox potentials. In some of these, the reaction tendency (i.e. direction) can be readily explained by the actual potentials of the interacting couples—that is, the actual values predicting electron flow in the direction observed. There are a number of other examples of such situations in the literature (see Gray and Gest, 1965), and it is worth pointing out at least one of relevance. Cytochrome c_3 ($E_0' = -205$ mV), when strongly reduced, mediates production of mole-cular hydrogen ($E_0' = -420$ mV) from dithionite by *Desulfovibrio desulfuricans* extracts (Ishimoto *et al.*, 1957). Substantially purified hydrogenase preparations from *D. desulfuricans* also catalyse the same reaction, and it is of interest that ferredoxin is unable to replace the electron transfer function of cytochrome c_3 (Yagi *et al.*, 1968). In the

latter connexion, it may also be noted that a particular flavoprotein ("flavodoxin") from *Cl. pasteurianum* can replace ferredoxin as an electron carrier in evolution of hydrogen from dithionite by extracts of this organism (Knight *et al.*, 1966). In principle, the actual redox potential of any system can be extremely low, no matter what the standard potential of the redox couple, and it seems very probable that strongly reduced micro-environments are created *in vivo* for particular physiological purposes. The difficulties involved in extrapolating from standard potential *in vitro* to actual potentials *in vivo* prompted Clark (1955) to admonish: "Beware when the attempt is made to apply such [thermodynamic] data too casually to the dynamic affairs of living cells and to such heterogeneous systems as are living cells".

Evidently, alternative pathways or mechanisms may result in a basically similar "reducing power potential" in respect to molecular hydrogen formation in different organisms. A superficial similarity in this regard, however, does not necessarily imply similarity of physiological significance. In fact, the overall energy metabolism and associated electron-transfer patterns in photosynthetic bacteria and clostridia are quite different and, in a sense, inverse to one another. One of the most obvious metabolic properties of growing clostridial cells is the catabolism of large quantities of organic substrates, i.e. relative to the amount of new cell material synthesized. Ordinarily, the bulk of the substrate is consumed for the purpose of obtaining energy (ATP). Formation of molecular hydrogen is an integral feature of the energy-yielding oxidative conversions (e.g. phosphoroclastic split of pyruvate), and presumably is a device enabling "disposal" of electrons from such reactions without the necessity for a terminal electron acceptor other than hydrogen ions (Gray and Gest, 1965). The yield of clostridial cells per unit weight of substrate fermented is always quite meagre, and it is clear that the net reducing power needed for biosynthesis is small relative to the electron flow "transactions" required to achieve an overall oxidation-reduction balance in the rather inefficient quest for ATP. Photosynthetic bacteria, on the other hand, are not "catabolic" organisms in the clostridial sense, and have a powerful ATP generator (cyclic photophosphorylation) that is essentially independent of carbon metabolism. Their "problem" is the generation of net reducing power for biosynthesis, and it seems that they solve it, in part at least, by using ATP as an "energy transmitter" to push electron flow so as to achieve a high reducing potential of the nicotinamide nucleotide couples. Production (light-dependent) of hydrogen by photosynthetic bacteria is not an inherent feature of the energy-conversion process but, rather, represents a control system for regulation of the reducing power and chemical energy fluxes so that these are properly integrated with the biosynthetic machinery. The meaning

of the ATP-dependent production of molecular hydrogen from dithionite by clostridial extracts (and *Azotobacter* preparations) is still uncertain, and it may be that a similar *in vivo* reaction—with a physiological electron donor (possibly reduced nicotinamide nucleotide)—has a control function comparable to that of light-dependent hydrogen evolution in photosynthetic bacteria.

VI. Regulatory Mechanisms

Regulation of light-dependent energy conversion and reducing power generation will doubtless be a major focus of research during the 1970s. Much of the earlier work in this connexion derived from concern with the actual mechanics of light-induced processes and, thus, was limited in scope and viewpoint. The various elegant cellular control mechanisms now known to operate in heterotrophic cells can be expected to have counterparts in photosynthetic cells, and relevant examination of photosynthetic bacteria is gaining momentum. Although at this point it is probably true that many more questions can be posed than answers given, it seems worth while to present a brief survey of regulation problems and possible directions of future research.

The overall control plan clearly has two major aspects: regulation of *activity* of the "photosystem," and control of its *synthesis*. Light-activated systems cannot be considered as completely isolated from "dark" metabolism, but as a starting-point statement of the control problem, this oversimplified position may have certain advantages. Activity control will be considered first.

It is immediately evident that, below the saturation levels, the rate of photophosphorylation may be affected by the concentrations of the reactants (ADP, inorganic phosphate, Mg^{2+} and light) available to the energy-conversion "centres". Aside from incident light intensity, which the individual cell cannot control, the concentrations of the other substrates at the active membrane sites *in vivo* must each be dependent on numerous intracellular factors. In respect to Mg^{2+}, the most significant circumstance may be the ambient levels of physiological chelating agents, among which we may particularly cite ATP, inorganic pyrophosphate, and organic acids.

Relatively little attention has been paid to inorganic pyrophosphate, which is an important direct product of numerous ATP-dependent biosynthetic reactions and, at the same time, an immediate precursor of inorganic phosphate, through pyrophosphatase action. At first glance it would seem that the pyrophosphate level should be related in some simple way to the "balance" between the overall average rate of the biosynthetic steps in question and pyrophosphatase activity. Previously,

it has been generally assumed that growing cells contain "excess" pyrophosphatase, and that this provides for a "pulling" of biosynthesis. Recent studies with photosynthetic bacteria (Klemme and Gest, 1971; Klemme et al., 1971), however, indicate a more complex regulatory picture. For example, R. rubrum contains a soluble "allosteric" pyrophosphatase whose in vitro activity under certain conditions is activated by Mg^{2+} and inhibited by 2-phosphoglyceric acid. In addition, the enzyme is subject to reversible inactivation by ATP and reduced nicotinamide nucleotides. Thus, it is possible that the recycling of inorganic phosphate to the energy-converting system may be regulated in part by a sophisticated control device that interlocks with other regulatory "loops" involving the adenylate and nicotinamide nucleotides.

Delineation of the complex control pattern that accounts for regulation of ADP concentration would be a major accomplishment. The temporal ADP level in growing cells must be affected by many enzymic activities. Adenosine diphosphate is regenerated through kinase reactions (including adenylate kinase), but is simultaneously drained from the "energy metabolism pool" for nucleic acid synthesis. Maintenance of a relatively constant pool of ADP consequently implies continuous net synthesis, at a properly controlled rate, from the simple carbon and nitrogen sources provided in the medium. It would seem that the overall rate of biosynthesis of cell material, which is some function of the ATP flux, eventually determines the concentration of ADP available to the energy-converting system for resynthesis of ATP. This is one way of depicting the remarkable self-regulating aspect of energy metabolism kinetics.

The concentrations of the adenylate nucleotides in the steady state "pools" of rapidly dividing cells are very small in comparison with the quantity of ATP required for appreciable growth. This indicates that there must ordinarily be a very close coupling between energy conversion and biosynthesis and, moreover, implies a sensitive mechanism through which these two phases of metabolism can be delicately balanced against each other. A sensible model for understanding how this may be achieved can be developed using Atkinson's "energy charge" concept (Atkinson and Walton, 1967; Atkinson, 1968) as a basis.

The adenylate and nicotinamide nucleotides link energy conversion processes and biosynthesis, and therefore would be plausible components of a signal system for integration of energy-yielding and anabolic metabolism. A quantitative formulation of the control effects exerted by the adenylate nucleotides was developed by Atkinson (see papers cited; also, Klungsøyr et al., 1968; and Shen et al., 1968) in terms of "Energy Charge", defined as being equal to:

$$\frac{[ATP] + 0 \cdot 5 \,[ADP]}{[ATP] + [ADP] + [AMP]}$$

In essence, energy charge is a measure of the number of anhydride-bound phosphates per adenosine moiety and thus is a quantity that reflects the energy level of a system. Of particular significance in respect to the overall regulation problem under discussion is the fact that the *in vitro* activities of certain enzymes involved in heterotrophic ATP regeneration ("R enzymes") are progressively inhibited as the energy charge increases (especially beyond about 0·7), while activities of a number of ATP-utilizing biosynthetic enzymes ("U enzymes") are stimulated. The reciprocal response pattern, which is illustrated in Fig. 9, would be expected to provide the cell with a dynamic means of poising the *in vivo* energy charge value, i.e., any "disturbance" that tends to alter energy charge would be "buffered" by opposing changes in the rates of R- and U-type enzyme activities. In other words, metabolic stability in respect to ATP turnover would be enhanced, and this is equivalent to saying

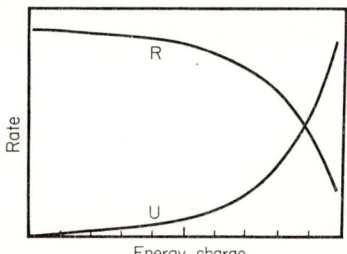

FIG. 9. Generalized representation of enzyme activity responses to the Energy Charge. R indicates enzymes involved in ATP regeneration; U, biosynthetic enzymes that utilize ATP. After Atkinson (1968).

that integrative control of energy-yielding and energy-utilizing processes is effected.

The energy charge concept of metabolic regulation was first applied to photosynthetic cells in investigations with *Rps. capsulata* (Sojka *et al.*, 1967; Sojka and Gest, 1968). Intermittent illumination, at certain frequencies, of growing cultures was found to depress photosynthetic growth rate and cause other changes suggesting a "syndrome" of energy stress, apparently caused by derangement of the integration between energy conversion and biosynthesis. Particularly noteworthy were effects on synthesis of bacteriochlorophyll, which may be taken as an index of the bacteriochlorophyll-membrane complex responsible for energy conversion (Sojka *et al.*, 1969; Steiner *et al.*, 1970). The experiments with intermittent light, and others with continuous light, were interpreted to indicate that high energy charge somehow suppresses synthesis of the bacteriochlorophyll-complex. This behaviour might seem contrary to expectations for biosynthetic processes regulated by energy charge, but it has been suggested (Zilinsky *et al.*, 1971; Gest, 1972) that the

unidentified enzymes affected may behave as R-type enzymes since they are concerned with synthesis of a system with R-type function (i.e. photophosphorylation). Additional evidence for this view was obtained from experiments in which exogenous ATP was found to inhibit synthesis of bacteriochlorophyll (and eventually growth) in cells dependent on photophosphorylation as a source of energy (Zilinsky *et al.*, 1971). The latter study also described the use of inorganic arsenate as a "probe" for disturbing the energy charge of growing cells of *Rps. capsulata*. Photosynthetic growth of this bacterium is severely inhibited by arsenate when this anion is present equimolar to orthophosphate in the culture medium. Studies (Zilinsky *et al.*, 1971; Lien *et al.*, 1971) on an arsenate-resistant mutant indicate that its resistance can be explained by a significantly increased photophosphorylation activity of the energy-converting machinery, which can compensate for the decrease in ATP regeneration rate caused by esterification of arsenate in place of phosphate. It is envisaged that the high photophosphorylation capacity of the mutant enables cells growing in the presence of arsenate to maintain the energy charge within the range necessary for orderly biosynthesis. Repeated subculture of the mutant in the presence of arsenate, under certain conditions, yields cell populations which have strikingly elevated contents of membrane-bound reaction-centre bacteriochlorophyll, cytochromes, and photophosphorylation coupling factor. Membranes from such cells can be expected to provide useful experimental systems for more refined analysis of the energy-conversion process.

The great complexity of energy-converting membranes is rapidly becoming more apparent, and it can be anticipated that the mechanisms regulating fabrication of such membranes are of corresponding intricacy. In photosynthetic bacteria, light intensity and molecular oxygen (with non-sulphur purple bacteria) are prominent among the factors affecting synthesis. As implied above, it is reasonable to believe that in the case of light (anaerobically) the actual chemical signals involved must include ATP (generated by photophosphorylation) and, thereby, the energy charge. Molecular oxygen causes repression of synthesis of certain enzymes that participate in production of bacteriochlorophyll; mutants of *Rps. spheroides* that are insensitive to oxygen repression have been obtained (Lascelles and Wertlieb, 1971), and should be helpful for further studies. Oxygen also has other effects, and the possibility that vigorous aeration may influence the energy charge significantly should not be overlooked.

Returning to the question of energy charge and activity of "biosynthetic" enzymes, we may first briefly consider the photophosphorylation process itself. Clear-cut "feedback" controls have not been described

BACTERIAL PHOTOSYNTHESIS 277

as yet, but it seems likely that they must occur. Preliminary tests (J.-H. Klemme and H. Gest, unpublished observations) for adenylate nucleotide energy charge effects with *R. rubrum* particles have shown about 20% inhibition of photophosphorylation rate at very high energy charge, which is suggestive. It seems possible that the activity of the photophosphorylation system could be subject to control by a more complex energy charge "quantity" that includes terms for other kinds of effectors such as the nicotinamide nucleotides. In fact, $NADH_2$ is known to be an effective agent for adjusting the redox state of isolated chromatophores to the range optimal for photophosphorylation (Geller and Lipmann, 1960). In this connexion, it is worth noting that the possibility of "$NADPH_2$ charge", i.e.

$$\frac{[NADPH_2]}{[NADP] + [NADPH_2]}$$

functioning as a regulatory element in reducing-power regeneration control has been suggested by Atkinson (1968). There are, in fact, experimental indications that such controls exist (Duggleby and Dennis, 1970). Perhaps of even greater interest is the possibility that *both* the adenylate and nicotinamide nucleotides contribute to interlocking feedback control of phosphorylation, in which case the relevant "charge expression" might be formidable indeed. It seems to me that this kind of regulation is strongly implied by the fact that the activities of a very considerable number of enzymes are affected by both adenylate and nicotinamide nucleotides (e.g. see Blackkolb and Schlegel, 1968; Rindt and Ohmann, 1969; Shen and Atkinson, 1970; Klemme and Gest, 1971). The reader is referred to a useful review by Preiss and Kosuge (1970) that details recent literature dealing with the effects of nucleotides on enzymes from photosynthetic and other kinds of autotrophic cells.

The point of view developed in this essay stresses the importance of "ATP energy pressure" in generation of reduced nicotinamide nucleotides from the "accessory" exogenous substrates always required in bacterial photosynthesis. The reactions, *per se*, which provide net reducing power for biosynthesis are regarded as "dark" processes. Reduced nicotinamide nucleotides, in turn, are very probably involved, one way or another, in control systems that influence the conversion of light energy to ATP. Accordingly, I envisage the possibility of an extraordinarily intricate cyclical control system, which could well be a major aspect of regulation of metabolism in photosynthetic cells.

VII. Epilogue

Despite temporary diversions into unproductive channels, the inexorable forward motion of research on bacterial photosynthesis has

reached the point where we can define the central issues with a considerable degree of confidence. There seems to be an increasing acceptance of the view that the bacterial and green-plant photosyntheses, although sharing a number of basic features, are distinctly different in respect to others. Without doubt, the search for a "unitary" mechanism has contributed to the fitful history of research on the comparative biochemistry of photosynthetic processes, which has witnessed protagonists shifting their positions up and back and frequently being right for the wrong reasons. Perhaps this was due in large measure to the fact that there are numerous possible levels of correlation in comparative biochemical reasoning. It is not always evident or true that the proper feature has been seized upon for comparison. Historically, close examination of apparent disunities has led to many of the great advances in biochemistry. Some remarks of Lipmann (1956) are pertinent: "It is astonishing to realize that the more one proceeds with the understanding of the working of the organism, the more one becomes concerned with methodological problems. Strangely, the prying into the mystery of life reduces more and more to an unravelling of a sometimes rather unusual and unexpected methodology of the cell".

VIII. Acknowledgements

Recent work from the author's laboratory, discussed here, was supported by Grant GB-7333X from the U.S. National Science Foundation. I am indebted to the John Simon Guggenheim Memorial Foundation for a fellowship, which made possible the writing of this article (completed in January, 1971).

Addendum: A recent report [Shanmugam, K. T. and Arnon, D. I. (1971) *Fed. Proc.* **30**, number 3, abstract 486] indicates the presence of two types of bound ferredoxin in *R. rubrum*, which can be "released" by treatment of cells with 5% (w/v) Triton X-100. Hopefully, the function(s) of these proteins will be defined in future studies.

REFERENCES

Abeles, F. B. (1964). *Pl. Physiol., Lancaster* **39**, 169.
Atkinson, D. E. (1968). *Biochemistry, N.Y.* **7**, 4030.
Atkinson, D. E. and Walton, G. M. (1967). *J. biol. Chem.* **242**, 3239.
Baccarini-Melandri, A., Gest, H. and San Pietro, A. (1970). *J. biol. Chem.* **245**, 1224.
Bachofen, R. and Arnon, D. I. (1966). *Biochim. biophys. Acta* **120**, 259.
Baltscheffsky, M. (1967a). *Nature, Lond.* **216**, 241.

BACTERIAL PHOTOSYNTHESIS 279

Baltscheffsky, M. (1967b). *Biochem. biophys. Res. Commun.* **28**, 270.

Baltscheffsky, M. (1969). *Archs Biochem. Biophys.* **133**, 46.

Baltscheffsky, H. and Arwidsson, B. (1962). *Biochim. biophys. Acta* **65**, 425.

Baltscheffsky, M. and Baltscheffsky, H. (1971). *In* "Wenner-Gren Symposium on Structure and Function of Oxidation Reduction Enzymes", in press.

Baltscheffsky, H., Baltscheffsky, M. and Thore, A. (1971). *In* "Current Topics in Bioenergetics" (D. R. Sanadi, ed.), Vol. 4, p. 273. Academic Press, New York.

Baltscheffsky, H., von Stedingk, L.-V., Heldt, H.-W. and Klingenberg, M. (1966). *Science, N.Y.* **153**, 1120.

Bennett, R., Rigopoulos, N. and Fuller, R. C. (1964). *Proc. natn. Acad. Sci. U.S.A.* **52**, 762.

Blackkolb, F. and Schlegel, H. G. (1968). *Arch. Mikrobiol.* **63**, 177.

Bose, S. K. and Gest, H. (1962). *Nature, Lond.* **195**, 1168.

Bose, S. K. and Gest, H. (1963a). *Proc. natn. Acad. Sci. U.S.A.* **49**, 337.

Bose, S. K. and Gest, H. (1963b). *In* "Energy-Linked Functions of Mitochondria" (B. Chance, ed.), p. 207. Academic Press, New York.

Bregoff, H. M. and Kamen, M. D. (1952). *Archs Biochem. Biophys.* **36**, 202.

Buchanan, B. B. and Bachofen, R. (1968). *Biochim. biophys. Acta* **162**, 607.

Buchanan, B. B. and Evans, M. C. W. (1969). *Biochim. biophys. Acta* **180**, 123.

Buchanan, B. B., Evans, M. C. W. and Arnon, D. I. (1967). *Arch. Mikrobiol.* **59**, 32.

Buchanan, B. B., Matsubara, H. and Evans, M. C. W. (1969). *Biochim. biophys. Acta* **189**, 46.

Burns, R. C. and Bulen, W. A. (1966). *Archs Biochem. Biophys.* **113**, 461.

Chance, B. and Nishimura, M. (1960). *Proc. natn. Acad. Sci. U.S.A.* **46**, 19.

Chance, B. and Olson, J. M. (1960). *Archs Biochem. Biophys.* **88**, 54.

Clark, W. M. (1955). *Bact. Rev.* **19**, 234.

Clayton, R. K. (1965). "Molecular Physics in Photosynthesis". Blaisdell Publishing Co., New York, Toronto and London.

Cohen-Bazire, G., Sistrom, W. R. and Stanier, R. Y. (1957). *J. Cell. Comp. Physiol.* **49**, 25.

Cruden, D. L. and Stanier, R. Y. (1970). *Arch. Mikrobiol.* **72**, 115.

Cusanovich, M. A., Bartsch, R. G. and Kamen, M. D. (1968). *Biochim. biophys. Acta* **153**, 397.

Cusanovich, M. A. and Kamen, M. D. (1968). *Biochim. biophys. Acta* **153**, 418.

Danielson, L. and Ernster, L. (1963). *Biochem. Z.* **338**, 188.

Duggleby, R. G. and Dennis, D. T. (1970). *J. biol. Chem.* **245**, 3751.

Duysens, L. N. M. (1964). *In* "Progress in Biophysics and Molecular Biology" (J. A. V. Butler and H. E. Huxley, eds.), Vol. 14, p. 1. Pergamon Press, Oxford, London.

Ernster, L. and Lee, C.-P. (1964). *A. Rev. Biochem.* **33**, 729.

Evans, M. C. W. (1969). *In* "Progress in Photosynthesis Research" (H. Metzner, ed.), Vol. III, p. 1474. H. Laupp, Jr., Tübingen.

Evans, M. C. W. and Buchanan, B. B. (1965). *Proc. natn. Acad. Sci. U.S.A.* **53**, 1420.

Fisher, R. R. and Guillory, R. J. (1969a). *J. biol. Chem.* **244**, 1078.

Fisher, R. R. and Guillory, R. J. (1969b). *FEBS Letters* **3**, 27.

Fowler, C. F. and Sybesma, C. (1970). *Biochim. biophys. Acta* **197**, 276.

Frenkel, A. W. (1954). *J. Am. Chem. Soc.* **76**, 5568.

Frenkel, A. W. (1956). *J. biol. Chem.* **222**, 823.

Frenkel, A. W. (1958). *J. Am. Chem. Soc.* **80**, 3479.
Frenkel, A. W. (1959). *Brookhaven Symp. Biol.* **11**, 276.
Frenkel, A. W. (1970). *Biol. Rev.* **45**, 569.
Frenkel, A. W. and Cost, K. (1962). *Nature, Lond.* **195**, 1171.
Fuller, R. C. (1969). *In* "Progress in Photosynthesis Research" (H. Metzner, ed.), Vol. III, p. 1579. H. Laupp, Jr., Tübingen.
Fuller, R. C., Conti, S. F. and Mellin, D. B. (1963). *In* "Bacterial Photosynthesis" (H. Gest, A. San Pietro and L. P. Vernon, eds.), p. 71. Antioch Press, Yellow Springs, Ohio.
Fuller, R. C. and Nugent, N. A. (1969). *Proc. natn. Acad. Sci. U.S.A.* **63**, 1311.
Gaffron, H. (1944). *Biol. Rev.* **19**, 1.
Gaffron, H. and Rubin, J. (1942). *J. Gen. Physiol.* **26**, 219.
Geller, D. M. (1962). *J. biol. Chem.* **237**, 2947.
Geller, D. M. and Lipmann, F. (1960). *J. biol. Chem.* **235**, 2478.
Gest, H. (1951). *Bact. Rev.* **15**, 183.
Gest, H. (1954). *Bact. Rev.* **18**, 43.
Gest, H. (1963). *In* "Bacterial Photosynthesis" (H. Gest, A. San Pietro and L. P. Vernon, eds.), p. 129. Antioch Press, Yellow Springs, Ohio.
Gest, H. (1966). *Nature, Lond.* **209**, 879.
Gest, H. (1972). *In* "First International Symposium on Genetics of Industrial Microorganisms," Czechoslovak Acad. Sci., in press.
Gest, H. and Kamen, M. D. (1948). *J. biol. Chem.* **176**, 299.
Gest, H. and Kamen, M. D. (1949). *Science, N.Y.* **109**, 558.
Gest, H. and Kamen, M. D. (1960). *In* "Encyclopedia of Plant Physiology" (W. Ruhland, ed.), Vol. V/2, p. 568. Springer Verlag, Berlin.
Gest, H., Kamen, M. D. and Bregoff, H. M. (1950). *J. biol. Chem.* **182**, 153.
Gest, H., Ormerod, J. G. and Ormerod, K. S. (1962). *Archs Biochem. Biophys.* **97**, 21.
Gottschalk, G. and Chowdhury, A. A. (1969). *FEBS Letters* **2**, 342.
Gray, C. T. and Gest, H. (1965). *Science, N.Y.* **148**, 186.
Healey, F. P. (1970). *Pl. Physiol., Lancaster* **45**, 153.
Hill, R. and Bendall, F. (1960). *Nature, Lond.* **186**, 136.
Hinkson, J. W. (1965). *Archs Biochem. Biophys.* **112**, 478.
Hood, S. L. (1964). *Biochim. biophys. Acta* **88**, 461.
Horio, T. and Kamen, M. D. (1962). *Biochemistry, N.Y.* **1**, 144.
Horio, T. and Kamen, M. D. (1970). *A. Rev. Microbiol.* **24**, 399.
Horio, T., Yamashita, J. and Nishikawa, K. (1963). *Biochim. biophys. Acta* **66**, 37.
Isaev, P. I., Liberman, E. A., Samuilov, V. D., Skulachev, V. P. and Tsofina, L. M. (1970). *Biochim. biophys. Acta* **216**, 22.
Ishimoto, M., Yagi, T. and Shiraki, M. (1957). *J. Biochem., Tokyo* **44**, 707.
Jackson, J. B. and Crofts, A. R. (1968). *Biochem. biophys. Res. Commun.* **32**, 908.
Jones, C. W. and Vernon, L. P. (1969). *Biochim. biophys. Acta* **180**, 149.
Jungermann, K., Thauer, R. K., Rupprecht, E., Ohrloff, C. and Decker, K. (1969). *FEBS Letters* **3**, 144.
Kamen, M. D. (1963). "Primary Processes in Photosynthesis". Academic Press, New York.
Kamen, M. D. and Gest, H. (1949). *Science, N.Y.* **109**, 560.
Keister, D. L. and Minton, N. J. (1969a). *Biochemistry, N.Y.* **8**, 167.
Keister, D. L. and Minton, N. J. (1969b). *In* "Progress in Photosynthesis Research" (H. Metzner, ed.), Vol. III, p. 1299. H. Laupp, Jr., Tübingen.

Keister, D. L. and Yike, N. J. (1966). *Biochem. biophys. Res. Commun.* **24**, 519.

Keister, D. L. and Yike, N. J. (1967a). *Archs Biochem. Biophys.* **121**, 415.

Keister, D. L. and Yike, N. J. (1967b). *Biochemistry, N.Y.* **6**, 3847.

Ketchum, P. A. and Holt, S. C. (1970). *Biochem. biophys. Acta* **196**, 141.

Klemme, J.-H. (1969a). *Z. Naturf.* **24B**, 67.

Klemme, J.-H. (1969b). *In* "Progress in Photosynthesis Research" (H. Metzner, ed.), Vol. III, p. 1492. H. Laupp, Jr., Tübingen.

Klemme, J.-H. and Gest, H. (1971). *Proc. natn. Acad. Sci. U.S.A.* **68**, 721.

Klemme, J.-H., Klemme, B. and Gest, H. (1971). *J. Bact.* in press.

Klemme, J.-H. and Schlegel, H. G. (1967a). *Z. Naturf.* **22B**, 899.

Klemme, J.-H. and Schlegel, H. G. (1967b). *Arch. Mikrobiol.* **59**, 185.

Klungsøyr, L., Hagemen, J. H., Fall, L. and Atkinson, D. E. (1968). *Biochemistry, N.Y.* **7**, 4035.

Knight, E., Jr., D'Eustachio, A. J. and Hardy, R. W. F. (1966). *Biochim. biophys. Acta* **113**, 626.

Kohlmiller, E. F., Jr. and Gest, H. (1951). *J. Bact.* **61**, 269.

Krebs, H. A. and Veech, R. L. (1969). *In* "The Energy Level and Metabolic Control in Mitochondria" (S. Papa, J. M. Tager, E. Quagliariello and E. C. Slater, eds.), p. 329. Adriatica Editrice, Bari.

Krebs, H. A. and Veech, R. L. (1970). *In* "Pyridine Nucleotide-Dependent Dehydrogenases" (H. Sund, ed.), p. 413. Springer Verlag, Berlin.

Lascelles, J. and Wertlieb, D. (1971). *Biochim. biophys. Acta* **226**, 328.

Lien, S., San Pietro, A. and Gest, H. (1971). *Proc. natn. Acad. Sci. U.S.A.* **68**, 1912.

Lipmann, F. (1956). *In* "Currents in Biochemical Research 1956" (D. E. Green, ed.), p. 241. Interscience Publishers, Inc., New York.

Losada, M., Nozaki, M. and Arnon, D. I. (1961). *In* "Light and Life" (W. D. McElroy and B. Glass, eds.), p. 570. Johns Hopkins University Press, Baltimore.

Löw, H. and Alm, B. (1964). *Abst. First Meet. Feder. Europ. Biochem. Soc.*, p. 68.

Medawar, P. B. (1969). "Induction and Intuition in Scientific Thought", p. 34. Methuen and Co., Ltd., London.

Melandri, B. A., Baccarini-Melandri, A., Gest, H. and San Pietro, A. (1971). *In* "Colloquium on Bioenergetics: Energy Transduction in Respiration and Photosynthesis". Pugnochiuso, Italy, 1970, in press.

Melandri, B. A., Baccarini-Melandri, A., San Pietro, A. and Gest, H. (1970). *Proc. natn. Acad. Sci. U.S.A.* **67**, 477.

Nozaki, M., Tagawa, K. and Arnon, D. I. (1961). *Proc. natn. Acad. Sci. U.S.A.* **47**, 1334.

Orlando, J. A., Sabo, D. and Curnyn, C. (1966). *Pl. Physiol., Lancaster* **41**, 937.

Ormerod, J. G. (1956). *Biochem. J.* **64**, 373.

Ormerod, J. G. and Gest, H. (1962). *Bact. Rev.* **26**, 51.

Ormerod, J. G., Ormerod, K. S. and Gest, H. (1961). *Archs Biochem. Biophys.* **94**, 449.

Peck, H. D., Jr. (1968). *A. Rev. Microbiol.* **22**, 489.

Preiss, J. and Kosuge, T. (1970). *A. Rev. Pl. Physiol.* **21**, 433.

Qadri, S. M. H. and Hoare, D. S. (1967). *Biochim. biophys. Acta* **148**, 304.

Rindt, K.-P. and Ohmann, E. (1969). *Biochem. biophys. Res. Commun.* **36**, 357.

Scholes, P., Mitchell, P. and Moyle, J. (1969). *Eur. J. Biochem.* **8**, 450.

Shen, L. C. and Atkinson, D. E. (1970). *J. biol. Chem.* **245**, 3996.

Shen, L. C., Fall, L., Walton, G. M. and Atkinson, D. E. (1968). *Biochemistry, N.Y.* **7**, 4041.

Sirevåg, R. and Ormerod, J. G. (1970). *Biochem. J.* **120**, 399.

Sistrom, W. R. (1965). *J. Bact.* **89**, 403.

Sojka, G. A., Baccarini, A. and Gest, H. (1969). *Science, N.Y.* **166**, 113.

Sojka, G. A., Din, G. A. and Gest, H. (1967). *Nature, Lond.* **216**, 1021.

Sojka, G. A. and Gest, H. (1968). *Proc. natn. Acad. Sci. U.S.A.* **61**, 1486.

Steiner, S., Sojka, G. A., Conti, S. F., Gest, H. and Lester, R. L. (1970). *Biochim. biophys. Acta* **203**, 571.

Tagawa, K. and Arnon, D. I. (1962). *Nature, Lond.* **195**, 537.

Thauer, R. K., Jungermann, K., Rupprecht, E. and Decker, K. (1969). *FEBS Letters* **4**, 108.

Tuttle, A. L. and Gest, H. (1959). *Proc. natn. Acad. Sci. U.S.A.* **45**, 1261.

Uffen, R. L. and Wolfe, R. S. (1970). *J. Bact.* **104**, 462.

van Niel, C. B. (1941). *In* "Advances in Enzymology" (F. F. Nord and C. H. Werkman, eds.), Vol. I, p. 263. Interscience Publishers, Inc., New York.

Vernon, L. P. (1968). *Bact. Rev.* **32**, 243.

von Stedingk, L.-V. (1964). M.S. Thesis: University of Wisconsin.

von Stedingk, L.-V. and Baltscheffsky, H. (1966). *Archs Biochem. Biophys.* **117**, 400.

Wassink, E. C. (1947). *Antonie van Leeuwenhoek* **12**, 281.

Weaver, P., Tinker, K. and Valentine, R.C. (1965). *Biochem. biophys. Res. Commun.* **21**, 195.

Yagi, T., Honya, M. and Tamiya, N. (1968). *Biochim. biophys. Acta* **153**, 699.

Yamamoto, N., Hatakeyama, H., Nishikawa, K. and Horio, T. (1970). *J. Biochem., Tokyo* **67**, 587.

Yamanaka, T. and Kamen, M. D. (1965). *Biochem. biophys. Res. Commun.* **18**, 611.

Yamanaka, T. and Kamen, M. D. (1967). *Biochim. biophys. Acta* **131**, 317.

Zilinsky, J. W., Sojka, G. A. and Gest, H. (1971). *Biochem. biophys. Res. Commun.* **42**, 955.

AUTHOR INDEX

Numbers in italics refer to the pages on which references are listed at the end of each article.

A

Abell, M. A., *72*
Abeles, F. B., 267, *278*
Abram, D., 18, 26, *71*
Abrams, A., 19, 21, 27, *71*, *76*
Adam, A., 36, 37, *78*, *81*, 103, *116*
Adams, G. A., 58, 59, *71*, *72*
Adams, J. B., *72*
Adinolfi, A., 186, *202*
Ahrens, J., 208, 226, *240*
Akamatsu, Y., 61, *76*
Albro, P. W., 61, 62, *78*
Alexander, M., 160, *200*
Allam, A. M., 163, 168, *203*
Alm, B., 260, *281*
Ames, G. F., 61, *72*
Anderson, J. C., 109, 110, *115*
Anderson, L., 174, *197*
Andreesen, J. R., 176, *198*
Anthony, C., 135, 145, 155, 156, 157, 161, *198*
Archibald, A. R., 32, 33, 34, 42, 43, 44, 45, 72, *73*, *75*, 96, 99, 109, 110, 113, 114, *115*, *116*
Argaman, M., 26, *79*
Arima, K., 25, *81*
Arkad'eva, Z. A., 131, *200*
Armstrong, J. J., 32, *72*, 89, 92, *115*
Arnon, D. I., 171, 172, 173, 174, 175, *198*, *199*, 250, 253, 254, 264, 269, *278*, *279*, *281*, *282*
Aronson, A., 19, *72*
Arwidsson, B., 249, *279*
Asbell, M. A., 64, 65, *72*
Ashman, D. F., 64, *78*
Asselineau, J., 62, *80*
Atkinson, D. E., 206, *240*, 274, 275, 277, *278*, *281*
Azuma, I., 37, *81*

B

Baccarini, A., 275, *282*
Baccarini-Melandri, A., 250, *278*, *281*

Bachofen, R., 171, *198*, 256, 262, 269, *278*, *279*
Baddiley, J., 32, 33, 34, 42, 43, 44, 45, 72, *73*, *75*, *80*, 86, 89, 92, 95, 96, 99, 102, 109, 110, 113, 114, *115*, *116*
Bakay, B., 20, *80*
Ball, E. G., 158, *202*
Baltscheffsky, H., 249, 250, *279*, *282*
Baltscheffsky, M., 249, 257, *278*, *279*
Balyuzi, H. H. M., 70, 71, *72*
Banerjee, A. K., 239, *240*
Barker, H. A., 120, 133, 135, 140, 141, 162, 192, *198*, *200*, *202*, *203*
Barkulis, S. S., 44, 46, *72*, *75*
Baron, C., 26, 27, *71*
Bartha, R., 207, 223, *240*, *242*
Bartsch, R. G., 269, *279*
Bass, J. A., 46, *72*
Bassalik, K., 133, 135, *198*
Bassham, J. A., 169, *198*
Bayer, M. E., 2, 47, 51, *72*
Bayley, S. T., 53, *80*
Bazil, S. L., 70, 71, *73*
Becker, G. E., 160, *201*
Beckwith, J. R., 211, *241*
Beebe, J. L., 24, *72*
Begg, K. J., 115, *115*
Beijerinck, M. W., 138, *198*
Bendall, F., 252, *280*
Bennett, E. O., 25, *80*
Bennett, R., 270, *279*
Benson, A. A., 61, *80*, 169, *198*
Bergeron, G., 13, 42, *78*
Berman, M. F., 35, 38, 40, *80*
Bernal, J. D., 196, *198*
Bernstein, S., *240*
Bertsch, L. L., 23, *72*
Betz, J. V., 32, *72*
Bezer, A., 42, *80*
Bhat, J. V., 133, 135, *198*, *200*
Birdsell, D. C., 4, 48, *72*
Bishop, D. J., 19, 21, 24, 25, *72*
Blackkolb, F., 221, 223, 224, 225, 226, *240*, *242*, 277, *279*
Blackmore, M. A., 170, 180, *198*

Bladen, H. A., 49, 59, *72*
Blair, P. V., 17, *74*
Blakley, R. L., 120, 164, 181, *198*
Blaylock, B. A., 162, 163, 164, 165, *198*
Bobo, R. A., 61, *72*
Bock, E., 210, *240*
Bodman, H., 18, *72*
Boltralik, J. J., 46, *72*
Bonaly, R., 36, 37, *77*
Bond, E. C., 5, *72*
Bone, D. H., *240*
Bongers, L., 209, *240*
Bonson, P. P. M., 23, *72*
Bose, S. K., 249, 251, 252, 253, 256, 258, 264, *279*
Bovell, C., 211, *240*
Bowien, B., 221, *240*
Boylan, R. J., *73*
Boylen, C. W., 17, 31, *72*
Branton, D., 13, 18, 51, *73*, *74*
Braun, V., 55, 56, *72*
Bray, J. P., 46, *72*
Breed, R. S., 133, *198*
Bregoff, H. M., 264, 265, *279*, *280*
Brenner, S., 4, *72*
Bricas, E., 36, 38, 41, 54, *73*, *75*, *77*
Bridgeland, E. S., 185, *200*
Brill, W. J., 163, 164, 168, *198*, *203*
Brinton, C. C., 33, *72*
Brock, T. D., 45, *73*
Brooks, D., 43, *72*
Brot, N., 164, *198*
Brown, A. D., 47, 52, *72*, 112, *115*
Brown, C. M., 102, *117*
Brown, L. R., 122, 123, 143, 145, 146, 147, *198*, 219, *240*
Brown, M. R. W., 65, *72*
Brundish, D. E., 43, *72*
Bryan-Jones, G., 144, 149, 196, *198*
Bryant, M. P., 142, 143, 162, 163, *198*, 201
Buchanan, B. B., 171, 172, 173, 174, 175, *198*, *199*, 253, 254, 256, 269, 270, *279*
Buchanan, J. G., 89, 92, *115*
Buckmire, F. L. A., 4, 51, 65, *72*, *73*
Budd, J. A., 150, *198*
Bulen, W. A., 270, *279*
Burde, R., 14, 71, *80*
Burdett, I. D. J., 4, 6, 7, 10, 12, 17, 18, 20, 21, 23, *72*, *79*
Burge, R. E., 47, 49, 50, 59, 69, 70, *72*
Burge, W. D., 192, *201*
Burger, M. M., 32, *72*, *73*, 88, 91, 95, *115*
Burns, R. C., 270, *279*
Burns, R. O., 236, *240*
Burton, A. J., 59, *73*
Burton, K., 161, *198*

Butler, T. F., 17, 26, *73*, *75*
Button, D., 42, 44, *72*, *73*

C

Callow, D. S., 98, *115*
Calvin, M., 169, *198*
Calvo, I. M., 236, *240*
Campbell, J. N., 41, *73*
Cardini, G., 153, *198*
Carito, S. L., 70, 71, *73*
Carnahan, J., 33, *72*
Carroll, K. K., 17, 18, 21, *74*
Carson, K. J., 65, *73*
Carter, H. E., 59, *73*
Card, G. L., *73*
Cesari, I. M., 7, *79*
Chaix, P., 11, 15, *74*
Chan, S. I., 29, 30, *75*
Chance, B., 252, *279*
Chang, J. P., 230, *240*
Chapman, G. B., 7, 10, *78*
Chapman, H. M., 123, *198*
Chappelle, E. W., 161, *198*
Charalampous, F. C., 192, *198*
Chatterjee, A. N., 5, 45, *73*
Cherry, W. B., 24, *77*
Chin, T., 42, *73*, 88, 91, 95, *115*
Cho, K. Y., 24, 25, 60, 61, 62, 68, *73*
Chou, J., 66, *78*
Chowdhury, A. A., 271, *280*
Chung, K. L., 32, *73*
Clarke, K., 52, 53, 61, 63, 67, *73*
Clark, W. M., 272, *279*
Clayton, R. K., 248, *279*
Clive, D., 5, *73*
Coapes, H. E., 72
Cohen-Bazire, G., 247, *279*
Cohn, F., 2, *73*
Colby, J., 134, 151, 158, *199*
Cole, R. M., 10, 11, 15, 17, 46, *73*, *75*, *78*, *80*
Coleman, G., 19, 20, *73*
Coleman, R., 71, *74*
Coles, E., 241, *80*
Collins, F. M., 85, *115*
Collins, V. G., 138, *199*
Conover, M. J., 20, *80*
Conti, S. F., 15, *73*, 134, 135, 136, 138, *199*, 247, 275, *280*, *282*
Cook, D. W., 219, *240*
Coon, M. J., 144, *201*
Corner, T. R., 3, *73*
Corpe, W. A., 53, *81*
Cost, K., 258, *280*
Costerton, J. W., 4, 47, 50, *73*, *74*
Cota-Robles, E. H., 4, 13, 48, *72*, *73*
Coty, V. F., 120, 122, *199*

Cox, R. H., 601, *81*
Coyette, J. 35, 41, 44, *73*
Crane, F. L., 30, *73*
Crofts, A. R., 120, *203*, 252, *280*
Cronan, J. E., 61, 63, *73*
Crough, D. J., 222, *240*
Cruden, D. L., 249, *279*
Cullen, J., 49, 64, *76*, 113, *116*
Curnyn, C., 259, 263, *281*
Curtis, M. J., 109, 110, *115*
Cusanovich, M. A., 250, 269, *279*
Cutinelli, C., 3, 44, *73*

D

Danielson, K., 258, *279*
Dandl, R., 41, *76*
Danielli, J. F., 30, *73*
Daniels, M. J., 13, *80*
Dans, C. L., 207, *241*
Dark, F. A., 4, *72*
Daron, H. H., 25, *73*
Das, B. C., 37, *81*
Da Silva, P. P., 18, *73*
Davey, J. F., 124, 126, 143, *203*
Davey, N. B., 109, *115*
Davie, J. M., 45, *73*
Davies, H. C., 113, *115*
Davies, S. L., 6, *73*, 124, 126, 143, *203*
Davis, B., 19, *81*
Davis, D. H., 206, 219, 221, *240*
Davis, J. B., 120, 122, *199*
Davis, R. H., 120, 123, 143, 146, 147, *199*
Davis, S. L., *199*
Davison, A. L., 32, 33, 34, *72, 75*, 113, 114, *116*
Davson, H., 30, *73*
Decker, K., 270, 271, *280, 282*
De Cicco, B. T., 211, 216, 217, 219, 239, *240, 242*
Dekegel, D., 49, 53, 59, 67, *74*
Delwiche, C. C., 160, *199*, 201
Dennis, D. T., 277, *279*
Deus, B., 185, 191, *200*
D'Eustachio, A. J., 272, *281*
De Petris, S., 49, 50, 51, 55, 59, 67, *73*
De Voe, I. W., 4, *73*
Dezélée, P., 36, 54, *73*
Dickens, F., 192, *199*
DiGiacomo, G., 70, 71, *73*
Din, G. A., 248, 275, *282*
Dische, Z., 192, *199*
Dittmer, J. C., 61, 62, *78*
Donachie, W. D., 115, *115*
Done, J., 67, *73*
den Dooren de Jong, L. E., 134, 154, *199*
Doudoroff, M., 13, *202*, 206, 219, 221, *240*

Dowell, V. R., 24, *77*
Doy, C. H., 68, *73*
Drägert, W., 231, *241*
Draper, J., 47, 49, 50, 59, 69, *72*
Dröge, W., 57, *73*
Drucker, H., 234, *242*
Drummond, D. G., 47, *72*
Duggleby, R. G., 277, *279*
Dunnick, K. K., 63, *73*
Duysens, L. N. M., 248, *279*
Dvorak, H. F., 67, *73, 81*
Dworkin, M., 53, *81*, 120, 121, 124, 143, 146, 147, *199*
Dyer, W. J., 150, *199*

E

Eady, R. R., 134, 151, 152, 153, *199, 200, 201*
Eagon, R. G., 61, 64, 65, *72, 73*
Eberhardt, U., 206, 207, 211, 213, 216, 217, 218, 219, 220, 223, 239, *240*
Edwards, M. R., 3, *73*
Eisenberg, R. C., 26, 28, *73, 74*
Elizarova, T. N., 131, *199*
Ellar, D. J., 3, 10, 11, 12, 13, 18, 19, 25, 33, *74, 77, 79*
Ellwood, D. C., 31, 44, 45, 58, *74*, 84, 85, 90, 91, 92, 94, 95, 97, 102, 112, 114, *115, 116*
Elson, H. E., 49, *78*
Engel, H., 210, *240*
Engelman, D. M., 26, 29, *74*
Ensign, J. C., 31, 35, 36, 53, 55, *72, 76, 80*
Erebo, L., 131, *199*
Ernster, L., 251, 258, *279*
Eroshin, V. K., 196, *199*
Evans, C. G. T., 85, *115*
Evans, M. C. W., 171, 172, 173, 174, 175, *198, 199*, 253, 254, 262, 269, 270, *279*

F

Fall, L., 274, *281*
Farr, L., *77*
Farshtchi, D., 24, *77*
Fensom, A. H., 58, 59, *74*
Fernández-Moran, H., 17, *74*
Ferrandes, B., 11, 15, *74, 79*
Fewson, C. A., 212, *240*
Fiil, A., 13, 18, 51, *74*
Finch, J. T., 33, *74*
Finean, J. B., 71, *74*
Finstein, M. S., 160, *199*
Fischer, A., 2, *74*
Fischman, D. A., 51, 53, 64, *74, 81*

286 AUTHOR INDEX

Fisher, R. R., 259, *279*
Fitch, W. M., 155, *199*
Fitz-James, P. C., 6, 7, 9, 10, 11, 15, 18, 19, *74*
Fleck, J., 35, 40, *76*
Fleischer, B., 17, *74*
Fleischer, S., 17, *74*
Foote, J. L., 24, 25, *80*
Forman, A., 5, *76*
Forrest, H. S., 150, 157, *203*
Forsberg, C. W., 4, 47, 50, *73, 74*
Foster, J. W., 120, 121, 122, 123, 124, 126, 132, 143, 144, 146, 147, 150, 189, 196, *199, 201*
Fowler, C. F., 269, *279*
Fox, E. N., 46, *74*
Frank, H., 49, 53, 54, 55, 59, 67, *74, 77, 81*
Freer, J. H., 10, 11, 18, 19, 20, 25, 28, *74, 77, 78, 79*
Fréhel, C., *5*, 10, *76, 79*
Frenkel, A. W., 248, 252, 253, 258, 260, 261, *279, 280*
Frerman, F. E., 24, *74*
Frings, W., 211, 213, 214, 218, 230, *240*
Fuller, R. C., 174, *197, 199*, 244, 247, 250, 270, *279, 280*

G

Gaffron, H., 267, *280*
Galdiero, F., 3, 44, *73*
Gale, E. F., 225, *240*
Galindo, B., 34, *76*
Ganguli, B. N., 153, *200*
Garrett, A. J., 31, 43, *78, 79*
Gaughran, E. R. L., 113, *116*
Geller, D. M., 261, 277, *280*
Gel'man, N. S., 6, 14, *74*
Georgi, C. E., *73*
Gerharot, P., 3, 4, 33, *72, 74*
Gerisch, G., 59, *77*
Gest, H., 247, 248, 249, 250, 251, 252, 253, 256, 258, 264, 265, 270, 271, 272, 274, 275, 276, 277, *278, 279, 280, 281, 282*
Ghosh, B. K., 3, 10, 11, 15, 16, 17, 18, 19, 21, *74, 79*
Ghuysen, J. M., 30, 35, 36, 37, 38, 39, 41, 42, 43, 44, 52, 53, 54, *73, 74, 75, 77, 78*, 92, 93, 102, 103, 104, *116*
Gibbins, L. N., 195, *199*
Gibbs, H., 193, *199*
Gibbs, M., 120, 173, *199, 201*
Gibson, F., *240*
Giesbrecht, P., 32, *75*
Gillespie, D., 19, 20, *78*

Girard, A., 57, 59, 68, *79*
Glaser, L., 42, 66, *73, 75, 77*, 88, 91, 92, 95, *115, 116*
Glaser, M., 29, 30, 31, *75*
Glauert, A. M., 2, 17, 31, 32, 34, 47, 49, 51, *75, 80*
Glover, J., 174, *199*
Gmeiner, J., 57, 58, 59, *75*
Goldstein, D. B., 155, *199*
Gooder, H., 3, 4, 5, *75, 76*
Goodman, N. S., 211, 239, *241*
Gordon, R. C., 61, *75*
Gotschlich, E. C., 44, *77*
Gottlieb, J. A., 134, 179, *201*
Gotto, A. M., 186, *200*
Gottschalk, G., 176, *198*, 213, 218, 219, 220, 223, 224, 239, *240, 242*, 271, 280
Gould, G. W., 84, 86, *116*
Goundry, J., 32, 33, 34, *75*, 113, 114, *116*
Graham, J. A., 59, *80*
Grant, W. D., 38, 44, *75*
Gray, C. T., 15, 47, *73*, 265, 271, 272, *280*
Gray, G. W., 52, 53, 58, 59, 60, 61, 63, 65, 67, *73, 74, 75, 76*
Green, D. E., 17, 30, *74, 75, 80*
Greenawalt, J. W., 14, *81*
Griffiths, H., 34, *81*
Gross, S. R., 236, *242*
Grula, E. A., 26, 37, 53, *73, 75, 81*
Grula, M. M., 17, 53, *75*
Guillory, R. J., 259, *279*
Guinand, M., 41, *75*
Gunsalus, I. C., 153, *200*

H

Hagemen, J. H., 274, *281*
Haggstrom, L., 146, *199*
Hahn, J. J., 46, *75*
Hall, E. A., 44, *75, 76*
Hall, J. O., 30, *73*
Hammerling, G., 57, *75*
Hancock, I. C., 60, 61, 63, *75*, 113, *116*
Hansen, R. G., 230, *240*
Hard, G. C., 34, *75*
Harder, W., 180, 183, 184, 185, *199*
Hardy, R. F. W., 272, *281*
Harrington, A. A., 121, 133, 135, 155, 158, *199*
Harris, A. Z., 169, *198*
Harrison, J. E., 120, 148, 157, *202*
Hartsell, S. E., 4, *78*
Harwood, J. H., 143, 145, 146, 147, 148, 196, *199*
Hatakeyama, H., *282*
Hay, J. B., 32, *72*

Healey, F. P., 267, *280*
Heldt, H.-W., 249, *279*
Henning, U., 230, *240*
Heppel, L. A., 66, 67, *73, 75, 78, 81*
Heptinstall, S., 42, 44, 45, *72, 75,* 96, 99, *116,* 153, 156, 159, 180, 181, 183, *199*
Herbert, D., 83, 84, 85, 86, 98, *115, 116*
Heymann, H., 36, 38, 44, *75, 77*
Hift, H., *199*
Higgins, I. J., 143, 144, 146, 149, *199*
Higgins, M. L., 7, 12, 17, 20, 32, *75*
Highton, P. J., 7, 12, 17, *75*
Hill, F., 236, *240*
Hill, R., 252, *280*
Hinkson, J. W., 261, *280*
Hippe, H., 208, *240*
Hirota, Y., 13, *75*
Hirsch, P., 134, 135, 136, 138, 140, *199*
Hoare, D. S., 140, 168, 170, 174, 175, 196, *200, 201, 202, 242,* 270, *281*
Hoeksema, H., 195, *202*
Hofschneider, P. H., 49, *75*
Holden, J. T., 10, 12, 13, 14, 15, 17, *76*
Hollinshead, J. A., 181, *200*
Hollis, D. G., 62, *77*
Holmwood, K. J., 44, *76*
Holt, S. C., 2, 3, 18, 32, 33, 34, *75, 77,* 247, *281*
Holzer, H., 185, 191, *200,* 231, *241*
Homann, H. R., 218, *241*
Honya, M., 271, *282*
Hood, S. L., 261, *280*
Horio, T., 249, 250, 261, *280, 282*
Horne, R. W., 34, 59, 68, *79, 80*
Hornig, D., 154, 157, 188, *200*
Hough, L., 195, *200*
Houtsmuller, U. M. T., 24, *75,* 100, *116*
Howes, W. V., 222, *241*
Huennekens, F. M., 157, *200*
Hughes, R. C., 32, 35, 37, 38, 43, 44, *75, 76,* 88, 91, 101, 106, 108, 113, *116*
Hungate, R. E., 141, 142, *201, 202*
Hungerer, K. D., 35, 38, 40, *76*
Hunter, J. R., 84, 110, *116, 117*
Hurst, A., 2, *76*
Huss, L., 31, 38, 39, *79,* 103, *116*

I

Ibbott, F. A., 21, *76*
Ida, S., 160, *200*
Imaeda, T., 7, 34, *76, 79*
Ingraham, J. L., 63, *80,* 113, *116*
Inniss, W. E., 59, 60, *77*
Irion, E., 176, 177, *201*
Isaev, P. I., 255, 261, *280*
Ishida, M., 26, 27, 28, *76, 77*

Ishimoto, M., 271, *280*
Iyer, S. N., 134, 154, *200*

J

Jackson, J. B., 252, *280*
Jacob, F., 13, *75, 79*
Jacobs, N. J., 15, *73*
Jaenicke, L., 191, *200*
Janczura, E., 88, 91, *116*
Jann, K., 58, 59, *77*
Janota, L., 133, *200*
Jarman, T. R., 152, *200*
Jensen, R. A., 237, *240*
Jeynes, M. H., 4, *72*
Johnson, E. J., 215, 216, *241*
Johnson, M. J., 146, 147, *200, 203*
Johnson, M. K., 215, 216, *241*
Johnson, P. A., 144, 155, 156, 158, 159, 189, 190, *200*
Johnston, J. H., 58, *76*
Johnston, R. J., 58, *76*
Jones, M. F., 46, *72*
Jones-Mortimer, M. C., 159, *200*
Jones, C. W., 261, 263, *280*
Jones, K. M., 185, *200*
Joseph, A. A., 230, 231, *240*
Judge, J. A., 3, *74*
Jungermann, K., 270, 271, *280, 282*
Jurtshuk, P., 153, *198*
Jušić, D., 53, *76*

K

Kabat, E. A., 42, *80*
Kakefuda, T., 10, 12, 13, 14, 15, 17, *76*
Kallio, R. E., 121, 133, 134, 135, 154, 155, 158, *199, 200*
Kaltwasser, H., 209, 224, 231, 232, 233, 234, *240, 241*
Kamen, M. D., 174, *199,* 245, 246, 248, 249, 250, 254, 264, 265, 269, *279, 280, 282*
Kandler, O., 4, 31, 37, 38, 39, 41, *72, 75, 76, 79,* 103, *116*
Kaneda, T., 25, *76,* 135, 159, 179, *200*
Kanemasa, Y. Y., 61, *76*
Kaneshiro, T., 62, *76*
Kanetsuna, F., 34, *76*
Karakawa, W. W., 46, *76*
Karlsson, J. L., 140, *200*
Karush, F., 113, *115*
Kaserer, H., 138, *200,* 205, *241*
Katagiri, M., 153, *200*
Kates, M., 59, *71*
Kato, K., 37, *76*
Katz, W., 54, 55, *76*

Kaufman, S., 150, *200*
Kay, L. D., 169, *198*
Kazuo, I., 38, *76*
Keech, D. B., 170. *202*
Keister, D. L., 259, 260. 261, 263, *280, 281*
Keleman, M. V., 69, *76*
Kellenberger, E., 2, 4, *72, 76*
Kelly, D. P., 120, *200*
Kelly, H., 173, *202*
Kemp, M. B., 179, 189, 190, 191, 193, *200, 201*
Kemper, S., 53, *77*
Kersten, D. K., 130, *200*
Ketchum, P. A., 247, *281*
Key, B. A., 58, 65, *76*
Khambata, S. R., 135, *200*
Kickhöfen, B., 59, *77*
Kiesow, L. A., 120, *200*
Kikuchi, G., 185, *200*
King, J. R., 5, *76*
King, R. D., 17, 26, *75*
Kistner, A., 138, *200*
Klein, S. M., 185, *200*
Klemme, B., 274, *281*
Klemme, J.-H., 254, 256, 258, 261, 262, 263, 274, 277, *281*
Klieneberger-Nobel, E., 4, *72*
Klingenberg, M., 249, *279*
Klug, A., 33, *74*
Klungsøyr, L., 274, *281*
Kluyver, A. J., 141, 142, *200*
Knight, E., Jr., 272, *281*
Knivett, V. A., 113, *116*
Knox, K. W., 44, 49, 64, *75, 76*
Knutton, S., 71, *74*
Koch, J., 120, 191, *200, 202*
Kochi, H., 185, *200*
Kocun, E. J., 24, *76*
Koffler, H., 212, 239, *242*
Kohlhaw, G., 185, 191, *200*, 231, *241*
Kohlmiller, E. F., Jr., 270, *281*
Kohn, A., 4, *76*
Kolb, J. J., 20, *80*
Kolenbrander, P. E., 53, 55, *76*
Kondrat'eva, E. N., 139, 170, *200*
König, Chr., 220, 231, 232, *241*
Korczynski, M. S., 49, 59, *81*
Korn, E. D., 17, *76*
Kornberg, A., 23, *72*
Kornberg, H. L., 174, 186, *199, 200*
Kortstee, G. J. J., 154, 155, *200*
Kosuge, T., 277, *281*
Kotani, S., 37, *76*
Kozlova, E. I., 131, *200*
Krämer, J., 230, 231, 232, 234, *240, 241*
Krause, R. M., *76*
Krauss, I., 208, *242*
Krebs, H. A., 186, *200*, 263, *281*

Krulwich, T. A., 35, 36, *76*
Kuehn, G. D., 216, 217, 218, *241*
Kung, H., 135, 151, 152, 154, 158, 187, 188, *200, 203*
Kuratomi, K., 176, *201*
Kusunose, E., 144, *201*
Kusunose, M., 144, *201*

L

Lacave, C., 62, *80*
Lacey, B. W., 112, *116*
Lache, M., 41, *75*
Ladner, A., 135, 156, *201*
Lafferty, R., 208, *242*
Lambert, R., *80*
Lampen, J. O., 16, *74, 76, 79*
Landman, O. E., 4, 5, *72, 73, 76, 77*
Landsberg, E., 192, *199*
Lang, D. R., 46, *76*
Lange, C. F., *76*
Langenberg, K. F., 143, *201*
Large, P. J., 134, 151, 152, 153, 179, 180, 181, *199, 200, 201*
Larsen, H., 68, 69, *76, 80*
Lascelles, J., *241*, 276, *281*
Latzko, E., 173, 193, *201*
Laudelot, H., 134, 160, *203*
Law, J., 62, *76*
Lawrence, A. J., 179, 189, 190, 192, 193, 195, *201*
Leadbetter, E. R., 2, 3, 18, 32, 33, 34, *75, 77*, 121, 122, 126, 132, 134, 143, 144, 146, 147, 150, 179, 189, 196, *201*
Leavitt, R., 236, *241*
Lebedeff, A. F., 207, 209, *241*
Lederer, E., 36, 37, *78, 81*, 103, *116*
Lee, C.-P., 251, *279*
Lee, R., 46, *76*
Lee, S. B., 213, *241*
Leech, D. B., *202*
Leene, W., 14, *77*
Lefrancier, P., 36, 38, *77*
Lehmann, J., 57, *73*
Leive, L., 65, 66, *77*
Lenard, J., 71, *77*
Lennarz, W. J., 20, 61, *78*
Lessie, T., 229, *241*
Lester, R. L., 247, 275, *282*
Leutgeb, W., 53, 54, *77, 81*
Levenberg, B., 150, *200*
Levintow, L., 188, *201*
Levy, S. B., 65, *77*
Lewis, V. J., 62, *77*
Leyh-Bouille, M., 35, 36, 37, 38, 39, 41, *73, 74, 75, 77*
Liberman, E. A., 255, 261, *280*
Lien, S., 276, *281*

AUTHOR INDEX

Limbrick, A. R., 71, *74*
Linday, E. M., 212, *241*
Linke, H. A. B., 207, 209, 210, *241*
Lindstrom, E. S., 140, 169, 170, 175, *202, 203*
Linke, H. A. B., 177, *201*
Lipmann, F., 277, 278, *280, 281*
Litwack, G., 93, *116*
Ljungdahl, L. G., 175, 176, 177, *201, 203*
Loh, V., 46, *72*
Lohmann, K., 192, *201*
Loke, J. P., 67, *73*
London, J., 196, *202*, 229, *242*
Lopes, J., 59, 60, *77*
Loriya, Zh. K., 134, *202*
Losada, M., 264, *281*
Löw, H., 260, *281*
Lowry, O. H., *77*
Lüderitz, O., 56, 57, 58, 59, *73, 75, 77*
Lui, T. Y., 44, *77*
Lukoyanova, M. A., 6, 14, *74*
Lundgren, D. G., 12, 13, 19, 33, 53, 59, *74, 76, 78, 80, 81*
Lusty, S. M., 154, *203*

M

McBride, B. C., 142, 163, *198*
McCarty, M., 46, 47, 49, *77*
McCleskey, C. S., 120, 122, 123, 133, 134, 143, 145, 146, 147, *198, 202*
McConnell, M., 12, 17, 23, *79*
McElhaney, R. N., 29, *80*
MacElroy, R. D., 215, 216, *241*
McFadden, B., 206, *240*
McFadden, B. A., 215, 216, 217, 218, 222, *241*
McIntosh, E. N., *201*
MacLeod, R. A., 4, 47, 50, 61, 65, *72, 73, 74, 75*
McNary, J. C., 33, *72*
McQuillen, K., 4, *72, 77*
Maass, D., 53, *77*
Malchow, D., *59, 77*
Magasanik, B., 225, *241*
Mahler, H. R., *199*
Malavolta, E., 159, *201*
Mandell, M., 206, 221, *240*
Mandelstam, J., 15, 53, *77*
Manniello, J. M., 44, *75*
Margolin, P., 236, *240*
Marquis, R. E., 3, *73, 78*
Marr, A. G., 113, *116*
Martin, H. H., 47, 49, 53, 54, 55, 67, *75, 76, 77, 80*
Martinetti, G. V., 23, *80*

Matin, A., 229, *241*
Matsubara, H., 269, *279*
Matsushima, K., 195, *201*
Matthew, D. D., 45, *77*
Matula, T. I., 4, *73*
Mauck, J., 31, *77*, 92, *116*
Meadow, P. M., 58, 60, 61, 63, *74, 75*, 113, *116*
Medawar, P. B., 243, *281*
Meers, J. L., 97, 102, *116, 117*
Meister, A., 188, *200*
Melanori, B. A., 250, *281*
Mellin, D. B., 247, *280*
Melling, J., 65, *72*
Melo, A., 66, *75*
Mendelson, N. H., 17, *73*
Mercer, E. H., 68, *73*
Merdinger, E., 46, *76*
Mergenhagen, S. E., 49, 59, 65, 66, *72, 77*
Metz, H., 235, *242*
Mevius, W., 136, *201*
Meyerhof, K., 192, *201*
Michalka, J., 181, *201*
Michalover, J. L., 149, *202*
Militzer, W. E., *73*
Miller, I. L., 5, *77*
Minton, N. J., 259, 260, 261, *280*
Mirelman, D., 36, 37, 45, *73, 77*
Mirsky, R., 20, 22, *77*
Mitchell, P., 2, *77*, 249, *281*
Miura, T., 26, 27, 28, 48, *77*
Mizushima, S., 26, 27, 28, 48, *76, 77*
Montague, M. D., *77*
Morowitz, H. J., 26, *74*
Morris, J. G., 230, *240*
Morse, S. I., 46, 47, *77*
Moss, C. W., 24, *77*
Moulds, J. D., *77*
Moyle, J., 2, *77*, 249, *281*
Mueller, G. C., 192, *201*
Müldner, H., 28, *78*
Munoz, E., 18, 26, 27, 36, 38, 39, 44, *77, 78*
Murden, D. J., 153, *201*
Murray, E. G. D., 133, *198*
Murray, R. G. E., 3, 10, 11, 15, 19, 32, 33, 34, 49, 51, *72, 74, 78*
Myerholtz, L. E., 4, *78*
Myers, P. A., 134, 152, *201*

N

Nachbar, M. S., 26, *77*
Nakane, K., 57, *81*
Nanninga, N., 2, 7, 18, 19, 51, *78*
Nason, A., 160, *201*
Nasser, D. S., 237, *240*

Nath, K., 59, *80*
Neale, E. K., 7, 10, *78*
Neidhardt, F. C., 83, *116*, 229, *241*
Nermut, M. V., 32, 33, 34, *74*, *78*
Nesbitt, J. A., 61, *78*
Nester, E. W., 237, *240*
Netschey, A., 25, *79*
Nev, H. C., 64, 66, 67, *78*, *81*
Ng, M. H., 27, *78*
Nicholas, D. J. D., 212, *240*
Nielsen, L., 19, *71*
Nikaido, H., 57, *78*, *81*
Nishikawa, K., 261, *280*, *282*
Nishimura, M., 252, *279*
Nisonson, I., 67, *78*
Nixon, R., 19, 20, *78*
Nojima, S., 61, *76*
Norris, J. R., 126, *201*
North, R. J., 47, *72*
Nossal, N. G., 66, *78*
Novick, A., 84, 94, *116*
Nozaki, M., 250, 264, *281*
Nugent, N. A., 250, *280*

O

Oda, T., 17, *74*
Ohmann, E., 277, *281*
Ohrloff, C., 270, *280*
O'Kelley, J. C., 160, *201*
Okoyama, H., 113, *116*
Okuda, S., 64, *81*
Okuyama, H., 63, *78*
O'Leary, W. M., 63, *73*
Olson, J. M., 252, *279*
Ooyama, J., 123, 196, *201*
Op Den Kamp, J. A. F., 4, 11, 24, *78*, 100, *116*
Orla-Jensen, S., 138, *201*
Orlando, J. A., 259, 263, *281*
Ormerod, J. G., 173, *202*, 260, 264, 265, *280*, *281*
Ormerod, K. S., 244, 264, *280*, *281*
Oro, J., 24
Osborn, M. J., 30, 35, 38, 56, 58, *78*, 111, *116*
Ostrovskii, D. N., 6, 14, *74*
Ou, Li-Tse, 3, *78*
Owen, P., 28, *78*
Oxender, D. L., 66, *78*

P

Packer, L., 223, *241*
Painter, H. A., 138, *203*
Palleroni, N. J., 133, *202*

Pardee, A. B., 66, *78*
Pardee, A. D., 211, *241*
Park, J. T., 45, *73*
Parlik, J. G., 106, 108, *116*
Patel, R., 196, *201*
Patterson, D., 19, 20, *78*
Patterson, P. H., 20, *78*
Paul, R., 66, *75*
Pavlik, J. G., 38, *76*
Paynter, M. J. B., 141, 142, *201*
Pearlman-Kothencz, M., 68, *79*
Peck, H. D., Jr., 120, *201*, 206, 210, *241*, 245, 258, *281*
Peel, D., 134, 155, 179, 180, *201*
Pelizza, G., 186, *202*
Pelzer, H., 53, 54, *81*
Perdue, J. F., 30, *75*
Perkins, H. R., 3, 5, 6, 21, 24, 27, 31, 36, 41 43, 47, *73*, *78*, *79*, *81*, 88, 91, 93, *116*
Perry, J. J., 123, 196, *201*
Peterkofsky, A., 172, *201*
Peterson, J. A., 144, *201*
Petit, J. F., 36, 37, 38, 39, *77*, *78*, *81*, 103, *116*
Pfennig, N., 120, *201*
Pfister, R. M., 19, *78*
Pfizner, J., 206, 207, 209, 210, 213, 230, *241*
Phillips, K. C., 121, 124, 126, 131, 132, 136, 145, 146, 147, 148, 195, *203*
Phipps, P. J., 85, 86, 98, *116*
Pichinoty, F., 230, *241*
Pierce, W. A., 46, *78*
Piperno, J. R., 66, *78*
Pirt, S. J., 98, *115*, 196, *199*
Pittard, J., *240*
Pizer, L. I., 181, 185, *201*
Planta, R. J., 19, *80*
Plapp, R., 37, 39, 41, *79*
Pollak, J. K., 67, *73*
Ponce-de-Leon, M., 181, *201*
Pontefract, R. D., 13, *78*
Popkin, T. J., 10, 11, 15, 17, *73*, *78*, *80*
Poston, M. J., 176, *201*
Potochny, M. L., 181, 185, *201*
Powelson, D., 212, 239, *242*
Prasad, A., 93, *116*
Preiss, J., 277, *281*
Price, T. D., 64, *78*
Proctor, H. M., 126, *201*
Purko, M., 201

Q

Qadri, S. M. H., 140, 168, 175, *201*, 270, *281*
Quadling, C., 58, *72*

Quayle, J. R., 134, 143, 144, 146, 149, 153, 155, 156, 158, 159, 170, 177, 179, 180, 181, 183, 184, 185, 189, 190, 191, 193, 195, *198, 199, 200, 201, 202*
Quinn, E. M., 192, *201*

R

Racker, E., 158, *202*
Racker, E. R., 172, *201*
Rader, R. L., 29, *80*
Raines, L. T., 24, *77*
RajBhandary, U. L., 92, *115*
Ramsey, H. H., 221, *240, 241*
Randall, R. J., *77*
Randle, C. L., 61, 63, *78*
Razin, S., 26, 29, *78, 79*
Reaveley, D. A., 4, 10, 11, 15, 18, 19, 20, 21, 28, 52, 53, 61, 63, 67, *73, 78, 79*
Redai, I., 24, *78*
Redwood, W. R., 28, *78*
Reh, M., 235, *241, 242*
Rehn, K., 55, *72*
Reinert, J. C., 29, *78, 80*
Remsen, C. C., 6, 7, 49, 51, 52, *72, 78, 79, 81*
Repaske, R., 4, 64, *79*, 206, 207, *241, 241*
Rheinheimer, G., 136, *199*
Ribbons, D. W., 120, 122, 123, 126, 127, 128, 129, 130, 148, 149, 157, *198, 201, 202*
Rich, R., 53, *81*
Rieber, M., 7, *79*
Rigopoulos, N., 173, *202*, 270, *279*
Rindt, K.-P., 277, *281*
Rittenberg, S., 229, *241*
Rittenberg, S. C., 120, 196, *202*, 206, 210, 211, 212, 213, 217, 229, 239, *241*
Rizza, V., 60, 62, *79*
Roberton, A. M., 166, 167, 168, *202*
Robinow, C. F., 17, *79*
Rodwell, A. W., 26, *79*
Rogers, H. J., 2, 4, 6, 7, 10, 11, 12, 15, 17, 18, 19, 20, 21, 23, 27, 28, 30, 31, 38, 43, 47, 69, *72, 76, 78, 79*, 88, 91, 106, 108, *116*
Rogers, N. L., 154, *203*
Rolls, J. P., 140, *202*
Romeo, D., 57, 59, 68, *79*
Rose, Z. B., 158, *202*
Rosenblum, E. D., 45, *77*
Rosebrough, N. J., *77*
Rothfield, L., 57, 59. 67, 68, *79, 81*
Rotta, J., 76
Rottem, S., 26, *79*
Roxburgh, J. M., 135, 159, 179, *200*
Rowen, R., 68, 69, *80*

Roy, C., 53, *76*
Rozanova, L. I., 131, *200*
Rubin, J., 267, *280*
Rubio-Hurios, M., 4, *72*
Ruch, F., 63, *80*
Rudd, J. H., 113, *115*
Rueckert, R. R., 192, *201*
Ruffo, A., 186, *202*
Ruhland, W., 207, 209, *241*
Rupprecht, E., 270, 271, *280, 282*
Ruska, C. H., 32, *75*
Rutberg, L., 19, 21, 24, 25, *72*
Ryter, A., 2, 5, 6, 7, 10, 11, 12, 13, 15, *74, 75, 76, 79*

S

Sabo, D., 259, 263, *281*
Sagers, R. D., 185, *200*
Saito, Y., *80*
Salton, M. R. J., 3, 4, 6, 18, 19, 20, 21, 24, 25, 26, 27, 31, 32, 47, 52, 60, 61, 62, *72, 73, 77, 78, 79*, 93, *116*
Sammler, I., 220, *241*
Samuelsson, B., 19, 21, 24, 25, *72*
Samuilov, V. D., 255, 261, *280*
Sanderson, A. R., 92, *115*
San Pietro, A., 250, 276, *278, 281*
Sargent, M. G., 16, *74*
Schaechter, M., 13, *80*, 85, *116*
Schindler, J., 222, 229, *241*
Schlegel, H. G., 176, *198*, 206, 207, 208, 209, 210, 214, 215, 218, 219, 220, 221, 222, 223, 224, 225, 226, 229, 231, 232, 235, 236, 239, *240, 241, 242*, 254, 277, *279, 281*
Schleifer, K. H., 31, 37, 38, 39, 40, 41, *75, 76, 79*, 103, *116*
Schmitt, M. D., 20, 26, *79*
Schnaitman, C. A., 48, 64, *79, 80*
Schnellen, C. G. T. P., 141, 142, *200, 202*
Schocher, A. J., 53, *80*
Scholes, P., 249, *281*
Schön, G., 160, *202*
Schor, M. T., 26, 27, *77, 78*
Schrauzer, G. N., 164, *202*
Schroeder, W., 195, *202*
Schuster, E., 214, 215, 235, *242*
Schuster, P., *201*
Sedar, A., 14, *80*
Shands, T. W., 59, 67, *80*
Shapiro, B., 188, *203*
Shaposhnikov, V. N., 134, *202*
Sharon, N., 36, 37, *77*
Shaw, D. H., 59, *71*
Shaw, M. K., 63, *80*, 113, *116*

Shaw, N., *80*
Shaw, W. V., 135, 153, 154, 187, 188, *202*
Sheetz, M., 29, 30, *75*
Shen, L. C., 274, 277, *281*
Shen, P. Y., 24, 25, *80*
Shewan, J. M., 150, *202*
Shieh, H. S., 155, *202*
Shiraki, M., 271, *280*
Shively, J. M., 14, 61, *80, 81*
Shockman, G. D., 7, 12, 17, 20, 32, *75, 80*
Shon, M., 96, *116*
Shorey, C. D., 67, *73*
Shovlin, V. K., 65, 66, *77*
Sieglin, U., 55, 56, *72*
Silbert, D. F., 63, *80*
Silva, M. T., 10, 17, *80*
Silver, W. S., 159, *202*
Simmons, D. A. R., *76*
Simmons, G. P., 58, 65, *73*
Simpson, F. J., 195, *199, 201*
Simpkins, H., 29, 30, *75*
Singer, H. J., 45, *73*
Singer, S. J., 29, 30, 71, *75, 77*
Singh, P. P., 58, 59, *71*
Sirevåg, R., 173, *202*, 244, *281*
Sisler, E. C., 174, *199*
Sistrom, W. R., 247, 261, *279, 282*
Siu, P. M. L., 181, *203*
Skulachev, V. P., 255, 261, *280*
Slepecky, R. A., 12, 13, *74*
Smillie, R. M., 173, 174, *199, 202*
Smith, A. J., 196, *202*, 229, *242*
Smith, C. E., 46, *72*
Smith, D. S., 126, 129, 130, 149, *202*
Smith, G. L., 17, 26, 53, *73, 75*
Smith, N. R., 133, *198*
Smith, P. H., 141, 142, *202*
Smith, U., 126, 127, 128, 129, 130, 149, *202*
Söhngen, N. L., 120, 121, *202*
Sojka, G. A., 247, 248, 275, 276, *282*
Sorokin, Y. I., 134, 170, *202*
Speck, J. F., 188, *202*
Spencer, C. P., 150, *198*
Spicer, S. S., 67, *81*
Stacey, B. E., 195, *200*
Stadtman, E. R., 135, 153, 154, 176, 187, 188, *201, 202, 203*
Stadtman, T. C., 141, 162, 163, 164, 165, 167, *198, 202*
Stafford, G. H., 42, *72*
Stanier, R. Y., 133, 196, *202*, 206, 219, 221, 229, *240, 242*, 247, 249, *279*
Stanley, J. P., 120, 122, *199*
Steed, P., 49, *78*
Steensland, H., 68, 69, *80*
Steim, J. M., *78, 80*
Stein, O., 26, 29, *79*
Steiner, S., 247, 275, *282*

Stenesh, J., 24, *80*
Stevens, R. W., 3, *73*
Stocks, P. K., 120, 123, 133, 134, 143, *202*
Stoeckenius, W., 17, 68, 69, *74, 80*
Stokes, E., 32, 38, *76*
Stokes, J. E., 170, 175, *202*
Stokstad, E. L. R., 120, *202*
Stoof, T. J., 19, *80*
Stout, H. A., 212, 239, *242*
Stouthamer, A. H., 146, 167, *202*
Strange, R. E., 4, *72*, 96, *116*
Strawinski, R. J., 122, 123, 143, 145, 146, 147, *198*
Strittmatter, P., 158, *202*
Strominger, J. L., 35, 36, 37, 38, 42, 43, *74, 75, 76, 77, 80, 81*, 92, *117*
Stubbs, J. M., 2, *76*
Stukus, P. E., 211, 216, 217, 219, 239, *240*
Sud, I. J., *85, 116*
Suganuma, A., 17, *80*
Sugimori, T., 195, *202*
Suhadolinik, R. J., 195, *202*
Sun, A. Y., 176, *203*
Sundaralingam, M., 30, *80*
Sybesma, C., 269, *279*
Sykes, J., 84, *117*
Syrett, P. J., 212, *241*
Szilard, L., 84, 94, *116*

T

Tagawa, K., 250, 253, *281, 282*
Takebe, I., 54, *80*
Tamiya, N., 161, *203*, 271, *282*
Tamura, G., 25, *81*
Tannenbaum, M., 67, *78*
Tanner, P. J., 32, 38, *76*, 106, 108, *116*
Taylor, B. F., *201*
Taylor, R. T., 120, 196, *203*
Tempest, D. W., 31, 45, *74*, 84, 85, 86, 90, 92, 94, 96, 97, 98, 102, 110, 112, 114, *115, 116, 117*
Terry, T. M., 26, *74*
Testa, E., 186, *202*
Thaemert, J., 19, *71*
Thatcher, F. S., 13, *78*
Thayer, R. K., 270, 271, *280, 282*
Theodore, T. S., 10, 11, 15, *78, 80*
Thiele, O. W., 62, *80*
Thompson, J., 4, *73*
Thompson, J. E., 71, *74*
Thompson, T. E., 28, *78*
Thore, A., *279*
Thornley, M. J., 3, 32, 34, 47, 49, 51, *75, 80*
Thurman, P. F., 43, 47, 60, *75, 76*, 91, *116*
Tinelli, R., 36, 37, *77*

Tinker, K., 255, *282*
Tipper, D. J., 35, 36, 37, 38, 39, 40, 42, 43, 53, *74, 75, 76, 77, 80, 81*, 92, *117*
Tischer, R. G., 219, *240*
Toennies, G., 20, *80*
Tomasz, A., 45, *80*
Tomcsik, J., 4, *72*
Tori, M., 42, *80*
Tornabene, T. G., 25, 58, *71, 72, 80*
Tourtellotte, M. E., 29, *80*
Trefts, P. E., 20, *79*
Trembley, G. Y., 13, *80*
Trijbels, F., 233, *242*
Trotsenko, Yu. A., 139, 170, *200*
Trudinger, P. A., 206, 210, 211, *242*
Trüper, H. G., 6, 51, *79*, 222, 223, 224, 225, *242*
Tsai, L., 135, 153, 154, 187, 188, *202*
Tsofina, L. M., 255, 261, *280*
Tu Ch, L., 215, 216, 217, 218, *241*
Tucker, A. N., 60, 62, 64, 66, *79, 80*
Tuttle, A. L., 249, *282*

U

Uffen, R. L., 243, *282*
Umbarger, H. E., 181, *203*, 236, *240, 241*
Umbarger, M. A., 181, *203*
Urushibara, T., 150, 157, *203*
Utech, N. M., 10, 12, 13, 14, 15, 17, *76*

V

Vagelos, P. R., 63, *80*
Valentine, R. C., 234, *242*, 255, *282*
Valois, F. W., 6, *78*
Van Deenen, L. L. M., 4, 24, *75, 78*, 100, *116*
van Delden, A., 138, *198*
Vanderkooi, G., 30, *80*
Van Dijk-Salkinoja, M. S., 19, *80*
Van Genderen, H., 174, *199*
van Gool, A., 134, 160, *203*
van Iterson, W., 4, 11, 13, 14, *78, 80*
van Niel, C. B., 139, *203*, 251, *282*
Vary, P. S., 146, *203*
Veech, R. L., 263, *281*
Veerkamp, J. H., 25, 38, *80*
Verma, J. P., 53, *80*
Vernon, L. P., 250, 261, 263, *280, 282*
Vishniac, W., 139, *203*, 206, 210, 211, 223, *240, 241, 242*
Vogels, G. D., 233, 234, *242*
Volcani, B. E., 140, *200*
Volesky, B., 131, *203*
Volk, W. A., 58, *80*
von Stedingk, L.-V., 249, 250, *279, 282*
Vorbeck, M. L., 23, *80*

Vorob'eva, L. I., 131, *200*
Voss, J. G., 64, *80*

W

Wadzinski, A. M., 120, 148, 157, *202*
Wagner, C., 135, 151, 152, 154, 157, 158, 188, *200, 203*
Walker, D. A., 120, *203*
Walker, I. O., 170, *198*
Walton, G. M., 274, *278, 281*
Wang, W. S., 49, 53, 59, *80, 81*
Ward, J. B., 5, 21, 24, *73, 81*
Wardlaw, A. C., 63, 64, *81*
Ware, G. C., 138, *203*
Warth, A. D., 38, *81*
Wassink, E. C., 252, *282*
Waterbury, J. B., 6, *79*
Watson, R. W., 53, *76, 80*
Watson, S. W., 6, 49, 51, 52, *78, 79, 81*
Wawszkiewicz, E. J., 192, *203*
Weaver, P., 255, *282*
Weaver, R. E., 62, *77*
Webster, R. E., 236, *242*
Weibull, C., 3, 4, 10, 13, 18, *72, 81*
Weidel, W., 53, 54, *77, 81*
Weigand, R. A., 14, *81*
Weinbaum, G., 51, 53, 64, *74, 81*
Weinstein, M., 19, 20, *78*
Weiser, M., 67, 68, *81*
Weissbach, H., 120, 164, *198, 203*
Weitzerbin-Falszpan, J., 36, 37, *81*
Welker, N. E., 18, *72*
Wellman, A., 131, *203*
Wertlieb, D., 139, *203*, 276, *281*
Westphal, O., 57, 58, 59, *73, 75, 77*
Wetzel, B. K., 67, *73, 81*
Whatley, F. R., 174, *199*
Wheat, R., 58, 59, *77*
White, D., 53, *81*
White, D. C., 24, 60, 62, 64, 66, *74, 79, 81*
Whiteside, T. L., 53, *81*
Whitney, J. G., 37, *81*
Whittenbury, R., 6, *73*, 121, 124, 126, 131, 132, 136, 143, 145, 146, 147, 148, 161, 195, *199, 203*
Wicken, A. J., 38, 42, 44, *75, 81*
Wieringa, K. T., 140, 175, *203*
Wietzerbin-Falszpan, J., *78*
Wilde, E., 219, 220, 223, *241, 242*
Wilkinson, J. F., 121, 124, 126, 131, 132, 136, 144, 145, 146, 147, 148, 149, 195, 196, *198, 203*
Wilkinson, S. G., 58, 61, 65, *75, 76, 81*
Williamson, D. H., 192, *199*
Wilson, A. T., 169, *198*
Wilson, E., 212, 239, *242*
Wilson, P. W., 213, *241*

Winshell, E. B., 66, *81*
Wittenberger, C. L., 206, 207, *242*
Wittner, M. K., 46, *74*
Wixom, R. L., 230, 231, *240*
Wolfe, R. S., 142, 143, 162, 163, 164, 165, 166, 167, 168, *198, 201, 202, 203*, 234, *242*, 243, *282*
Wolff, H., 56, *72*
Wolin, E. A., 142, 162, 163, 164, 168, *198, 203*
Wolin, M. J., 142, 162, 163, 164, *198, 203*
Wolin, M. J., 25, 26, 28, *73, 74, 81*
Wood, H. G., 175, 176, 177, *201, 203*
Wood, J. M., 163, 164, 165, 166, 168, *203*
Wood, W. A., *201*
Woolfolk, C. A., 188, *203*
Work, E., 34, 49, 64, *76, 81*
Wutzerbin-Falszpan, J., 103, *116*

Y

Yagi, T., 161, *203*, 271, *280, 282*
Yaguchi, M., 58, 59, *71, 72*

Yamaguchi, T., 25, *81*
Yamamoto, N., 250, *282*
Yamanaka, T., 254, *282*
Yamashita, J., 261, *280*
Yike, N. J., 259, 260, 261, 263, *281*
Yoch, D. C., 169, 170, 175, *203*
Young, F. E., 31, 43, 45, *81*, 92, 59, 103, 108, *117*
Yu, L., 25, 26, 28, *73, 74, 81*
Yuasha, R., 57, *81*
Yudkin, M. D., 19, 23, 24, *81*

Z

Zajic, J. E., 131, *203*
Zalkin, H., 62, *76*
Zatman, L. J., 134, 135, 151, 152, 155, 156, 157, 158, *198, 199, 201*
Zeikus, J. C., 32, *72*
Zhilina, T. A., 239, *242*
Zilinsky, J. W., 275, 276, *282*
Zsigray, R. M., 5, *77*

SUBJECT INDEX

A

Abequose in bacterial lipopolysaccharides, 57

Acetate kinase activity of hydrogenomonads, 222

Acetate-metabolizing enzymes, effect of hydrogen on synthesis of, 225

Acetate, repression of synthesis of ribulose diphosphate carboxylase synthesis by, in knallgasbacteria, 216

Acetohydroxy acid synthetase activity of hydrogenomonads, 236

Acetyl-CoA cycle, 171

Acetyl-CoA synthetase in bacteria, 172

Acid phosphatase, release of, from Gram-negative bacteria by osmotic shock, 66

Aconitate hydratase in bacteria, 172

Active transport, involvement of adenosine triphosphatase in, 28

Acyloin condensation in methane-utilizing bacteria, 191

Adenine utilization by hydrogenomonads, 233

Adenosine diphosphate, as a control factor in oxidation of hydrogen by bacteria, 208

Adenosine triphosphatase in bacterial membranes, 27

Adenosine triphosphate, capacity of, to drive nicotinamide nucleotide reduction, 263

intracellular level of, in hydrogen-oxidizing bacteria, 208

involvement in methane formation, 165

regeneration of, in bacterial photosynthesis, 245

requirement for, in carbon dioxide fixation in knallgasbacteria, 215

Adenylate control of glucose 6-phosphate dehydrogenase, 229

Aerobacter aerogenes, allose metabolism in, 195

effect of environment on composition of wall of, 112

effect of environment on wall content of, 85

Aerobic micro-organisms that utilize one-carbon compounds, 120

Aerobic oxidation of one-carbon compounds, 143

Aerobiosis, effect of, on phospholipid composition of bacteria, 24

Alanine, stimulation of ribulose diphosphate carboxylase synthesis by, in knallgasbacteria, 216

Alanylation of bacterial teichoic acids, 98

Alcaligenes spp., taxonomy of, 206

Algae, ability of, to oxidize formate, 161

photoproduction of molecular hydrogen by, 267

Alkylamines, metabolism of, 150

Allantoic acid production by hydrogenomonads, 233

Allantoin utilization by hydrogenomonads, 233

Allose metabolism in *Aerobacter aerogenes*, 195

Allose phosphate as a product of formaldehyde fixation in bacteria, 191

Allose 6-phosphate ketol isomerase in *Aerobacter aerogenes*, 195

Allosteric inhibition of adenosine triphosphate in hydrogenomonads, 227

Allulose, in 6-amino-9-D-psicofuranosyl-purine, 195

metabolism of, in bacteria, 195

phosphate as a possible product of formaldehyde fixation in bacteria, 191

Altrose phosphate as a possible product of formaldehyde fixation in bacteria, 191

Amines, methylated, production of, in the biosphere, 119

Amino-acid composition of wall peptidoglycans, effect of growth conditions on, 31

Amino acids, biosynthesis of, by hydrogenomonads, 234

growth of knallgasbacteria on, 230

in bacterial cell-wall peptidoglycans, 36

para-Aminobenzoic acid, effect of, on growth of methane bacteria, 181

6-Amino-9-D-psicofuranosylpurine, allulose in, 195

295

SUBJECT INDEX

Ammonia, growth of knallgasbacteria on, 230
limitation, effect of, on peptidoglycan of *Bacillus subtilis*, 107
effect of, on teichoic acid synthesis by bacilli, 89
Anaerobes, non-photosynthetic, net fixation of carbon dioxide by, 175
Anaerobic dismutation of one-carbon compounds, 161
Anaerobic growth of *Chlorobium thiosulphatophilum*, 171
Anaerobic growth of photosynthetic bacteria, 243
Anaerobic methanogenic bacteria, 140
Anaplerotic pathways in hydrogen bacteria, 230
Anionic polymers in bacterial walls, 90
Anteiso acids in bacteria, 24
Antigenicity of Gram-negative bacteria, effect of environment on, 113
Antigens of bacterial walls, 46
Antimycin as an uncoupler of bacterial photophosphorylation, 254
Appearance of intracytoplasmic membranes in bacteria, 5
Aromatic amino-acid pathway, regulation of, in hydrogenomonads, 237
Arthrobacter crystallopoietes, peptidoglycans in walls of, 35
Arthrobacter globiformis, metabolism of glycine by, 185
Asparagine, growth of knallgasbacteria on, 224
Aspartic acid cross bridges in bacterial peptidoglycans, 41
Assembly of lipopolysaccharide units in Gram-negative bacteria, 60
Assimilation of carbon dioxide by bacteria, 169
Assimilation of cations by bacteria, 97
Assimilation of reduced one-carbon compounds, 177
ATPase, *see* Adenosine triphosphatase
Autolysis of bacterial walls, 44
Autotrophic growth, effect of, on enzyme synthesis in hydrogenomonads, 219
Azotobacter tumefaciens, phospholipid composition of, 62

B

Bacilli, composition of mesosomes from, 11
structure of surfaces of, 34
Bacillus licheniformis, cytoplasmic membranes in, 17
effect of nutrient limitations on phosphorus contents of walls of, 89

Bacillus licheniformis—continued
electrophoresis of proteins from cytoplasmic membranes from, 23
mesosomes in, 6, 9
teichoic acids in walls of, 42
teichuronic acids in walls of, 101
Bacillus megaterium, anionic polymers in walls of, 92
effect of nutrient limitations on phosphorus contents of walls of, 89
protoplasts of, 4
Bacillus methanicus, isolation of, 120
Bacillus polymyxa, structure of walls of, 33
Bacillus sphaericus, oxidation of N-methylurea by, 154
peptidoglycan composition in, 37
peptidoglycan in wall of, 40
utilization of one-carbon compounds by, 134
Bacillus stearothermophilus, teichoic acids in walls of, 42
Bacillus subtilis, anionic polymers in walls of, 91
comparative effects of lysozyme on, 105
effect of nutrient limitations on phosphorus contents of walls of, 89
L forms, 5
var. *niger*, anionic polymers in walls of, 90
changes in teichoic acid content of wall of, with growth rate, 95
effect of environment on wall content of, 85
effect of pH value on wall composition of, 99
Bacitracin, effect of, on bacterial mesosomes, 7
Backbone structure of bacterial cell-wall peptidoglycan, 103
Bacterial cytoplasmic membrane, structure of, 28
Bacterial cytoplasmic membranes, association of, with ribosomes, 19
enzymes of, 26
Bacterial membranes, 1
Bacterial photophosphorylation, 248
Bacterial photosynthesis, energy conversion in, 243
Bacterial protoplasts, 3
Bacterial sphaeroplasts, 3
Bacterial wall content, 84
Bacterial walls, 1
composition of, 83
mucopeptide in, 102
Bacteriochlorophylls, 249
types of, 249
Bacterium 4B6, utilization of trimethylamine by, 151

Bacterium formicum, utilization of one-carbon compounds by, 134

Bacterium formoxidans, growth of, on formate, 170

utilization of one-carbon compounds by, 134

Batch cultures, limitations of, in studies on effect of environment on bacterial walls, 84

Benzoquinone as an electron mediator in fumarate reduction by photosynthetic bacteria, 256

Benzyl viologen as an electron mediator in fumarate reduction by photosynthetic bacteria, 256

Betaine, metabolism of, 150

Binding of cations by bacterial walls, 45

Binding of magnesium ions to bacterial walls, 96

Biosynthesis of amino acids by hydrogenomonads, 234

Bizarre shapes of bacterial protoplasts, 4

Branched chain fatty acids in bacteria, 25

Buffering of culture media, 98

Butyribacterium rettgeri, peptidoglycan in walls of, 37

C

Calvin cycle, possible operation of, in knallgasbacteria, 219

Camphor methylene hydroxylase, 153

Capsule formation among methane-utilizing bacteria, 132

Carbon assimilation, by methane bacteria, 169

by the serine pathway, 177

pathways in methane-utilizing bacteria, 195

Carbon balances for methane oxidation by bacteria, 147

Carbon dioxide, assimilation of, by bacteria, 169

fixation by knallgasbacteria, 214

net fixation of, by non-photosynthetic anaerobes, 175

utilization of, by bacteria, 142

Carbon monoxide, oxidation of, by bacteria, 160

utilization by anaerobic bacteria, 142

utilization of, 138

Carbonyl cyanide-*m*-chlorophenylhydrazone, effect of, on methane production, 168

effect of, on oxidation of hydrogen by bacteria, 208

Carboxydomonas oligocarbophila, utilization of carbon monoxide by, 138

Carboxypeptidases in bacterial walls, 40

Catalytic amounts of adenosine triphosphate, need for, in methanogenesis, 167

Cationic detergents, action of, on bacterial membranes, 25

Cations, assimilation of, by bacteria, 97

binding of, by bacterial walls, 44

Cell cycle in bacteria, changes in teichoic acid content of walls during, 95

Cell envelope of halophilic bacteria, 68

Cell envelopes, of bacteria, 3

of Gram-negative bacteria, 47

Cell-free extracts in studies on methane oxidation, 149

Cell-wall polymers in bacteria, 46

Cell walls of Gram-positive bacteria, 30

Chain length, hydrocarbon, role of, in microbial utilization, 123

Chain lengths of bacterial teichoic acids, 43

Changes in teichoic and teichuronic acid contents of walls of *Bacillus subtilis* var. *niger*, 94

Chemical composition of bacterial mesosomes, 11

Chemolithotrophic bacteria, facultative, 205

Chemolithotrophic metabolism, 206

Chemo-organotrophic metabolism in knallgasbacteria, 219

Chemostat cultures, use of, in studies of environment on bacterial walls, 84

Chilling, effect of, on bacterial mesosomes, 7

Chlamydomonas eugametos, photoproduction of molecular hydrogen by, 267

Chlamydomonas moewusii, photoproduction of molecular hydrogen by, 267

Chlorobium thiosulphatophilum, anaerobic growth of, 171

enzymes of acetyl-CoA cycle in, 172

enzymes of ribulose diphosphate cycle in, 173

hydrogenase activity of, 255

Chloropseudomonas spp., formate utilization by, 139

metabolism of formate by, 170

Choline, metabolism of, 150

Chromatium spp., photophosphorylation in, 250

Chromatophores of bacteria, 249

Chromatophores of *Rhodopseudomonas capsulata*, rates of redox reactions catalysed by, 262

Chromobacterium violaceum, phospholipid composition of, 62

wall structure in, 53

Citrate, growth of knallgasbacteria on, 224

lyase in bacteria, 172

Classification of methane-utilizing bacteria, 126

Clostridia, comparison of energy metabolism in photosynthetic bacteria and, 269

Clostridial ferredoxins, 254

Clostridium aceticum, growth of, on hydrogen and carbon dioxide, 175

Clostridium formico-aceticum, reductive carboxylation of acetyl-CoA by, 177

Clostridium kluyveri, molecular hydrogen production by, 271

Clostridium thermo-aceticum, homoacetate fermentation in, 176

Clostridium welchii, cross bridges in wall peptidoglycan of, 39

Cold shock in Gram-negative bacteria, 65

Colony colour of methane-utilizing bacteria, 132

Comparative effects of lysozyme on *Bacillus subtilis*, 105

Comparison of energy metabolism in photosynthetic bacteria and clostridia, 269

Composition of bacterial cytoplasmic membranes, 20

Composition of peptidoglycans in bacterial walls, 35

Concentration of cations in bacterial walls, 96

Connections between cell wall and membrane in bacteria, 3

Continuous culture, use of, in studies of effect of environment on bacterial walls, 84

Control of teichoic acid synthesis in bacteria, 100

Conversion of energy in bacterial photosynthesis, 243

Co-ordinate synthesis of hydrogenases in hydrogenomonads, 219

Core region of lipopolysaccharides of Gram-negative bacteria, 56

Corrinoids, in carbon dioxide fixation by clostridia, 176

involvement in methane production, 164

Corynebacterium, octane hydroxylase of, 152

Corynebacterium ovis, wall structure of, 34

Cross bridges, in peptidoglycans in walls of Gram-negative bacteria, 54

in bacterial wall peptidoglycans, 38

Cross bridging in bacterial lipopolysaccharides, 59

Cyanide utilization by micro-organisms, 138

Cyclic flow of electrons in bacterial photosynthesis, 249

Cyclopropane acids in Gram-negative bacteria, 63

Cycloserine, effect of on bacteria, mesosomes, 7

Cysts in methane-utilizing bacteria, 131

Cytidine diphosphate glycerol in synthesis of teichoic acids, 101

Cytidine diphosphate ribitol in synthesis of teichoic acids, 101

Cytochrome function in bacterial photosynthesis, 250

Cytochromes, in bacterial photosynthesis, 257

involvement of, in oxidation of one-carbon compounds, 161

solubilization of, from bacterial membranes, 25

Cytoplasmic membranes, bacterial, enzymes of, 26

bacterial, isolation of, 18

in bacteria, 2

of Gram-positive bacteria, 16

Cytosine deaminase of hydrogenomonads, 234

Cytosine, utilization by hydrogenomonads, 234

D

Dark electron flow in bacteria, 252

O-Demethylase, possible role of in methane oxidation by bacteria, 149

Demethylation by bacteria, 150

Density-gradient centrifugation of bacterial cytoplasmic membranes, 19

Description of micro-organisms that utilize one-carbon compounds, 120

Desulfovibrio desulfuricans, ability of, to oxidize carbon monoxide, 161

production of molecular hydrogen from dithionite by, 271

Detergents, action of, on bacterial cytoplasmic membranes, 20

action of, on bacterial membranes, 25

Dimethylamine, metabolism of, 150

mono-oxygenase, 151

Dimethyl ether, as a possible intermediate in methane oxidation by bacteria, 149

as a substrate for methane-utilizing bacteria, 121

Dimethylglycine, metabolism of, 150

2,4-Dinitrophenol, effect of, on methane production, 168

Diplococcus spp., serine pathway in, 179

Disaggregation of bacterial membranes, 25

Dismutation, anaerobic, of one-carbon compounds, 161

SUBJECT INDEX

Dithionite, production of molecular hydrogen from, by *Desulfovibrio desulfuricans*, 271

Diversity of assimilation pathways in methane-utilizing bacteria, 195

Drifts in pH value in batch culture, 98

E

Early products of carbon dioxide fixation in bacteria, 175

Ectothiorhodospira mobilis, membranes in, 6

wall structure in, 52

Effect of hydrogen on enzyme formation in hydrogenomonads, 223

Effect of hydrogen on enzyme function, 226

Effect of hydrogen on hexose degradation in hydrogenomonads, 222

Effect of pH value on bacterial wall composition, 98

Effects of environment on bacterial wall composition, 83

Efficiency of conversion of methane carbon to cell carbon by bacteria, 146

Electron acceptor, molecular hydrogen as an, in bacteria, 253

Electron transport, relationship to synthesis of ribulose diphosphate carboxylase in hydrogenomonads, 217

Electrophoresis of proteins from bacterial cytoplasmic membranes, 20

Embden-Meyerhof pathway in knallgasbacteria, 228

Energy charge, concept of, in control of bacterial photosynthesis, 274

Energy charge in hydrogenomonads, 222

Energy conversion in bacterial photosynthesis, 243

Energy generation by anaerobic dismutation, 167

Energy metabolism of micro-organisms growing on one-carbon compounds, 143

Enterobacter aerogenes, phospholipid composition of, 62

Entner-Doudoroff pathway, in chemolithotrophic bacteria, 229

operation of, in hydrogenomonads, 219

Environment, effect of, on composition of bacterial cell-wall peptidoglycans, 103

effects of, on bacterial wall composition, 83

ionic, effect of, on bacterial wall composition, 95

Enzyme formation, effect of hydrogen on, in hydrogenomonads, 223

Enzyme function, effect of hydrogen on, 226

Enzymes of acetyl-CoA cycle, 172

Enzymes of bacterial cytoplasmic membranes, 26

Enzymes of serine pathway in bacteria, 180

Eras of photosynthesis research, 245

Eructation by cattle, methane production during, 119

Erwinia spp., wall structure in, 53

Erythrulose phosphate, synthesis of, 192

Escherichia coli, intracytoplasmic membranes in, 13

lipoprotein of rigid layer of, 55

mini cells of, 67

peptidoglycan in, 54

serine biosynthesis in, 181

wall structure in, 48

Ethanolamine in Gram-negative lipopolysaccharides, 56

Ethylenediamine tetra-acetic acid, effect of, on Gram-negative bacteria, 64

use of, to prepare bacterial protoplasts, 4

Exospores in methane-utilizing bacteria, 131

Extrusion of mesosomal vesicles in bacteria, 10

F

Facultative chemolithotrophic bacteria, 205

Fatty acids in bacterial lipids, 24

Ferredoxin, involvement of, in acetyl-CoA cycle, 171

Ferredoxin-linked formate dehydrogenase in bacteria, 164

Ferredoxin, photoreduction of, 270

Ferredoxins in photosynthetic bacteria, 253

Ferric ions, effect of, on synthesis of ribulose diphosphate carboxylase, 217

Fibres in bacterial walls, 34

Fish, occurrence of methylated amines on, 150

Fixation of carbon dioxide by knallgasbacteria, 214

Flagellation in methane-utilizing bacteria, 132

Flow of hydrogen in photosynthetic bacteria, 252

Flow of ions across bacterial walls, 96

Folate antagonists, effect of, on pseudomonads, 181

Folate derivatives in metabolism of one-carbon compounds, 163

Folates, involvement of, in carbon dioxide fixation, 177
Formaldehyde dehydrogenase, 158
Formaldehyde fixation, ribose phosphate cycle and, 189
Formaldehyde, formation of, from methane, 190
oxidation of, 156, 158
use of, as a carbon source, 151
Formate dehydrogenase, 159
of *Rhodopseudomonas palustris*, 168
Formate, growth of *Bacterium formoxidans* on, 170
growth of *Pseudomonas oxalaticus* on, 170
metabolism of, by *Rhodopseudomonas palustris*, 168
oxidation of, by bacteria, 159
utilization, by anaerobic bacteria, 141
by photosynthetic bacteria, 139
Formate-utilizing bacteria, 131
Formation and function of tricarboxylic acid cycle enzymes in knallgasbacteria, 229
Formic acid, production of, in the biosphere, 119
Fractionation of methyltransferase activity, 165
Freeze-etching of bacterial walls, 51
Freezing of fatty acids in bacterial membranes, 29
Fructose, effect of, on synthesis of hydrogenase, 212
Fructose-limited growth of hydrogenomonads, 220
Fructose utilization by hydrogenomonads, enzymes of, 220
Fumarate hydratase in bacteria, 172
Fumarate, reduction of, by photosynthetic bacteria, 256
Function of teichoic acids in bacteria, 44
Functioning of enzymes, effect of hydrogen on, 226

G

Generation of reducing power in bacterial photosynthesis, 243
Generation times of methane-utilizing bacteria, 132
Gluconate utilization, by hydrogenomonads, effect of hydrogen on, 227
by knallgasbacteria, enzymes involved in, 219
Gluconokinase activity of knallgasbacteria, 221

Glucose limitation, effect of, on synthesis of teichoic acids by bacilli, 89
effects of, on peptidoglycan of *Bacillus subtilis*, 107
Glucose 6-phosphate dehydrogenase activity in hydrogenomonads, 219
Glucosaminylphosphatidylglycerol in bacterial membranes, 4
Glutamate, as a nitrogen source for photosynthetic bacteria, 253
dehydrogenases of hydrogenomonads, 230
effect of, on synthesis of hydrogenase, 212
stimulation of ribulose diphosphate carboxylase synthesis by, in knallgasbacteria, 216
Glutamic acid cross bridges in bacterial wall peptidoglycans, 41
Glutamine synthetase, use of methylamine by, 188
γ-Glutamylmethylamide, synthesis of, by pseudomonads, 188
Glutathione as a cofactor of formaldehyde dehydrogenase, 158
Glycan chains in bacterial cell-wall peptidoglycans, 35
Glycerate dehydrogenase activity in hydrogenomonads, 233
Glycerol limitation, effect of, on composition of walls of *Aerobacter aerogenes*, 112
Glycerol teichoic acid, structure of, 87
Glycerol teichoic acids in bacterial walls, 42
Glycine bridges in bacterial cell-wall peptidoglycans, 38
Glycine skeleton, net synthesis of, 184
Glycolipid composition of bacterial cytoplasmic membranes, 21
Glycosaminopeptide, in bacterial walls, 102
Glyoxylate, as a precursor of glycine in pseudomonads, 184
carboligase activity in *Pseudomonas acidovorans*, 233
cycle in hydrogenomonads, 222
G-polysaccharide from walls of *Streptococcus pyogenes*, 44
Gram-negative bacteria, cell envelopes of, 47
effect of ethylaminediaminetetraacetic acid on, 64
lipoprotein of rigid layer of, 55
osmotic shock in, 66
peptidoglycans of, 52
protoplasts of, 4
Gram-positive bacteria, cytoplasmic membranes of, 16

SUBJECT INDEX

301

Gram-positive bacteria—*continued*
  peptidoglycan contents of wall of, 3
  septum formation in, 12
  walls of, 30
Growth conditions, effect of, on hydrogen-
    ase synthesis in bacteria, 213
  effects of on bacterial cell-wall peptido-
    glycans, 106
Growth of *Bacterium formoxidans* on
    formate, 170
Growth phase, effect of, on bacterial wall
    composition, 31
Growth rate, effect of, on bacterial wall
    composition, 95
  effect of, on composition of walls of
    Aerobacter aerogenes, 112
    on hydrogenase synthesis by *Hydro-
      genomonas* H 16, 214
Growth temperature, effect of, on fatty-
    acid composition of Gram-negative
    bacteria, 63
  effect of, on lipid composition of bac-
    teria, 113
Growth yields, of bacteria on methane, 146
  of methane bacteria, 167
Guanine utilization by hydrogenomonads,
    233

H

Haemophilus influenzae, serine biosyn-
    thesis in, 181
Halobacterium cutirubrum, cell envelope of,
    68
Halobacterium halobium, cell envelope of, 68
Halobacterium salinarium, cell envelope
    of, 68
Halophilic bacteria, cell envelope of, 68
Hamamelose phosphate as a possible
    product of formaldehyde fixation in
    bacteria, 191
Heptose in bacterial lipopolysaccharides,
    57
Heterotrophic growth, effect of, on enzyme
    synthesis in hydrogenomonads, 219
  effect of, on hydrogenase synthesis, 211
Hexosamine residues in bacterial cell walls,
    35
Hexose degradation in hydrogenomonads,
    effect of hydrogen in, 222
Hexose phosphate synthetase in bacteria,
    193
Hexose utilization by knallgasbacteria,
    enzymes involved in, 219
Histidine, growth of knallgasbacteria on,
    224
Homoacetate fermentation in *Clostridium
    thermo-aceticum*, 176

Hydrogen, ability of, to induce synthesis
    of hydrogenase, 211
  effect of, on enzyme function, 226
    on hexose degradation in hydrogeno-
      monads, 222
  utilization of, by bacteria, 142
Hydrogen dehydrogenase synthesis by
    bacteria, 210
Hydrogen effect, operation of, in knallgas-
    bacteria, 223
  similarity of, to the glucose effect, 225
Hydrogen flow in photosynthetic bacteria,
    252
Hydrogen ion concentration, effect of, on
    bacterial wall composition, 98
  effect of, on synthesis of bacterial
    peptidoglycans, 108
Hydrogen ion uptake, light stimulation of,
    in bacteria, 250
Hydrogen limitation, effect of, on hydro-
    genase synthesis by bacteria, 214
Hydrogen oxidation by hydrogen bacteria,
    207
Hydrogen-oxidizing bacteria, 205
Hydrogenase activity of *Chlorobium thio-
    sulfatophilum*, 255
Hydrogenase, relationship to nitrate re-
    ductase, 230
  synthesis in bacteria, 210
Hydrogenases, membrane-bound, of bac-
    teria, 207
Hydrogenomonads, effect of growth rate
    on synthesis of hydrogenase by, 214
  operation of hydrogen effect in, 223
Hydrogenomonas carboxydovorans, utiliza-
    tion of carbon monoxide by, 138
Hydrogenomonas eutropha, hydrogenase
    synthesis by, 210
  taxonomy of, 206
Hydrogenomonas facilis, taxonomy of, 206
Hydrogenomonas ruhlandii, taxonomy of,
    206
Hydrogenomonas spp., taxonomy of, 206
Hydrophobic interactions in bacterial
    membranes, 30
β-Hydroxybutyrate, utilization of, by
    hydrogenomonads, 223
Hydroxymethylation of glycine in bac-
    teria, 177
β-Hydroxymyristic acid in lipid A of
    Gram-negative bacteria, 59
Hydroxypyruvate reductase, in methane-
    utilizing bacteria, 195
  mutants of pseudomonads, 182
Hyphomicrobia, properties of, 136
Hyphomicrobium spp., ecology of methane-
    utilizing strains of, 136
  photomicrographs of, 137

Hyphomicrobium spp.—*continued*
utilization of one-carbon compounds by, 134
Hyphomicrobium vulgare, growth of, on one-carbon compounds, 138
serine pathway in, 179
Hypoxanthine utilization by hydrogenomonads, 233

I

Incorporation of methylamine into N-methylglutamate, 187
Incubation temperature, effect of, on mucopeptide composition of *Bacillus subtilis*, 109
Induction of hydrogenase synthesis in bacteria, 212
Induction of urease in hydrogenomonads, 231
Inhibition of fructose utilization by hydrogen, 226
Intracellular teichoic acids in bacteria, 102
Intracytoplasmic membranes in bacteria, 6
Intrinsic factor as an inhibitor of methane production, 164
Involvement of cytochromes in oxidation of one-carbon compounds, 161
Iodoacetate, effect of, on methane oxidation by bacteria, 145
Ion exchange resins, bacterial walls as, 3
Ion flow across bacterial walls, 96
Ionic environment, effect of, on wall composition in bacteria, 95
Ionic interactions in bacterial walls, 50
Iso acids in bacteria, 24
Isocitrate dehydrogenase in bacteria, 172
Isoleucine, growth of knallgasbacteria on, 224
Isopropylmalate synthetase activity of hydrogenomonads, 236
Isovalerate, growth of knallgasbacteria on, 224

J

Janus green B as an electron mediator in fumarate reduction by photosynthetic bacteria, 256

K

2-Keto-3-deoxyoctonic acid in bacterial lipopolysaccharides, 57
2-Keto-3-deoxy-6-phosphogluconate dehydrogenase activity in hydrogenomonads, 219

2-Ketogluconate, growth of hydrogenomonads on, 221
Knallgasbacteria, regulatory phenomena in metabolism of, 205
utilization of nitrogenous compounds by, 230

L

Lactate, effect of, on hydrogenase synthesis, 212
Lactobacilli, cross bridges in wall peptidoglycans of, 39
Lactobacillus acidophilus, thickness of walls in, 31
Lactobacillus casei, peptidoglycans in walls of, 35
Lactobacillus plantarum, mesosomes in, 10
Lamellar mesosomal membranes in bacteria, 7
Latent adenosine triphosphatase in bacterial membranes, 27
Layering, in bacterial walls, 32
in walls of Gram-negative bacteria, 49
Leerlauf oxidation in knallgasbacteria, 208
Lengths of peptidoglycans in bacterial walls, 35
Leucine biosynthetic pathway, regulation of, in hydrogenomonads, 236
Leucine, growth of knallgasbacteria on, 224
L-forms, of bacteria, 5
wall of, 49
Light-dependent cyclic electron transfer in purple bacteria, 250
Linkage of peptidoglycan to teichoic acids in bacterial walls, 43
Lipid A in bacterial lipopolysaccharides, 57
Lipid bilayers in bacterial membranes, 30
Lipid-depleted bacterial membranes, 26
Lipid layers in bacterial walls, 34
Lipids, bacterial, effect of environment on synthesis of, 113
of Gram-negative bacteria, 60
Lipopolysaccharides, bacterial, effects of environment on, 110
of Gram-negative bacteria, 56
Lipoprotein of rigid layer of Gram-negative bacteria, 55
Listeria monocytogenes, cytoplasmic membranes in, 17
mesosomes in, 10
Localization of respiratory enzymes in bacteria, 14
Low-potential ferredoxins in bacteria, 254
Lysis of Gram-negative bacteria by ethylenediaminetetraacetic acid, 65
Lysozyme, comparative effects of, on *Bacillus subtilis*, 105

SUBJECT INDEX 303

Lysozyme, use of, to prepare protoplasts of bacteria, 3
Lytic enzymes in bacteria, 45

M

Magnesium ion binding to bacterial walls, 96
Magnesium ions, and bacterial cytoplasmic membranes, 26
binding of, by bacterial walls, 45
effect of, on bacteria, mesosomes, 11
Magnesium limitation, effect of, on composition of walls of *Aerobacter aerogenes*, 112
effect of, on mucopeptide composition of *Micrococcus lysodeikticus*, 109
on peptidoglycan of *Bacillus subtilis*, 107
on wall content of bacteria, 85
Malate dehydrogenase in bacteria, 172
Marine pseudomonad, protoplasts of, 4
Mechanism of photoproduction of molecular hydrogen, 266
Mechanism of the hydrogen effect, 226
Melting of fatty acids in bacterial membranes, 29
Membrane-bound hydrogenases of bacteria, 207
Membrane structure and metabolic pathways in methane-utilizing bacteria, 195
Membranes, in bacteria, 1
in methane-utilizing bacteria, classification of, 126
of methanogenic bacteria, 143
of *Methylococcus capsulatus*, electron micrograph showing membranes in, 130
Menadione as an electron mediator in fumarate reduction by photosynthetic bacteria, 256
Mesosomes, functions of, 15
Mesophilic bacteria, fatty acids in, 25
Mesosomal lipids in bacteria, 23
Mesosomal sac in bacteria, 7
Mesosomal tubules in bacteria, 7
Mesosomes, in L forms, 5
relation of, to nucleus in bacteria, 12
Metabolic interrelationships of the serine pathway, 178
Metabolic regulation, of synthesis of intracellular teichoic acids, 102
of teichoic acid synthesis, 101
Metabolism, chemolithotrophic, 206
of one-carbon compounds by microorganisms, 119
regulation of, in knallgasbacteria, 205

meta-Chlorocarbonylcyanide phenylhydrazone as an uncoupler of bacterial photophosphorylation, 254
Methane bacteria, carbon assimilation by, 169
serine pathway in, 179
Methane, oxidation of by micro-organisms, 119
pathway for oxidation of, 143
production of, in the biosphere, 119
reduction of methyl groups to, 164
utilization by photosynthetic bacteria, 139
Methane-utilizing micro-organisms, 120
cytochromes in, 161
Methanobacterium formicum, utilization of carbon monoxide by, 142
utilization of formate by, 141
Methanobacterium methano-oxidans, involvement of an oxygenase in methane oxidation by, 144
Methanobacterium mobilis, utilization of formate by, 141
Methanobacterium omelianskii, ferredoxin-linked formate dehydrogenase in, 164
mechanism of methanogenesis in, 166
resolution of, 141
Methanobacterium ruminantium, utilization of formate by, 141
Methanococcus vannieli, utilization of formate by, 141
Methanogenesis, adenosine triphosphate requirement during, 166
Methanogenic bacteria, carbon assimilation by, 175
Methanol, as an oxidation product of methane, 143
dehydrogenase in bacteria, 155
oxidation of, 155
production of, in the biosphere, 119
reduction of one-carbon compounds to level of, 163
utilization by anaerobic bacteria, 141
Methanol-utilizing micro-organisms, 120
Methanomonas methano-oxidans, carbon balance for growing on methane, 147
electron micrograph of, 127
isolation of, 120
properties of, 122
Methanosarcina barkeri, utilization of carbon monoxide by, 142
utilization of methanol by, 141
Methods for isolating methane-utilizing bacteria, 124
N-Methylalanine, synthesis of, by pseudomonads, 188
Methylamine dehydrogenase of bacteria, 153

SUBJECT INDEX

Methylamine, metabolism of, by bacteria, 186

Methylamine-utilizing bacteria, 131

Methylated amines, metabolism of, 150
production of, in the biosphere, 119

N-Methylated amino acids, metabolism of, by bacteria, 186

Methylated thiols, metabolism of, 150

Methylation of homocysteine, 164

2-Methylbutane, utilization of, by microorganisms, 123

Methylcobalamin, reduction of methyl groups of, to methane, 164

Methyl cobamides as substrates in methane production, 164

Methylene tetrahydrofolate in serine pathway in bacteria, 181

N-Methylglutamate, incorporation of methylamine into, 187
synthetase, 188

Methyl groups, oxidation of, by bacteria, 152
reduction of, to methane, 164

Methylobacter bovis, photomicrograph of, 152

Methylobacter chroococcum, photomicrograph of, 125

Methylococcus capsulatus, carbon balance for growing on methane, 147
cell-free extracts of, 149
electron micrograph of, 129
incorporation of methane carbon in, 190
isolation of, 120
photomicrograph of, 125
properties of, 123

Methylococcus minimus, photomicrograph of, 125

Methylocystis parvus, photomicrograph of, 125

Methylomonas albus, carbon assimilation in, 196

Methylomonas methanica, photomicrograph of, 125

Methylosinus sporium, photomicrograph of, 125

Methylosinus trichosporium, carbon assimilation in, 196
photomicrograph of, 125

Methylotrophy in bacteria, 196

Methylsulphonium salts, oxidation of, by bacteria, 154

Methyltetrahydrofolate in metabolism of one-carbon compounds, 163

Methyltransferase activity, fractionation of, 165

Michaelis constant of glucose 6-phosphate dehydrogenase in knallgasbacteria, 228

Microbial utilization, of carbon monoxide, 138
of cyanides, 138

Micrococcus denitrificans, hydrogenase, synthesis in, 212
urease activity of, 232

Micrococcus freudenreichii, cross bridges in wall peptidoglycan of, 39

Micrococcus lysodeikticus, adenosine triphosphatase in membranes of, 27
anionic polymers from walls of, 93
effect of environment on wall content of, 86
effect of growth conditions on mucopeptide composition of, 109
effect of nutrient limitations on phosphorus contents of walls of, 89
mesosomes in, 10
teichuronic acids in walls of, 43

Micrococcus radiodurans, wall structure in, 34

Micrococcus roseus, bridges in peptidoglycan of, 39

Micro-organisms that utilize one-carbon compounds, description of, 120

Mini cells of *Escherichia coli*, 67

Mixed function oxidase, possible involvement of, in methane oxidation, 144

Mixotrophic growth, enzymes of the tricarboxylic acid cycle during, 229

Mixotrophic metabolism in knallgasbacteria, 219

Molar growth yields of bacteria on methane, 148

Molecular hydrogen, photoproduction of, by algae, 267
photoproduction of, by photosynthetic bacteria, 264
production of, in bacterial photosynthesis, 252

Molecular sieves, bacterial walls as, 3

Monomethylamine, metabolism of, 150

N-Monomethylurea, metabolism of, 150

Mono-oxygenase, possible involvement of, in methane oxidation, 144

Moraxella spp., lipopolysaccharides in, 58

Morphology, of bacterial walls, 31
of Gram-negative bacterial walls, 49

M protein, of streptococcal walls, 46
of streptococci, effect of environment on synthesis of, 113

Mucopeptide, *see* Peptidoglycan

Mucopeptides, in bacterial walls, 102
of *Bacillus subtilis*, effects of substrate limitations on, 107

Muramic acid in bacterial walls, 35

Murein, in bacterial walls, 102

SUBJECT INDEX

305

Mutants, of methane bacteria, 181
on the serine pathway in bacteria, 178
Mycobacteria, wall structure in, 34
Mycobacterium smegmatis, peptidoglycan in walls of, 36
Myxococcus xanthus, wall structure in, 53
Mycoplasmas, membranes in, 29

N

Natural gas environments, isolation of methane bacteria from, 122
Neisseria catarrhalis, lipopolysaccharides of, 58
Neopterin, isolation of, from methane bacteria, 150
Net fixation of carbon dioxide by non-photosynthetic anaerobes, 175
Net synthesis of the glycine skeleton, 184
Neutral lipid composition of bacterial cytoplasmic membranes, 21
Nicotinamide adenine dinucleotide reducing system in hydrogen bacteria, 207
Nicotinamide nucleotide reduction, energy linked in bacteria, 252
Nile blue A as an electron mediator in fumarate reduction by photosynthetic bacteria, 256
Nitrate, effect of, on hydrogenase synthesis, 213
growth of knallgasbacteria on, 230
reductase of hydrogenomonads, 230
reduction with hydrogen in bacteria, 213
Nitrifying bacteria, ability of, to oxidize formate, 159
Nitrite oxidase in bacteria, 160
Nitrobacter spp., ability of, to oxidize formate, 159
generation of reducing power in, 258
Nitrobacter winogradskyi, coupling of electron transport and phosphorylation in, 210
utilization of one-carbon compounds by, 134
Nitrocystis oceanus, wall structure in, 52
Nitrogen source, effect of, on hydrogenase synthesis in bacteria, 212
Nitrogenous compounds, utilization of, by knallgasbacteria, 230
Nitrosocystis oceanus, membranes in, 6
Non-cyclic electron flow mechanisms in bacteria, 268
Non-cyclic photophosphorylation in bacteria, 250
Nonidet, action of, on bacterial membranes, 25
Non-photosynthetic anaerobes, net fixation of carbon dioxide by, 175

Non-photosynthetic bacteria, ability of, to utilize one-carbon compounds anearobically, 140
Norvaline, effect of, on hydrogenomonads, 237
Nucleoside triphosphates, effect of, on activity of glucose 6-phosphate dehydrogenase activity in knallgasbacteria, 228
3'-Nucleotidase, release of, from Gram-negative bacteria by osmotic shock, 66
Nucleotides, release of, from Gram-negative bacteria by osmotic shock, 66
Nucleus, relation of, to mesosomes in Gram-positive bacteria, 12
Nutrient limitations, effect of, on teichoic acid synthesis in bacteria, 88

O

Obligate methylotrophy in bacteria, 196
Occurrence of intracytoplasmic membranes in bacteria, 6
Octane hydroxylase, 153
One-carbon compounds, assimilation of, 177
description of micro-organisms that utilize, 120
metabolism of, by micro-organisms, 119
photometabolism of, 168
reduction of, to methane by bacteria, 162
One-carbon transfer in microbial metabolism, 120
Organic compounds, ability of, to repress synthesis of hydrogenase, 211
Organic substrates, effect of hydrogen on utilization of by hydrogenomonads, 223
Organism JOB5, properties of, 123
Oscillations of urease activity in hydrogenomonads, 232
Osmotic shock in Gram-negative bacteria, 66
Outer membrane, structure of, in Gram-negative bacteria, 67
Oxidation, of betaine, 154
of choline, 154
of formaldehyde, 158
of hydrogen by knallgasbacteria, 207
of methane, pathway for, 143
of methanol, 155
of methyl groups by bacteria, 152
of methylamine by bacteria, 153
of N-methylurea by bacteria, 154
of sarcosine, 154

Oxygen tension, effect of, on synthesis of ribulose diphosphate carboxylase synthesis, 217
Oxygenase, possible involvement of, in methane oxidation, 144

P

Penicillin treatment, use of, to prepare sphaeroplasts, 4
Penicillinase secretion, and mesosomes in bacteria, 16
Pentachlorophenol, effect of, on methane production, 168
Pentaglycine bridges in bacterial wall peptidoglycans, 38
Peptide units in bacterial walls, 36
Peptidoglycan, in bacterial walls, 3, 31, 35, 102
  synthesis in L forms, 5
Peptidoglycans of Gram-negative bacteria, 52
Peptococcus glycinophilus, metabolism of glycine by, 185
Peripheric vesicles of photosynthetic bacteria, 249
Periplasmic enzymes, release of, from Gram-negative bacteria by osmotic shock, 66
Phage adsorption, role of teichoic acids in, 45
Phage-resistant mutants of bacteria, 45
Phenol extraction of bacterial walls, 53
Phenol, use of, to extract bacterial envelopes, 67
Phenotypic changes in bacterial cell-wall composition, 94
Phenylalanine, growth of knallgasbacteria on, 224
Phenylhydrazine, use of, to extract teichoic acids, 44
Phosphate limitation, effect of, on composition of walls of *Aerobacter aerogenes*, 112
  effect of, on mucopeptide composition of *Micrococcus lysodeikticus*, 109
  on peptidoglycan of *Bacillus subtilis*, 107
  on wall composition of *Micrococcus lysodeikticus*, 93
  on wall content of bacteria, 85
Phosphatidylglycerol, content of, in bacteria, 24
Phosphoenolpyruvate carboxylase in bacteria, 172
Phosphoenolpyruvate synthase in bacteria, 172
Phosphoglucoisomerase activity in hydrogenomonads, 219

6-Phosphogluconate dehydratase activity in hydrogenomonads, 219
Phosphoglycerate, production of, by the serine pathway, 178
Phospholipid composition of bacterial cytoplasmic membranes, 21
Phospholipids of Gram-negative bacteria, 61
Phosphoramidate bonds in bacterial walls, 43
Phosphoribulokinase in knallgasbacteria, 215
Phosphoserine phosphohydrolase mutants of pseudomonads, 182
Photometabolism of one-carbon compounds, 168
Photomicrographs of methane-utilizing bacteria, 124
Photophosphorylation in bacteria, 248
Photoproduction of molecular hydrogen by photosynthetic bacteria, 264
Photoreduction of nicotinamide nucleotides in bacteria, 252
Photosynthetic bacteria, ability of, to use carbon monoxide, 140
  formate utilization by, 139
  growth of on formate, 170
  metabolism of one-carbon compounds by, 168
  that utilize one-carbon compounds, 138
pH value, effect of, on bacterial mesosomes, 12
  effect of, on bacterial wall composition, 98
Pigment composition of photosynthetic bacteria, effect of nutrients on, 247
Pigments of methane bacteria, 121
Planococci, cross bridges in wall peptidoglycan of, 39
Plasma membranes in bacteria, 2
Plasmalemma in bacteria, 2
Plasmolysis, nature of, 2
Pneumococcal C substance, and teichoic acids, 42
Poly-β-hydroxybutyric acid, production of, by knallgasbacteria, 208
Polyphosphates, formation of, by hydrogen bacteria, 209
Polyribosomes in bacteria, 19
Potassium limitation, effect of, on teichoic acid synthesis by bacilli, 89
  effects of, on peptidoglycan of *Bacillus subtilis*, 107
Primary and secondary processes in bacterial photosynthesis, 245
Production of molecular hydrogen in bacterial photosynthesis, 252
Products of carbon dioxide fixation in bacteria, 175

SUBJECT INDEX

Profiles of walls of Gram-negative bacteria, 49

Proline, growth of knallgasbacteria on, 224

Propionibacteria, peptidoglycan structure in, 41

Propionibacterium pentosaceum, erythrulose phosphate synthesis in, 192

Protaminobacter alboflavus, utilization of one-carbon compounds by, 134

Protaminobacter ruber, effect of folate antagonists on, 181
formate dehydrogenase in, 159
methanol dehydrogenase in, 156
utilization of one-carbon compounds by, 133

Protein, cell-wall, effect of environment on bacterial, 113
composition of bacterial cytoplasmic membranes, 21
coupling factor in bacterial photophosphorylation, 250

Proteins, in bacterial cytoplasmic membranes, 20
in bacterial walls, 33
in walls of Gram-negative bacteria, 50
of the envelopes of Gram-negative bacteria, 63
release of, from Gram-negative bacteria by osmotic shock, 66

Proteolytic enzymes, action of, on *Escherichia coli*, 55

Proteus mirabilis, wall structure in, 49

Proteus. vulgaris, phospholipid composition of, 62
wall structure in, 50

Protoplast membranes in bacteria, 2

Protoplasts, bacterial, 3
bacterial, extrusion of mesosomal vesicles from, 10
definition of, 4

Pseudomonads, enzymes of serine pathway in, 180
glucose 6-phosphate dehydrogenase of, 229
metabolism of trimethylamine by, 151
mutants of, 182
oxidation of methylated amines by, 153
serine pathway in, 179

Pseudomonas acidovorans, allantoin metabolism in, 233

Pseudomonas aeruginosa, lipopolysaccharide, 58
wall structure in, 47

Pseudomonas aminovorans, oxidation of methylated amines by, 152
utilization of one-carbon compounds by, 134

Pseudomonas fluorescens, oxidation of choline by, 155

Pseudomonas methanica, carbon balance for growing on methane, 147
formate, dehydrogenase in, 159
incorporation of methane carbon into, 190
isolation of, 120
methanol dehydrogenase in, 156
oxidation of formaldehyde by, 158
oxidation of methanol by, 155
oxygenase involvement in methane oxidation by, 144
properties of, 121
ribose phosphate cycle in, 189

Pseudomonas methanitrificans, properties of, 122

Pseudomonas oxalaticus, formate dehydrogenase of, 159
growth of, on formate, 170
utilization of one-carbon compounds by, 135

Pseudomonas saccharophila, taxonomy of, 206

Pteridine derivatives, in methanol dehydrogenase, 157
isolation of, from methane bacteria, 150

Pteroyl derivative as a prosthetic group of methanol dehydrogenase, 157

Purine derivatives, growth of knallgasbacteria on, 230

Purple bacteria, light-dependent cyclic electron flow in, 250

Pyruvate oxidase, catalysis of glyoxylate cleavage by, 185

Pyruvate, repression of ribulose diphosphate carboxylase synthesis by, in knallgasbacteria, 216
synthetase in bacteria, 172

R

Rates of redox reactions catalysed by chromatophores of *Rhodopseudomonas capsulata*, 262

Reaction-centre chlorophylls in bacteria, 247

Re-aggregation of bacterial membranes, 25

Reduced nicotinamide nucleotides, production of, in bacterial photosynthesis, 244

Reducing power, effect of, on hydrogenase synthesis in bacteria, 214
generation of, during methane production, 168
in bacterial photosynthesis, 243
production in hydrogen bacteria, 229

Reduction of cytochromes in photosynthetic bacteria, 258
Reduction of fumarate by photosynthetic bacteria, 256
Reduction of methyl groups to methane, 164
Reduction of one-carbon compounds, to level of methanol, 163
to methane by bacteria, 162
Reductive carboxylation of acetyl-CoA by *Clostridium formico-aceticum*, 177
Reductive carboxylic acid cycle, 171
Reductive pentose phosphate cycle, in chemolithotrophic metabolism, 206
in knallgasbacteria, 215
Regulation, of amino-acid synthesis in hydrogenomonads, 235
of enzymes of the Embden-Meyerhof pathway in knallgasbacteria, 228
of fructose degradation in hydrogenomonads, 227
of hydrogenase synthesis in bacteria, 211
Regulatory mechanisms in photosynthetic bacteria, 273
Regulatory phenomena in metabolism of knallgasbacteria, 205
Relief of repression by hydrogen in hydrogenomonads, 224
Repression of enzyme synthesis by hydrogen in knallgasbacteria, 223
Repressors, nature of, in the hydrogen effect, 226
Respiratory enzymes in bacteria, localization of, 14
Reversion of L forms, 5
Rhodopseudomonas capsulata, energy charge in, 275
rates of redox reactions catalysed by chromatophores of, 262
Rhodopseudomonas gelatinosa, methane utilization by, 139
Rhodopseudomonas palustris, enzymes of ribulose diphosphate cycle in, 175
ferredoxins in, 254
growth of, on formate, 140
metabolism of formate by, 168
Rhodopseudomonas sp., ability of, to use carbon monoxide, 140
Rhodopseudomonas spheroides, transdehydrogenase activity of, 259
Rhodospirillum rubrum, enzymes of acetyl-CoA cycle in, 172
enzymes of ribulose diphosphate cycle in, 174
fatty-acid composition of, 61
photophosphorylation in, 249
Ribitol teichoic acids, in bacterial walls, 42
structure of, 87

Ribose phosphate, condensation of, with formaldehyde in methane utilization, 190
cycle of formaldehyde fixation, 189
Ribosomes, association of, with bacterial cytoplasmic membranes, 19
Ribulose diphosphate carboxylase in knallgasbacteria, 215
Ribulose diphosphate cycle, in methane bacteria, 169
relation of, to ribose phosphate cycle, 193
Rigid layer, lipoprotein of, in Gram-negative bacteria, 55
R Mutants of Gram-negative bacteria, 56
Rod-like forms of bacteria, 4

S

Saccharate, growth of hydrogenomonads on, 221
Salmonella spp., lipopolysaccharide of, 56
Salmonella typhimurium, lipopolysaccharide of, 57
serine biosynthesis in, 181
structure of lipopolysaccharide of, 111
Sarcina lutea, cross bridges in wall peptidoglycan of, 39
fatty acids in, 24
Sarcosine, metabolism of, 150
Saturation in bacterial fatty acids, effect of growth temperature on, 113
Secretion of enzymes by bacteria, and mesosomes, 16
Sensitivity of bacterial walls to environment, 83
Septum formation in Gram-positive bacteria, 12
Serine-glyoxylate aminotransferase mutants of pseudomonads, 184
Serine hydroxymethylase mutants of pseudomonads, 182
Serine pathway in bacteria, 177
Serratia marcescens, lipopolysaccharides of, 59
Sewage purification, methane production during, 119
Shape of bacterial walls, 3
Size of bacteria, effect of growth rate on, 84
Sodium chloride, effect of concentration of, on envelope of halophilic bacteria, 69
effect of, on bacterial wall composition, 97
Sodium dodecyl sulphate, action of, on bacteria, membrane, 25
Sodium ions, binding of, to bacterial walls, 96
Solubilization of bacterial membranes, 25

SUBJECT INDEX

Sphaeroplasts, bacterial, 3
 of Gram-negative bacteria, 48
O-Specific side chains of Gram-negative lipopolysaccharides, 56
Spherical proteins in bacterial membranes, 30
Spirillum serpens, wall structure in, 51
Stable L forms, 5
 of bacteria, 5
Stalked particles on bacterial cytoplasmic membranes in bacteria, 17
Staphylococcus aureus, anionic polymers in walls of, 92
 effect of environment on wall content of, 86
 effect of nutrient limitations on phosphorus contents of wall of, 89
 mesosomes, in, 10
 peptidoglycan in walls of, 103
 structure of peptidoglycan from wall of, 35
 structure of peptidoglycan of, 40
Staphylococcus epidermidis, peptidoglycans in, 35
Streptococcal walls, M protein of, 46
Streptococci, cross bridges in wall peptidoglycan of, 39
 haemolytic, effect of environment on synthesis of M protein of, 133
Streptococcus faecalis, cytoplasmic membranes in, 17
 cytoplasmic membranes of, 20
Streptococcus pyogenes, polysaccharides in walls of, 44
Streptomycin, effect of, on bacterial mesosomes, 7
Structure, of bacterial lipopolysaccharides, 59
 of outer membrane of Gram-negative bacteria, 67
 of the bacterial cytoplasmic membrane, 28
Structures, of bacterial teichoic acids, 87
 of peptidoglycans in bacterial walls, 35
Subunit nature of glucose 6-phosphate dehydrogenase, 228
Subunits in bacterial walls, 33
Succinate, as a hydrogen donor for photoreduction in photosynthetic bacteria, 260
 dehydrogenase, in bacteria, 172
 solubilization of, from bacterial membranes, 25
 effect of, on synthesis of ribulose diphosphate carboxylase in hydrogenomonads, 218
 production from fumarate by photosynthetic bacteria, 256

Succinyl-CoA synthase in bacteria, 712
Sugar phosphates, incorporation of carbon of methane into, 189
Sulphanilamide, effect of, on pseudomonads, 181
Sulphate limitation, effect of, on cell-wall composition of *Bacillus subtilis*, 114
 effect of, on peptidoglycan of *Bacillus subtilis*, 107
Surface patterns on bacteria, 32
Synthesis, of bacterial lipopolysaccharides, 56
 of hydrogenase in bacteria, 210
 of hydrogenase in *Micrococcus denitrificans*, 212
 of phosphoribulokinase in knallgasbacteria, 216
 of ribulose diphosphate carboxylase in knallgasbacteria, 216
 of the glycine skeleton, 184

T

Taxonomy, of hydrogen-oxidizing bacteria, 206
 of methane-utilizing bacteria, 133
Teichoic acid synthesis in bacteria, effect of nutrient limitations on, 88
Teichoic acids, bacterial, structures of, 87
 binding of cations by, in bacterial walls, 100
 effect of environment on wall content of, in bacteria, 86
 function of, in bacteria, 44
 in bacterial walls, 31, 42
 in walls in *Bacillus subtilis* var. *niger*, 91
 intracellular, in bacteria, 102
 location of, in bacterial walls, 32
Teichuronic acid synthesis, control of, in bacteria, 100
 in bacteria, 31
Teichuronic acids, effect of environment on wall content of, in bacteria, 86
 in bacterial walls, 42
 in walls of *Bacillus subtilis* var. *niger*, 91
Temperature growth, effect of, on fatty-acid composition of Gram-negative bacteria, 63
Temperature of growth, effect of, on lipid composition of bacteria, 113
Temperature of incubation, effect of, on mucopeptide composition of *Bacillus subtilis*, 109
Tetrahydrofolate in metabolism of one-carbon compounds, 163
Tetramethyl-*p*-phenylenediamine as an electron mediator in bacterial fumarate reduction, 257

Thermophilic bacilli, fatty acids in, 25
Thickness of bacterial walls, 31
    effect of environment on, 85
Thickness of cytoplasmic membranes in bacteria, 17
Thiobacillus intermedius, enzyme repression in, 229
Thiobacillus thiooxidans, phospholipids of, 61
Thionine as an electron mediator in fumarate reduction by photosynthetic bacteria, 256
Threonine deaminase activity of hydrogenomonads, 235
Threonine starvation, effect of, on bacterial mesosomes, 7
Thymine utilization by hydrogenomonads, 234
Tolulene blue as an electron mediator in fumarate reduction by photosynthetic bacteria, 256
Transdehydrogenase activity in photosynthetic bacteria, 258
Transport of fructose into hydrogenomonads, 220
Tricarboxylic acid cycle enzymes, formation and function of, in knallgasbacteria, 229
Trifluoroleucine, effect of, on hydrogenomonads, 236
Trimethylamine, metabolism of, 150
Trimethylamine-N-oxide, metabolism of, 150
Trimethylsulphonium chloride, metabolism of, 150
Triose phosphate, synthesis of, from formaldehyde, 193
Tris buffer, effect of, on Gram-negative bacteria, 64
Tryptophan, growth of knallgasbacteria on, 224
Types of peptide subunit in bacterial cell-wall peptidoglycans, 104
Tyrosine, growth of knallgasbacteria on, 224

U

Ubiquinones in photosynthetic bacteria, 250
Uptake of fructose by knallgasbacteria, 220
Uracil utilization by hydrogenomonads, 234
Urea, growth of knallgasbacteria on, 230
Urease activity of hydrogenomonads, 231
Ureidoglycolic acid production by hydrogenomonads, 233

Ureidoglycine production by hydrogenomonads, 233
Uric acid, growth of knallgasbacteria on, 224
    production of, by hydrogenomonads, 233
Uricase activity of hydrogenomonads, 233
Uridine diphosphate glucuronic acid in synthesis of teichuronic acid, 101
Utilization, of carbon monoxide, 138
    of nitrogenous compounds by knallgasbacteria, 230

V

Valine starvation, effect of, on bacterial mesosomes, 7
Variability, in peptidoglycan composition of bacterial walls, 102
    of bacterial lipopolysaccharides, 110
Veillonella parvula, wall structure in, 49
Vibrio extorquens, effect of folate antagonists on, 181
    formate dehydrogenase in, 159
    methanol dehydrogenase in, 156
    serine pathway in, 178
    utilization of one-carbon compounds by, 133
Vibrio oxaliticus, utilization of one-carbon compounds by, 135
Vitamin B_{12} and methane formation, 165

W

Wall composition in bacteria, effect of growth rate on, 95
    effect of ionic environment on, 95
Wall composition of bacteria, effects of environment on, 83
Wall contents of bacteria, 84
Walls in bacteria, 1

X

Xanthine utilization by hydrogenomonads, 233
X-Ray diffraction patterns of bacterial envelopes, 70

Y

Yeast, oxidation of formaldehyde by, 158
Yield of cytoplasmic membranes in bacteria, 20